Fundamentals of
Fibre Reinforced Composite Materials

Series in Materials Science and Engineering

Series Editors: **B Cantor**, University of York, UK
M J Goringe, School of Mechanical and Materials
Engineering, University of Surrey, UK
E Ma, Department of Materials Science and Engineering,
Johns Hopkins University, USA

Other titles in the series

Fundamentals of Ceramics
M W Barsoum
Department of Material Engineering, Drexel University, USA

Topics in the Theory of Solid Materials
J M Vail
University of Manitoba, Canada

Physical Methods for Materials Characterisation—Second Edition
P E J Flewitt, BNFL Magnox Generation and University of Bristol, UK, and
R K Wild, University of Bristol, UK

Metal and Ceramic Composites
B Cantor, F P E Dunne and I C Stone (eds)
University of York, UK, and University of Oxford, UK

High Pressure Surface Science and Engineering
Y Gogotsi and V Domnich (eds)
Department of Materials Engineering, Drexel University, USA

Computer Modelling of Heat, Fluid Flow and Mass Transfer in Materials
Processing
C-P Hong
Yonsei University, Korea

High-K Gate Dielectrics
M Houssa (ed)
University of Provence, France, and IMEC, Belgium

Novel Nanocrystalline Alloys and Magnetic Nanomaterials
B Cantor (ed)
University of York, UK

3D Nanoelectronic Computer Architecture and Implementation
D Crawley, K Nikolic and M Forshaw (eds)
University College London, UK

Series in Materials Science and Engineering

Fundamentals of Fibre Reinforced Composite Materials

A R Bunsell and J Renard

Materials Research Centre, Ecole des Mines de Paris, France

IoP

Institute of Physics Publishing
Bristol and Philadelphia

British Library Cataloguing-in-Publication Data

A catalogue record for this book is available from the British Library.

ISBN 0 7503 0689 0

Library of Congress Cataloging-in-Publication Data are available

Series Editors: **B Cantor, M J Goringe and E Ma**

Commissioning Editor: Tom Spicer
Editorial Assistant: Leah Fielding
Production Editor: Simon Laurenson
Production Control: Sarah Plenty
Cover Design: Victoria Le Billon
Marketing: Louise Higham, Kerry Hopkins and Ben Thomas

Published by Institute of Physics Publishing, wholly owned by The Institute of Physics, London

Institute of Physics Publishing, Dirac House, Temple Back, Bristol BS1 6BE, UK

US Office: Institute of Physics Publishing, The Public Ledger Building, Suite 929, 150 South Independence Mall West, Philadelphia, PA 19106, USA

Typeset by Academic + Technical, Bristol
Printed in the UK by MPG Books Ltd, Bodmin, Cornwall

Contents

Preface

The contents of this book on composite materials reflects courses that both authors have developed and presented over a period of more than 25 years to undergraduate and postgraduate students in France at the Ecole des Mines de Paris and the University of Paris Sud, and in Belgium at the University of Leuvan as well as numerous other courses at universities on three continents. The plan, however, is considerably influenced by workshops and courses both authors have developed separately for the Conservatoire National des Arts et Métiers in Paris which has attracted hundreds of professional people looking to broaden their experience so as to work in this exciting field. The authors have experience both in industrial and in academic research. The approach therefore has been to present composites not just as an academic subject but one which is increasingly entering into the day-to-day experience of all of us and having a growing influence on modern industry. Many of the challenges that society will face in this 21st century will require the use of composite materials. The synergistic combination of two very different materials to form these materials offers the means of overcoming many of the limitations of traditional structural materials. It is hoped that this book will encourage interest in and the development of fibre reinforced composites.

Acknowledgments

This book has benefited from contributions from a number of people. We should like to acknowledge the help given by colleagues at the Ecole des Mines de Paris, including the following people: S Lemercier, O Adam, Y Auriac, Y Favry, J-Ch Teissedre for help in the preparation of the manuscript, M-H Berger for discussions, particularly on the statistical analysis of fibre data, and A Thionnet for discussions on the viscoelastic nature of resin matrix systems. The book is the result of many years working in the composite field by the two authors and we wish to thank many of our students who have enabled us to explore many aspects of the subject. Some figures and micrographs have been taken from theses, especially in the case of studies on fibre reinforcements and although not all can be mentioned the theses of the following have been particularly useful in the preparation of the book, M-H Lafitte, J-Ch Veve, Ch Oudet, G Simon, Ph Bonniau, L Adamczak, B Devimille, D Laroche, C Baxevanakis, F Deléglise, A Poulon, A Marcellan, J Mercier and S Blassiau.

We have been fortunate that the glass fibre manufacturer, Vetrotex, has allowed reproduction of figures showing the manufacture of composite parts and also for the help and permission of S Di Dimizio of the company SNPE for the use of other figures concerning manufacturing processes.

The preparation of the book has benefited from discussion with numerous other people and we should like to thank them all for their help and contributions.

A R Bunsell and J Renard
January 2005

Chapter 1

Introduction

Fibre reinforced composite materials represent a radical approach to designing structural materials when compared with traditional materials such as metals and ceramics. Although the living world, both the structures of plants and animals, is based on the reinforcement of matrix materials by flexible filaments, it was not until the second half of the 20th century that fibre reinforced composites began to be really developed. This was possible as organic resins had begun to be produced in the 1930s and in the same decade the first synthetic fibres were made. Composite materials, today, find uses in all areas and in some have become the dominant form of structural material used against which other materials are judged. This is certainly the case for the aerospace industry and will become increasingly the case for other industrial sectors as the 21st century progresses.

Matter in the form of a filament can possess extraordinary mechanical properties of strength and stiffness. Composite materials, or at least most of them, exploit this fact. The reasons why fibres are so much stronger and often stiffer than the same material in bulk form are discussed in chapter 2 but for the present let us just remember that a glass fibre can be hundreds of times stronger than the same glass in bulk form. This is because the fine fibres contain far fewer defects than does the bulk material, so that an assembly of fibres making up a given volume will be much stronger than the same volume of the same glass in bulk form. What is true for glass fibres is also true for other materials in fibre form so that fibres are really an extraordinary form of matter. If we reconsider the assembly of fibres mentioned above it should be obvious that the strength we are considering is parallel to the fibres' axes. It is obvious that if we pull the bundle at right angles to the fibres the strength will be negligible. The fibre assembly is highly anisotropic. This would not be the case for the bulk material which has the same characteristics in all directions and is therefore isotropic. This immediately illustrates a difference between composites and most traditional materials. Composites, at the level of layers of fibres, are inherently anisotropic whereas most traditional structural materials are isotropic. This is dealt with in detail in chapter 7, which treats the micromechanics of reinforcement. In addition, composites are made

1

up of layers of fibres, similar to cloth, and many of the ways of handling them have their origins in the textile industry. Composite materials are fundamentally two-dimensional whereas traditional materials are usually made as a block in three dimensions and then formed into their final shape. This is an advantage in forming composites as, just as cloth can be draped and made to take the shape of complex structures, such as a body, so can composite materials be formed into complex shapes. The manufacturing of composite parts takes advantage of this, as is explained in chapter 4, and is able to integrate many parts in one operation so reducing the overall cost of the structure. This is important as often the initial outlay can appear costly. The raw material, generally glass fibre and polyester resin, is around three times the price of conventional materials but some carbon fibres and more exotic resins can be up to 30 times more expensive. However, the possibility of reducing operating steps in the manufacture of parts reduces considerably the overall final cost of a composite part.

While fibres are important in any composite part we do not simply knit or weave a wing of a plane, a hull of a boat or a body part for a car. The fibres are held together in a matrix. There are many resins which are available for composite materials and whilst most are organic polymers there are composites with other types of matrices. These can be rubber, as in tyres, or metals such as aluminium, magnesium and titanium, speciality glasses and even ceramics such as mullite and silicon carbide. There is a growing market for fibre reinforced cement although in this type of composite the fraction of fibres is small and their role is to prevent crack growth rather than to reinforce. The choice of fibre reinforcement and matrix is dictated by the end use to which the composite part will be put and will depend on the properties of the components used as well as the manufacturing route to be adopted. Many of the resins and some of the other matrix materials used in composites are discussed in chapter 3. The long-term behaviour and ageing of composites are intimately linked with the nature of the components used and this is considered in chapters 9, 10 and 11. The use of two types of very different materials to make up a composite part opens up many possibilities which are not available to the designer using conventional bulk materials. For instance, the tensile modulus can be decoupled from the shear modulus in a composite part which allows designs of structures which are not possible with conventional materials. This is exploited in some aircraft designs. However, in order to calculate the behaviour of such complex materials a greater number of variables have to be considered than is the case for bulk isotropic materials and this is explained in detail in chapters 5, 6 and 8.

The size of the overall composite market is expected to reach ten million tonnes towards the end of the first decade of the 21st century. Glass reinforced plastics (GRP) represent more than 90% of the overall world market and have been growing at rates which represent a mature market of around 4% per year although great regional differences exist.

The markets which have developed for these composites cover a very wide range of applications, including pleasure boats, wind turbines, storage tanks, body parts for cars, dental implants and many others. Glass fibres are a low value added product so that although they represent such a large part of the industry they represent a lower percentage of the value of the world's composite market. Advanced composites, based on higher performance fibres, account for a disproportionate part of the overall value of the market even though they account for only around 1% of market share. The rate of growth of advanced composites is considerably greater than that of GRP. The total composite market is composed of GRP and advanced composites with a developing contribution from natural fibre reinforced materials.

The greatest use of high performance composite materials, today, is in civil applications but initially it was military and in particular aerospace structures which used composites. The reason is obvious—weight saving. To go higher and farther it is necessary to have sufficient strength and rigidity for the aircraft, for example, but weight will limit performance. That is why there are baggage weight restrictions when boarding a plane and why some low cost carriers have greater weight restrictions than other, more expensive, carriers. We can quantify the effect of weight for a given material by dividing its ultimate strength and elastic modulus by its density to give what is known as the specific properties of the material. If this is done for the most common metals, steel, aluminium and titanium it is surprising to see that we find almost the same values if the Young's modulus is divided by the specific gravity, as shown in table 1.1.

Glass is an interesting borderline case as it is amorphous and even in the form of a fibre its microstructure does not become aligned as is the case for most other fibres. This results in glass fibres having the same elastic modulus as bulk glass and, as can be seen, there is no weight saving for stiffness to be

Table 1.1. Comparison of specific moduli of some common metals and fibres.

Material	Specific gravity	Young's modulus (GPa)	Specific modulus (GPa)
Steel	7.9	200	25.3
Aluminium	2.7	76	28
Titanium	4.5	116	25.7
Glass (bulk and fibre)	2.5	72	27.6
Carbon (high strength fibre)	1.8	295	164
Carbon (ultra high modulus fibre)	2.16	830	384
Kevlar 49 fibre	1.45	135	93
Zylon fibre	1.56	280	180

made by replacing the metals in the table by glass fibres. In addition the specific modulus of any glass fibre reinforced composite will be less than the value shown as the fibres have to be bound together with the resin matrix material and this has low stiffness. The other fibres, however, show much higher specific moduli than the conventional materials and even allowing for the need to combine them with a resin, with say a fibre volume fraction or 0.6, it is clear that there are considerable weight savings to be made by using composites. Even though glass fibres offer no weight savings they are the most widely used reinforcement for composites as their low cost and ease of use, which facilitates manufacturing, are often determining points in their favour. For high performance structures, however, it was the development of high performance fibres, such as carbon and Kevlar fibres, in the 1960s and 1970s which allowed composites to become the material of choice for an increasing number of applications.

1.1   Production and markets

The first applications for advanced composites were for military jets, both in the fuselage and wings as well as in the outer casing of their engines. The markets for these applications were primarily in those countries which had developed industrial military complexes so that the USA and the UK and then France became leading actors, with Germany and Italy also becoming involved. Japan has invested heavily in the development of advanced fibres but retained a much broader approach to markets so that from an early date the use of high performance composites in such areas as civil engineering was investigated. Japan was also very active in developing sports goods and in particular the ubiquitous carbon fibre composite golf club which is a necessity for every Japanese businessman. Very few types of sports goods are made today without a major part being in composite materials. Most tennis and squash racquets are made of composites although light metal alloys are also used. Taiwan has become the main and dominant producer of such composite racquets. Skis are almost inevitably made up of complex layers of composites, often with several types of fibre reinforcement used. India is strong in producing hockey sticks and China is producing advanced competition rowing boats and sculls.

From the first markets for composites others have developed so that the producers and the markets are truly world wide. Figure 1.1 shows that the distribution of composite markets is roughly divided evenly between North America, Europe and Asia.

As over 90% of composites produced are reinforced with glass fibres figure 1.1 is based on that market. Growth rates vary from one year to the next and between sources. However, this distribution has changed little since the last decade of the 20th century. Rates of growth suggest some

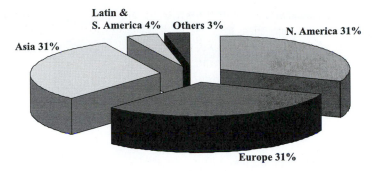

Figure 1.1. World wide composite market distribution.

saturation of the overall composite market in North America but impressive growth in Asia, principally in China and India.

The breakdown of markets for GRP is shown in figure 1.2. It can be seen that no mention is made of aerospace and military applications for GRP. This reflects the lack of weight advantage which is often at a premium for these applications, although the ease of manufacture coupled with the lower cost of GRP means that the interior cladding including baggage racks in civil planes are made in this composite. For the same reasons the bodies of civil helicopters are also produced in GRP. Glass fibres are also used in the Glare material, which is a sandwich of GRP and aluminium foil used in the construction of aircraft such as the A380.

Carbon fibres are the reinforcements of choice for many high performance composites. Although they represent only around 0.6% of the overall market they account for about 12% of the overall composite market value. Carbon fibre reinforced plastic (CFRP) composites have become a standard choice, not only for military aircraft but also, more importantly in terms of volume and value, for civil aircraft. The Airbus

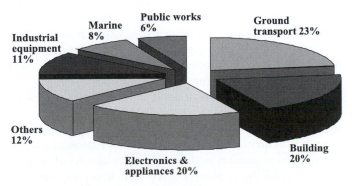

Figure 1.2. Breakdown of the overall GRP market.

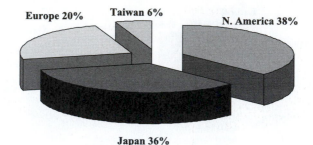

Figure 1.3. Carbon fibre production.

A380 and the Boeing 7E7 airliners make enormous use of advanced composites, reinforced by both carbon and aramid fibres. The wide use of high performance composites by the aerospace industry means that, even though the total volume of the material used is less than 1% of all composites, the sector represents 17% of the overall value of the composite market. The aeronautical industry, together with other fast growing industries such as gas pressure vessels, wind turbines, offshore oil applications as well as sports goods and a slowly emerging sports car sector, are stretching the world wide carbon fibre production capacity which is around 30 000 tonnes. The distribution of carbon fibre production is shown in figure 1.3.

Carbon fibre manufacture is overwhelmingly dominated by Japanese producers who have installed or bought production plants in both the USA and Europe. The Taiwanese carbon fibre production feeds its successful sports goods production industry. World wide growth rate of carbon fibre composite use is difficult to evaluate from one year to the next but is several times that of GRP, although for a much smaller market and the market is far from saturated. Indeed it could be expected that CFRP will mature, during the first part of the 21st century, into a large industry now that it is finding applications in sectors other than aerospace and sports goods. The broadening of the market offers opportunities to other producers who are producing fibres in very large tows, with up to 25–30 times the number of fibres generally produced. In this way production costs are being dramatically reduced. These large tows can then be separated into more manageable tows containing smaller numbers of fibres which then can be processed. As already mentioned carbon fibres represent around 0.6% of the fibre reinforcement market but 12% of its value.

Aramid fibres are another type of high performance fibre which is of major importance in advanced composites. They represent about 0.4% of the fibre reinforcement market and 5% of its value. As chapter 2 reveals, they are very different from glass and carbon fibres as they can deform plastically in compression whereas they are largely elastic in tension.

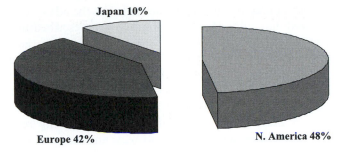

Figure 1.4. Production capacities of aramid fibres.

Although this is a limitation for their use in structures which can be subjected to compressive loads, the high energy absorption of composites made from these fibres finds them important markets. These fibres are used in many structures other than those which are the subject of this book, such as cables, aircraft tyres and cloths, but an approximate distribution of aramid fibre manufacture for use in composites is shown in figure 1.4. There are only two commercial producers. One is Du Pont in the USA which also produces Kevlar fibres in Europe and Japan, the latter as a joint venture with Toray. The other producer is Teijin-Twaron based in Holland. The Teijin group, which has its headquarters in Japan, produces two types of aramid fibre, the Twaron fibre and the Technora fibre.

The fibres are embedded in a matrix and in the overwhelming number of cases the matrix is a thermosetting resin or a thermoplastic. The first organic resins used were thermosets and they still represent around two-thirds of the overall composite market and about the same fraction of the overall market value (figure 1.5). Unsaturated polyester is by far the most important

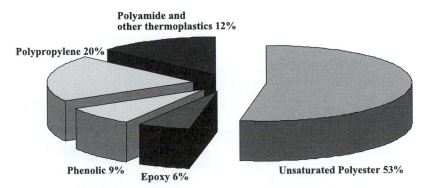

Figure 1.5. The overall composite market uses about two-thirds thermosetting resins and one-third thermoplastics as matrix materials.

thermosetting resin used in composite materials but the advanced end of the market uses predominantly epoxy resins which are described in chapter 3. Fibres can be easily impregnated with these resins and then the composite can be formed into its final shape before the thermosetting resin is crosslinked, as explained in chapters 3 and 4. Phenolics are used for specialist applications such as the interior of mass transport vehicles, e.g. trains and planes, because of their fire resistance, and in the facing of buildings. The finished product made of a fibre reinforced thermosetting resin cannot, however, be altered after manufacture and recycling off-cuts is not practicable. This is not the case for thermoplastic matrix composites. In addition, injection moulding of short fibre reinforced composites using thermoplastics is a developed industry, as is explained in chapter 4. Long or continuous fibre reinforced thermoplastics are a developing industry and produce high performance composite structures. Thermoplastic matrix composites represent around one-third of the overall market but generally have a higher growth rate. Polypropylene is the preferred matrix material for many markets because of its low cost and ease of recycling. However, polyamide 66 and saturated polyester are also widely used. These and other specialist thermoplastics are described in chapter 3.

A quick glance at the literature of fibre reinforced composite materials could easily give the impression that most are made from high performance fibres, usually carbon, embedded in epoxy resin or indeed in more exotic resins. This is not the case, as is hopefully clear from the above rough analysis of the overall market. Nevertheless, what is undoubtedly true is that the high performance components used in advanced composites represent a larger fraction of the overall value of the composite market than their volume share would indicate, as is shown in figure 1.6.

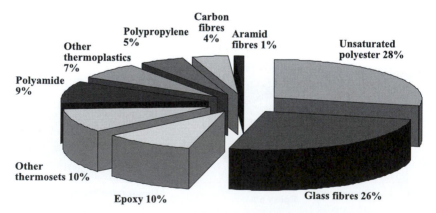

Figure 1.6. Percentages of the overall value of the fibre reinforced composite market of fibres and matrix materials.

1.2   Applications

1.2.1   Space

Advanced fibre reinforced composites may not represent the greatest part of the composite market but their applications are often startling for they are used where more conventional materials cannot match their properties. This means that they find applications at the frontiers of engineering. Space is a challenge to engineers and getting there is a competition between getting the rocket off the launch pad, the payload and gravity, which means that weight is all important. Carbon fibre reinforced resin accounts for 80% of the weight of the structure of a satellite because of the high specific properties of these composites. The stiffest carbon fibres can find a role in assuring the dimensional stability of large satellite or space station constructions. In addition to their stiffness and strength, carbon fibre composites have an almost negligible coefficient of thermal expansion in the direction of the fibres. In space, the stark contrasts in temperature between the sunlit side of a structure, up to $+150\,°C$, and the side turned away from the sun, which could be at $-150\,°C$, means that dimensional stability is a major issue. A space telescope would lose its function if its structure varied as the temperature changed. The low atmospheric pressures encountered in near space place other constraints on composites which are made on earth. Resin matrix composites absorb water which, in low pressure conditions, will leave the composite and could condense on lenses, mirrors or electronic circuits. For these reasons, resins such as isocyanates are used which absorb much less water than epoxies. The composites are usually in the form of panels bonded onto a honeycomb structure to form a lightweight structure which is stiff in bending. This type of structure is widely used, for example, in aircraft structures but there is a difference as the honeycomb in aircraft is made of organic nomex paper but more often it is of aluminium in satellite structures. This again is because of concerns about water absorption which justifies the use of the heavier metal honeycomb. Metal matrix composites are also considered for space applications for the same reasons and they also have the additional advantage of being able to conduct heat away from sensitive parts which the insulating resins could not do. In low orbits, atomic oxygen, which is very reactive, can cause the deterioration of resin systems so that this is another argument for the use of metal matrix composites in space.

1.2.2   On the seabed

The frontier of space is being conquered by the use of composite materials and in this area they have long since been considered as standard structural materials. A new frontier, however, is in deep water for offshore oil drilling. The world runs on oil and it is a diminishing resource so that we are obliged to look into increasingly inhospitable regions to uncover more. The offshore oil industry involves massive resources and installations and has amassed considerable experience in drilling in depths of water using steel piping and casings. As water depths increase, however, the weight of the steel structures used to bring the oil to the surface becomes a serious limitation to what is feasible. At depths greater than 1500 m the weight becomes a major problem, both for transporting the parts to the offshore site but also because the structure has to be supported by the floating platform. The density of steel is eight times that of water so that any buoyancy due to the Archimedes effect is negligible. Carbon fibre composites are obvious candidates for this frontier. Their low density means that the buoyancy effect of being in water is considerable. In addition they can match the strength and rigidity of steel and provide greater resistance to corrosion. As if that were not enough they also are better thermal insulators and this is not an insignificant property as the oil leaves the seabed at temperatures approaching 100 °C but the surrounding sea water is at 4 °C. As the oil cools its viscosity increases and flow can be stopped by the viscous oil blocking the riser, which is the name given to the pipe which brings it to the surface. Risers and other pipelines for transporting oil and gas from the seabed and also to land are complex structures which are required to have lifetimes of 20 years. Although the industry is by nature conservative and hesitant to adopt new materials with which they have no experience, advanced carbon fibre composites are set to make a big impact in this area. There is really no alternative for extracting oil in depths of water down to 3000 m. This is before drilling starts.

Above the sea, on the platform, composites are also finding uses despite initial fears because of fire safety. The first generation of offshore oil platforms did not use composites because of their resin matrices, which were considered to be a fire hazard. Later research showed that if the composites were of a greater thickness than 8 mm they performed better than steel in a major fire. The reason is the same as why it is better to have a wooden door between you and a fire than a metal door. The latter is a good conductor of heat and metals will soften and buckle if the temperature is great enough. Wood will first char and this will provide a protective layer to the underlying unburnt wood. Composites of sufficient thickness behave in the same way, with charring and out-gassing retarding the progress of a fire. Composites are now being used in offshore platforms and are used not only for piping but also stairways and walkways. As with the space applications, saving weight, even on an offshore oil platform which weighs thousands of tonnes, is still important. Even the mooring ropes of platforms are being made of organic fibres for the same reasons. Such ropes, made of polyester, aramid or high modulus polyethylene fibres, are required to have breaking loads of up to 2000 tonnes and be kilometres in length. They are replacing the much heavier chains and steel cables used in the past.

1.2.3 Aircraft

The aircraft industry was the first to use fibre reinforced composites and they have become, together with aluminium, the most important class of materials used in aerospace structures. Their adoption by the industry has been spectacular and rapid and all the more impressive as the industry has some of the toughest safety regulations for the use of materials of all industrial sectors. Originally, composites were introduced into military aircraft and since the 1960s each new generation of fighter aircraft and helicopters has seen an increase in their use so that the composites' share of these aircraft is approaching 50% of the structural weight. Original applications were for secondary structures such as wing extensions to increase range, nose cones and helicopter bodies but the use of carbon–carbon brakes on planes became standard and other advanced resin matrix composites found increasing numbers of applications. Very quickly the superior fatigue resistance of composites meant that they replaced the metal blades on helicopters which needed to be changed every 2000 flying hours because of the fatigue crack growth. Composite blades last as long as the body of the helicopter and are a remarkable demonstration of composite fatigue resistance compared with that of metals. The greatest part of the composite materials used in aircraft has been carbon or aramid fibre reinforced epoxy.

Composites in military aircraft and helicopters are now used in primary structures such as wings, fuselages and tail structures and so account for such a large percentage of their structural weight. When the low density of composites is considered, compared with even light alloys, it can be seen that their percentage volume in the aircraft is even greater than 50%.

The coming of age of advanced composites in the aircraft industry must, however, be seen in their adoption in civil aircraft. This trend started in the 1980s but matured into much wider use at the beginning of the 21st century. Not only are these aircraft so much bigger than military aircraft but safety standards are even stricter. The Airbus series of aircraft have seen a gradual but a continual, step by step, increase in the use of advanced composites. An aircraft such as the Airbus A380 uses composites in many structural parts such as the upper fuselage which is composed of a material called Glare, which is a laminate of glass reinforced epoxy resin and aluminium sheets. The use of Glare produces a 25% weight saving compared to other traditional materials. It is also used in the leading edges of the vertical and horizontal stabilizers, resulting in a 20% reduction in the weight of the 14 m high tail. This hybrid material is a lower cost variation of Aral, which was developed for Fokker by the University of Delft in Holland and is composed of layers of aramid reinforced epoxy resin and aluminium sheets. These materials have the deformability of aluminium but are lighter and have much superior fatigue properties. Glare represents 3% of the weight of an Airbus A380, not counting the weight of the engines and landing gear. Most composites, however, are reinforced by carbon fibres and are used in the rear pressure bulkhead separating the pressurized passenger cabin from the rear, unpressurized, section of the plane, floor beams, flight control surfaces such as flaps, spoilers and ailerons, landing gear doors, most of the tail and a 12 tonne centre wing box which links the two wings through the underbody of the fuselage. These advanced composite materials represent 22% of the weight of the plane without the engines and landing gear.

The other large plane maker, Boeing, is not absent in its use of composites although was slower in using them than Airbus. The large body Boeing 777 introduced in the last decade of the 20th century has around 10% by weight of advanced composites but the middle sized Boeing 7E7 aircraft capable of carrying 200–250 passengers is designed to have up to 50% of composites. This is to be compared with 20% aluminium, 15% titanium and 10% steel. It is claimed that this future generation of aircraft will have most of its fuselage and wings composed of carbon fibre laminate. It is clear that the use of advanced composites will dominate in 21st century aircraft over other types of materials.

Smaller planes, such as business jets, have been able to move faster in using advanced composites. It is interesting that their introduction was not solely because of superior mechanical properties linked to lower weight but also in some structures because the composite solution was cheaper than aluminium. The possibility of integrating several assembly steps which are necessary with metal structures reduces overall manufacturing costs. The possibility of using composite structures for more than just mechanical purposes is attractive. This multi-functionality allows the composite material also to be used for insulation, so reducing problems of condensation, noise reduction, fire resistance, corrosion resistance and greater design flexibility. Sensors can be incorporated into the composite material, turning it into what is known as an intelligent material and which can be used to monitor the state of the plane so as to reveal problems before they become acute. In addition composites offers better durability than aluminium for roughly equal cost. Nevertheless, there is a continuous evaluation of the materials used in aircraft and improvements will continue to be made, not only in composites but also light alloys of aluminium and titanium. The manufacturing processes used with composites will continue to evolve with the goal of reducing costs by using automated processes and the increased use of pre-forming technologies drawn from the textile industry.

1.2.4   Boats

Any visitor to a marina will see hundreds of sleek and shiny pleasure boats ranging from the small to the very big, up to 40 m. Overwhelmingly they will be made of glass reinforced polyester resin. The industry began with the introduction of room temperature curing unsaturated polyester resin in the late 1940s which could be reinforced with glass fibres. Over the years this material has become dominant in pleasure boat building. As is described in chapter 4, open mould manufacture of the hulls of these boats is the most common production process and has brought the cost of boats into the range of many more people than was the case before their introduction. The industry could evolve, however, to using more closed mould manufacturing because of the need to reduce styrene emissions.

Most high performance yachts and catamarans, used for races, contain carbon fibre reinforced vinyl or epoxy resin in the hulls and masts. The use of composites has enabled records to be continually broken but problems, linked usually to difficulties of manufacturing limited numbers of large structures and also the reduction of safety margins, have led to some high profile and unexpected failures of masts.

Very large military boats have been produced as mine sweepers. The non-metallic nature of the hulls is the prime attraction together with their resistance to fire, but also reduced maintenance over the lifetime of the boat is an important consideration.

1.2.5   Ground transport

Ground transport covers road vehicles, monorails, tramways and trains. Trains use a lot of composite materials, nearly always reinforced with glass fibres for reasons of cost, and they are found in the interior structures of the wagons. The need to reduce weight is important for high speed trains and this has encouraged consideration of the use of composite materials in their primary structures. Weight saving may not at first seem a major issue with trains which spend most of the time travelling at high speed with relatively little time spent in acceleration. However, the very fast trains in Europe and Japan run on specially laid track. In order to make the trains more economic it is desirable to increase the numbers of passengers transported by each train. Increased numbers of passengers means an increase in weight and this poses a problem for the track. If the track is not to be replaced, at great cost, so as to take heavier trains, the weight of the trains has to be reduced. The use of aluminium for wagon bodies was a first step but advanced composites are being investigated as an even better solution. Cost is the limiting factor but as the markets for advanced materials increase and material costs come down, trains will be constructed with large parts made of advanced composites. A major part which has been considered is the bogie, which is a swivelling framework with several axles connected to the wheels and springs, used to guide the locomotive. Several projects to make bogies out of carbon fibre reinforced plastic have shown the feasibility of this application. The use of carbon–carbon brakes could reduce the weight of a high speed train by up to 15 tonnes and could one day be fitted, but their efficiency increases with temperature so that although good for stopping a train travelling at 350 kph there would be a problem if the train were going at only 5 kph. A hybrid system could be the solution to this conundrum.

Subway trains are increasingly made with low cost composites making up an important part of the body parts of the carriages. Weight is an obvious disadvantage as the trains have to repeatedly accelerate and brake during normal use. The flexibility of manufacture that composites give is shown by the innovative designs which can be seen in many of the subway and tramway systems which have been installed in many European cities.

Cars and lorries have long attracted the attention of the composite community, above all because of the enormous volume of production which is involved. It is a well established industry, ruled by a conservative approach to material use and above all by cost. Cars are made with increasing numbers of extras such as more complicated seats, air-conditioning units and safety features, all of which increase weight at a time when fuel economy is increasingly important. The attraction of composites to the automobile industry is the possibility of making innovative body shapes as well as reducing weight to produce original vehicles with enhanced corrosion resistance. In addition, it has long been recognized that composites can absorb more energy in the case of a crash for the same weight of a metal structure. Sports cars have long used composites, whether for racing cars or sports cars for the road. The bodies of Formula 1 cars are almost all composites with an emphasis on carbon fibre composites for lightness and stiffness together with aramid fibre reinforced composites for toughness. However, these are small volume productions and composites have had a harder job being accepted for the average family car. The limitations are lack of experience and cost. The costs of retooling to make composite parts are high, but the tools and moulds are often cheaper than for metal parts as the forming pressures involved are much lower and they can even be made with other composite materials. Whether the composites solution is cost effective can often be a question of numbers of vehicles to be produced. Very large numbers of vehicles produced per day, say over 400, could mean that the advantages of cheaper tooling are lost as they have to be replaced frequently. Nevertheless, composites have made their mark in family cars. In 1972 the Renault 5 was the first car to use non-chromed bumpers and instead used a glass fibre polyester bumper. This was a big marketing gamble as up until then shiny bumpers were considered a must for cars. Today, although most bumpers are of a tough unreinforced plastic, it is interesting to note that this innovation has spread all around the car, protecting it from bumps and scrapes. Placing composites in car bodies has had mixed success. In Europe the approach has been to replace metal parts with composites. The bonnet was seen as a fairly straightforward structure to replace but early attempts underlined the importance of finish to the proud owner of the car sitting behind his or her steering wheel from where every imperfection could be seen. Getting the required class-A finish, which meant avoiding entrapped air or solvents in the composite from popping through the surface during the painting process, proved to be a difficult and costly exercise. Another unexpected problem was the absence of a Faraday's cage when the bonnet was made of glass reinforced resin. The result was that the radio picked up interference from the engine. The solution, which was far from ideal, was to put a metal grille under the bonnet and ground it to the rest of the car. The rear fifth door, however, was a much greater success as the composite solution proved to be less costly as the typical 27 assembly operations could be reduced to just four. An interesting innovation has been the development of the modified reinforced reaction injection moulding process, described in chapter 4, to produce major structural parts by including aligned fibrous parts in the mould rather than having a random fibre arrangement. This material is called 'structural RIM'. Cargo boxes and tailgates of pickup trucks have been produced in this way with greater resistance to damage, both mechanical and to corrosive products, and considerable weight savings. The manufacturing process has been successfully developed so as to meet the high production rates necessary in the car industry. In other attempts to introduce composites to cars, the composites are used to provide bolt-on body parts to a metal chassis which mean that designs can be changed much more quickly. Less obvious has been other innovations, often under the bonnet, with parts made of short glass fibre injected thermoplastics such polyamide 66 or saturated polyester. Thermoplastics are of particular interest to the automobile industry as they can be recycled at the end of the life of the car.

With the fall in price of carbon fibres they are finding applications in speciality, high cost and low volume sports cars, and it seems inevitable that these advanced composites will ultimately find wide use in family cars. If this happens it will revolutionize the carbon fibre production industry as, if only a few kilograms of CFRP were used in each car, the amount of carbon fibre which would be required would far outstrip present world wide production.

The use of carbon fibre reinforced pressure vessels for stocking natural gas used as an environmentally friendly carburant for buses is discussed in chapter 11. This is seen as a major developing market for advanced composites and if successful this technology could be extended to private vehicles.

1.2.6 Building and civil engineering

The building sector is one of the biggest for the use of composites which, in the past, were almost invariably reinforced with glass fibres; but a market for carbon fibre reinforced structures is developing. Façades in GRP for buildings allow architects to produce novel and interesting outward appearances of new and renovated buildings. Composite offer much more, however, and allow imaginative architects to make entirely novel roofing shapes with large spans which could not be produced in more conventional building materials. Protection from collapse in regions prone to earthquakes can be afforded by including in the original design or by retrofitting frameworks of advanced composites which would retain their integrity in the case of the building collapsing.

Much piping is also made of glass reinforced resin. For these applications, large pipes can be made by centrifugal casting as described in chapter 4 and are used to refit large diameter drainage and sewerage. Other techniques exist which allow existing degraded pipelines to be relined *in situ* using pre-impregnated glass felt which is crosslinked once the composite cladding has been put in place.

The ease of application of composites which, as pre-impregnated tapes or cloth, can be added to an existing structure and after crosslinking, act as a reinforcement, is exploited in many civil engineering structures. The renovation of subway systems, some of which have been in place for 150 years, and bridges, is facilitated by the use of high modulus carbon fibre composite bonded onto the steel beams which need supporting. Increasingly, new vehicular and pedestrian bridges are being made with composites incorporated into the structure with road decks, and struts being made from GRP. The struts are made by pultrusion, as described in chapter 4, while the decks are made in the form of lightweight sandwich or honeycomb structures.

Concrete is the main building material in civil engineering systems but, even in this type of material, fibre reinforcement is used. Historically, asbestos reinforced cement showed the way. In this type of material the role of the fibres is to hinder crack propagation in the cement. This they do by providing weak interfaces normal to the fibre crack propagation direction and also requiring pullout for the crack to develop further. Typically fibre volume fractions are low, around 5%, so that strength or stiffness reinforcement, as seen with other composites, is not the goal. Asbestos is no longer used because of health hazards but other fibres, polypropylene, cellulose, steel filaments, even aramid and carbon fibres, have replaced them. Such cement and concrete matrix composites show increased toughness and the result is that the thickness of say a prefabricated roof can be reduced. The short carbon fibres used in such materials are made from spinning isotropic pitch which is then converted into a carbon fibre felt. The fibres have low moduli but resist the alkaline environment of cement which is difficult for glass fibres. Pitch-based carbon fibres are discussed in chapter 2. High performance carbon fibres are also used to replace the steel rods in pre-stressed concrete for applications where their high resistance to corrosion or their transparency to electromagnetic fields are important.

An application which is of increasing importance is the use of composites in wind energy generation. These wind turbines are placed, often near the coast or offshore, where wind speeds are high and the wind blows almost continuously. Wind turbines can have composite blades up to 125 m in length, made usually of glass fibre cloth on a foam or other lightweight core, although advanced composites are also being used in the hybrid structures of the bigger turbines. They provide a valuable source of renewable energy with individual wind turbines able to generate more than 5 MW of power.

1.3   Conclusions

The above applications are given only as examples of the diversity of composite products. Look around and you will see many more, from fibre reinforced baths in the home to moulded bathroom units supplied as a finished product to hotels. All that is necessary is to connect it to the plumbing. Artists are using composites to make innovative sculptures and designing originally shaped furniture. Go to the local swimming pool and like as not the roof is made of laminated wooded beams. The use of hybrid beams of wood and aramid composites on the tension side allows less costly and more easily replaced wood to be used. Medical prostheses, surgical instruments and innovative filling materials for teeth all use composites. Take a plane and you can be sure to be within touching distance of advanced composites. From your mobile phone which contains composite printed circuit boards to under the bonnet of the car, composites are everywhere. If they are not there yet you can be pretty sure they are coming.

1.4   Bibliography

The markets and uses of composites are continually changing so to keep up to date the following journals and magazines are recommended.

1. JEC composites: www.jeccomposites.com
2. Composites technology: www.compositesworld.com
3. Reinforced plastics: www.reinforcedplastics.com

1.5   Revision exercises

1. Why are composite materials reinforced with fibres rather than any other form of matter?
2. Explain why a composite is intrinsically anisotropic.
3. Why do satellites contain so much composite material? Discuss some of the problems and advantages of using composites in near earth orbital space structures.

4. Why have composites replaced metals in helicopter blades?
5. Describe what is meant by Glare and Aral. Why are these materials used in aircraft?
6. To what do the specific properties of a material relate? Show that there is no advantage in the GRP specific properties compared with conventional metals such as aluminium.
7. Despite the poor specific properties of GRP, give reasons why it is used in civil aircraft.
8. What percentage of the whole composite market do advanced composites represent? Discuss the value related to the volume of composite markets and their values.
9. Discuss the issues limiting the use of carbon fibres as reinforcements for composites and explain how these are likely to be overcome.
10. What is the most widely used fibre reinforcements and matrix material?
11. Temperature variations in space can be large. Discuss how the use of carbon fibre composites can solve some of the problems that this causes.
12. Why is the choice of matrix important for a space application?
13. What is a honeycomb structure and why is it used?
14. Advanced composites are finding use under the seas. Discuss the reasons for this and why initially the offshore oil industry was reluctant to use them.
15. What percentage of the next generation of civil aircraft will be made of advanced composites?
16. Why is it important to reduce the weight of high speed trains?
17. Most pleasure boats are made of glass reinforced polyester made with open moulds but legislation is provoking greater use of closed moulds. Why is this?
18. Why is it necessary to reduce the weight of high speed trains and what technical difficulties must be overcome in the use of carbon–carbon brakes on trains even though they are used in aircraft?
19. The cost effectiveness of composites in the car industry can be related to the numbers of cars to be produced per day or year. Explain.

Chapter 2

Fibre reinforcements

2.1 Fibre development

Composite materials owe their remarkable characteristics to the fibres which are used to reinforce the matrix.

The reinforcement of a medium by a network of fibres has evolved in the living world to be the natural solution for structures which respond to and resist mechanical stresses. The combination of two very different phases, the matrix medium and the fibres, allows a great degree of flexibility in the development of microstructural arrangements. In addition, the long fine form of the fibres, coupled with their mechanical strength and stiffness, allows natural structures to be preferentially reinforced in areas of greater stress.

Although surrounded by naturally occurring fibre reinforced materials and the wide use of wood for construction, man, with very few exceptions, only began to create fibre reinforced structural materials at the beginning of the 20th century. Up to that time there were no artificial fibres produced but their development at the end of the 19th century coincided with the beginning of the automobile industry. Fibre reinforced rubber forms the basis of the tyre industry. Fibre reinforced resins, which form the basis of the composite industry, have developed in the second part of the 20th century and owe their existence to the production of organic resins and the beginning of the synthetic fibre industry in the 1930s.

Fibres can be divided into three groups, natural, regenerated and synthetic. Naturally occurring fibres can themselves be divided into three subgroups, vegetable, animal and mineral fibres. Vegetable fibres such as flax, cotton and hemp are short or at least discontinuous and based generally on cellulose. Mineral fibres such as asbestos are also short, being rarely longer than a few centimetres, and their fully crystalline structure places them apart from the others. Animal fibres which are based on animal proteins can be discontinuous, as in the case of wool, or they can be virtually continuous, as with silk. The natural fibres all share a common trait of possessing molecular structures which themselves are arranged in a fibrillar manner. A cotton fibre, for example, has a microfibrillar structure arranged in a helical fashion around the fibre axis. The helix reverses in direction at points along the fibre length and is reminiscent of the structure of cotton threads. Threads consisting of discontinuous fibres owe their integrity to the intertwining of the individual fibres and the frictional forces developed between them which lock the structure together and allow a continuous thread to be produced. The load transfer developed between the fibres within a thread is exactly analogous to the load transfer between matrix and fibre which accounts for the properties of fibre reinforced composites.

Regenerated fibres benefit from the fibrillar nature of the molecular structure of plants which is processed to form continuous filaments. In this way, this first class of artificial fibre to be produced avoids the difficulty of constructing its basic molecular structure.

Artificial fibres are only just over 100 years old and the earliest produced were based on the naturally occurring cellulose molecule. Towards the end of the 19th century workers in Great Britain and in France had found means of dissolving natural cellulose and then extruding it through holes to produce filaments. The first commercial exploitation of these regenerated fibres was in France, where a patent to produce the earliest form of rayon was awarded to Count Hilaire de Chardonnet in 1885. This was the beginning of a revolution in fibre technology. The regenerated fibres did not exploit the total potential of the cellulose molecules as they were broken up to shorter lengths during the fibre manufacturing process. The cellulose was obtained from wood, usually spruce, made up of molecules containing about 1000 glucose residues but these were broken down to about a fifth of their original length in the manufacture of rayon. The intention of the first fibre producers was to produce an artificial silk for textile purposes, but at the beginning of the 20th century they found a major application in the carcasses of tyres for the developing automobile industry. Rayon fibres have evolved significantly over the years and the descendants of these earliest artificial fibres are still to be found in some car tyres today. Rayon fibres were used as precursors for the early production of carbon fibres, mainly in the USA, and there are still some carbon fibres produced in this way. However, they have now mainly been superseded by other fibres for this purpose.

Truly synthetic organic fibres for which the long molecular structures, necessary for most fibre production, are made from relatively simple molecules have existed only since the late 1930s. The first organic truly synthetic fibres were developed independently in the USA and Germany in the late 1930s. These were polyamide or 'Nylon' fibres and these, together with polyester fibres developed in Great Britain in the 1940s, have become the most widely produced commodity synthetic fibres. Organic fibres have since been developed with elastic moduli rivalling that of steel but with only a fifth or less of the weight. Two approaches have been taken to achieve this remarkable behaviour. First, the molecular structures of organic fibres have been made more complex and consequently stiffer. The second approach has been to align the molecular structure of simple polymers to achieve greatly improved tensile properties.

Glass fibres, first commercially produced in the USA in the early 1930s were the first artificial fibres with a relatively high Young's modulus. The manufacture of glass fibres by drawing the molten glass through a spinneret is similar to the process used to produce fine thermoplastic fibres. However, the act of drawing aligns the molecular structure of the thermoplastic fibre producing an anisotropic structure. This is not the case with the glass fibres, which remain isotropic, but in both instances the act of drawing and plastically deforming the material during fibre production removes surface defects and accounts for the very high strength of glass fibres compared with bulk glass.

Although glass fibres possess Young's moduli considerably higher than those of conventional textile fibres, still greater tensile stiffness is required for structural composites in many applications and notably in the aerospace industry. Boron fibres produced in the early 1960s, first in the USA and then in France, allowed a leap from 70 GPa, which is the elastic modulus of glass, to over 400 GPa. These remarkable fibres were technically a success but their use has been limited by their extremely high cost.

Carbon fibres, first produced in Great Britain in the middle 1960s and produced commercially for the first time there in 1967, were also developed in parallel in Japan where commercial production begin in 1970. These carbon fibres, which are made by the pyrolysis of polyacrylonitrile precursor fibres, have revolutionized the aerospace industry and are finding increasing applications in other industrial areas. An alternative route for producing carbon fibres is from pitch, which is the residue from oil refining or the cokeing of coal process. Fibres with very high Young's moduli, approaching that of diamond, can be made from pitch.

Large diameter ceramic fibres of silicon carbide on a core have existed since the early production of boron fibres; however, fine ceramic fibres started to appear at the end of the 1970s and in the 1980s. One group of these fibres developed in Great Britain and in the USA is based on fine grained alumina, whereas others developed in Japan are silicon carbide or silicon nitride. This class of fibres allows light metal alloys to be reinforced and, probably more importantly, the production of fibre reinforced ceramics for high temperature structural applications. They are made by the pyrolysis of precursor fibres in an analogous way to the production of carbon fibres.

2.2 Organic fibres

It was not until around 1938 that the first truly synthetic organic fibres were produced by the polycondensation of short molecules. Polyamide 6.6 fibres were developed in the USA by E I du Pont de Nemours. Almost simultaneously with the work of Carothers at du Pont, polyamide 6 was produced in Germany by IG Farben.

The work that gave rise to polyamide 6.6 fibres, which were given the commercial name of Nylon, came from a wider ranging study of both polyamides and polyesters. The latter were made by the reaction between diacids and dialcohols, but this initial work found that polyesters were too easily hydrolysed and had low melting points so this avenue of research was abandoned. The polyamides were produced by the polycondensation of diacids and diamines and eventually gave rise to the first truly synthetic organic fibres.

Polyamide 6.6 fibres are produced by the condensation of hexamethylene diamine $(H_2N-(CH_2)_6-NH_2)$ and adipic acid $(HOOC-(CH_2)_4-COOH)$. The combination of these two products results in the elimination of water and the creation of long chain molecules with the repeat unit of $-NH-(CH_2)_6-NH-CO-(CH_2)_4-CO-$. This product is polyamide 6.6 with the numbers referring to the six carbon atoms found in the parent molecules and which are found in the repeat unit of the molecular structure which is produced.

Polyamide 6 fibres are produced from a single monomeric material which contains an acidic and amine function. The synthesis makes use of caprolactam which is produced from ε-aminocaproic acid which has the formula $H_2N-(CH_2)_5-COOH$. The reaction of this molecule with itself and the elimination of water gives the repeat unit $-(CH_2)_5-CO-NH-$. This is the repeat unit of polycaprolactam or polyamide 6, the number indicating the number of carbon atoms in the repeat unit.

Earlier man-made fibres made from regenerated cellulose were made by dissolving the cellulose in a solvent, then extruding the solution through fine holes of a spinneret and fibres were produced by the evaporation of the solvent. This technique is not suitable for the production of polyamide fibres which are spun from the molten material and, as there is no need for solvent recuperation, which is hardly ever complete, the inherent production costs are lower. Spinning from the melt also allows faster production rates. The fibres are spun through spinnerets with holes of around 0.025 mm in diameter. The melting point of polyamide 6.6 is around 260 °C when heated in nitrogen and the molecular weight is of the order of 15 000. The melting point of polyamide 6 is around 215 °C, the molecular weight is around 20 000 and the polymer is more stable and easier to spin than polyamide 6.6. These properties of relatively low melting points and high molecular weights makes the two types of polyamide very suitable for spinning into fibres. The polyamide 6.6 fibres are generally slightly stiffer and less extensible that the polyamide 6 fibres.

The polyamide fibres are first spun at speeds of the order of several thousand m min^{-1}, cooled to around 70 °C and wound onto bobbins. The

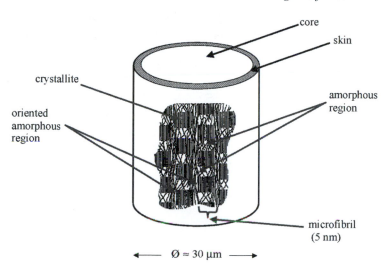

Figure 2.1. The molecular morphologies of PA and PET fibres, which are usually around 30 μm in diameter, are thought to be similar and to be composed of fine microfibrils which contain crystalline regions and amorphous regions. The fibre can have a skin–core macro structure, as illustrated.

fibres can be subsequently further drawn four to five times their initial lengths so as to increase mechanical properties by a better alignment of the molecular structure, although most modern processes include this in a one-step initial operation. Figure 2.1 shows a schematic diagram of the microfibrillar nature of the polyamide fibres as well as the possibility of a skin–core macro structure.

The fibres can be seen to be semi-crystalline, generally with a crystallinity of around 45%. The blocks of crystalline material are made up of folded molecules which can pass from one block to another. Within the fibrils and between the blocks the molecules form tie molecules which ensure the continuity of the fibrils which are themselves surrounded by other similarly oriented but not perfectly aligned molecules. It can be observed that the structure is anisotropic and as the molecules are not aligned continuously parallel to the fibre axis the properties of the fibres are far lower than what would be expected if the molecules were straight and aligned. Table 2.1 shows that these fibres possess initial moduli of up to 5 GPa but, as will be seen below, moduli of perhaps 50 times higher would be calculated if the molecules were straight and parallel to the axis.

Although the work on polyesters was initially abandoned because of the instability of the molecular structure it inspired work in England by J R Whinbfield and J T Dickson of the Calico Printers Association. These two chemists found that polyethylene terphthalate (PET) resisted hydrolysis and possessed a melting point of 260 °C. As can be seen from table 2.1, the

Table 2.1. The molecular structure of organic fibres has evolved to become increasingly more complex and rigid with higher melting or decomposition temperatures.

Fibre type and initial producer	Repeat unit in the macromolecule	Maximum elastic modulus (GPa)	Melting point or decomposition temperature (°C)
Polyamide 6 [Nylon 6] IG Farben	$-[NH-CH_2-CH_2-CH_2-CH_2-CH_2-CO]_n-$	4	230
Polyamide 6/6 [Nylon 6.6] du Pont	$-[NH-CH_2-CH_2-CH_2-CH_2-CH_2-CH_2-NH-CO-CH_2-CH_2-CH_2-CH_2-CO]_n-$	~5	260
Polyethylene terephthalate [Polyester] ICI	$-[O-CO-\phi-CO-O-CH_2-CH_2]_n-$	~15	260
Poly(*m*-phenylenediamine-isophthalamide) [Nomex] du Pont	(aromatic polyamide repeat unit structure)	17	400
Poly-paraphenylene/3,4-diphenylether terephthalamide [Technora] Teijin	(aromatic copolyamide repeat unit structure)	70	500

Fibre	Structure		
Poly(p-phenylene terephthalamide) [Kevlar] du Pont	—[HN— —NH—CO— —CO]—	135	550
Poly(p-phenylene benzobisoxqazole) PBO [Zylon] Toyobo		280	650
Poly{2,6-dimidazo [4.5-b:4'.5-e] pyridinylene-1,4 (2,5-dihydroxy) phenylene} (PIPD) M5 AKZO		330	650

molecule contains an aromatic ring which confers on it thermal and chemical stability as well as interesting mechanical properties. This modification to the linear macromolecules found in simpler polymers defined the route which chemists would take to produce the organic fibres with the highest tensile stiffness and which are used in high performance composites.

The polyester fibres which were developed from PET have become the most widely produced synthetic fibre. PET is obtained from two bifunctional reactants, ethyleneglycol

$$HO-(CH_2)_2-OH$$

and terephthalic acid

$$HOOC-\langle\bigcirc\rangle-COOH$$

Poly(ethylene terephthalate) is then obtained by the reaction between the alcohol and acid groups, with elimination of water. The long chain molecules which are produced have repeat units containing ester functions giving

$$-O-(CH_2)_2-O-CO-\langle\bigcirc\rangle-CO-$$

Predrying is essential before spinning if hydrolysis is to be avoided. The water content of the polymer must be inferior by 0.005% by weight and this is achieved through heating the polymer chips at 120 °C. Afterwards the PET fibres are spun in a similar manner to that of polyamide fibres at speeds of the order of several thousand $m\,min^{-1}$ and wound onto bobbins. The polymer is more viscous than polyamide because of the aromatic ring in the molecule and this leads to higher extrusion pressures. The fibres are subsequently further drawn, at a temperature of around 75 °C, four to five times their initial lengths so as to increase mechanical properties by a better alignment of the molecular structure. The molecular morphology of PET fibres is similar to that of polyamide fibres although crystallinity is lower. Table 2.1 shows that these fibres possess moduli of up to 18 GPa.

Polyamide (Nylon) and PET (polyester) fibres are extensively used for reinforcing rubber and these flexible composites are used in fan belts, hoses, flexible drives, moving walkways, tyres and other applications. Respectively, they account for around 25 and 50% of the total synthetic fibre production. Fibre reinforced rubber accounts for about the same volume of material as do all the other types of composites combined.

Neither fibre shows real elastic behaviour even at low strains, and plastic deformation as well as creep occurs when the fibre, either by itself or in an elastomeric composite, is loaded. The properties required of these fibres, for example in a tyre, are high strength, dimensional stability, fatigue resistance, thermal stability and good bonding to the matrix.

Figure 2.2. Polyamide and PET fibres fail in tension and creep from the surface by slow crack growth which opens the crack until rapid final failure occurs and leads to the morphology in the left image. In fatigue the crack initiates at the surface but runs along and into the fibre before it breaks, as shown in the right image.

Tensile failure in both Nylon and polyester fibres is similar, involving slow crack growth from the surface, in a radial direction. The crack opens as the material ahead of the crack deforms plastically. A point is reached at which the crack becomes unstable and rapid failure occurs. Almost identical fracture surfaces are obtained with both broken ends. The composite structures, for which organic fibres are used as reinforcements, are often subjected to dynamic loading and fatigue is a major concern. Many thermoplastic fibres can fail by a distinct failure process leading to a characteristic fracture morphology involving longitudinal crack growth which gradually reduces the load bearing fibre cross section.

This leads to the distinctive fatigue failure morphology. A comparison of the tensile and fatigue failure morphologies can be seen in figure 2.2. The loading conditions which lead to fatigue failure are unusual as a necessary criterion for fatigue failure is that the minimum cyclic load be nearly zero. Indeed it is possible to avoid fatigue by increasing the overall loading pattern on the fibre and, if this does not raise it into a regime of creep failure, the fibre will no longer break. The deformation of the molecular structures of a polyamide or polyester fibre during cyclic loading is a possible explanation of the fatigue behaviour. Upon loading, the tie molecules in the intrafibrillar amorphous zones are pulled slightly out of the crystalline zones so that on unloading they buckle, inducing shear stresses at the edge of the microfibril. These shear stresses lead to the breakage of molecules and a local deterioration of the fibre. Eventually the fatigue crack appears and its development is influenced greatly by the anisotropic structure of the fibre. The process is enhanced because the fast cooling rates experienced during manufacture of the fibres lead to differences in internal stresses at the surface and core. This may mean that on unloading the surface goes into compression below a certain minimum load, even if the overall load on the fibre remains positive. This initiates the fatigue crack at or near the fibre surface as it is known that polymers fail relatively more easily in compressive fatigue than in tension. The observation that the long fracture surface on one end of the fatigue fracture is curved always towards the fibre core lends support to the premise that the surface is in compression when the fibre is unloaded.

Since the production of the polyamide and polyester fibres other organic fibres have been produced and a trend towards the production of increasingly more rigid and complex molecules used in making the fibres is clearly discernible from table 2.1. The aramid family of fibres is made up of aromatic polyamides which possess remarkably high Young's moduli, more than 20 times that of conventional polyamide fibres. The term aramid comes from aromatic polyamide and covers the range of wholly aromatic polyamides so as to distinguish them from the linear aliphatic polyamides. The aramids consist of stiff molecules due to the aromatic rings which are not flexible. The result is a family of polymers which are more chemically resistant and thermally stable, and possess higher glass transition temperatures and melting points than aliphatic polyamides. Some of the aramids also possess much higher strengths and tensile stiffness than aliphatic polyamides due to the greater degree of alignment of the molecular structure which is possible. The properties of the most common classes of organic fibres are shown in table 2.2. An early type of aramid fibre which has found wide use is poly(*m*-phenylene isophthalamide), the form of which is shown in table 2.1. This fibre is sold by its makers, du Pont, under the name of Nomex. This polymer contains *meta*-oriented phenylene radicals which means that stresses through the molecule are not supported by straight covalent bonds and this results in a tensile stiffness which is still only about that of highly drawn PET. The fibre is, however, flame resistant, thermally stable to around 100 °C over the melting point of the aliphatic polyamides and has good dielectric properties. As a consequence the fibre is used in apparel for which resistance to heat is required such as firemen's uniforms and racing drivers' overalls. It is also used in chopped form to make a paper material which is widely used as a honeycomb in advanced composites structures. The low specific gravity of Nomex (1.38) compared with aluminium (2.7), which is also used for honeycomb structures, gives it an obvious advantage for applications for which weight saving is important such as in space or aircraft.

Table 2.2. Comparison of fibre properties.

Fibre	Polymer	Diameter (μm)	Density ρ (g/cm^3)	Young's modulus E (GPa)	Strength σ (GPa)	Strain to failure ε (%)	Specific modulus E/ρ (GPa/g/cm^3)	Specific strength σ/ρ (GPa/g/cm^3)
Polyamide 66	PA66	20	1.2	<5	1	20	4	0.8
Polyester	PET	15	1.38	<18	1	15	13	0.6
Nomex	MPD-I	15	1.38	17	0.64	22	12	0.5
Technora	PPDT	12	1.39	70	3	4.4	50	2.2
Kevlar 49	PPTA	12	1.45	135	3	4.5	93	2.1
Zylon	PBO	12	1.56	280	5.8	2.5	180	3.7
M5	PIPD	12	1.7	330	4	1.2	194	2.4
Polyethylene	PE	38	0.96	117	3	3.5	120	3.1

In order to obtain high strength and above all high stiffness the aramid polymers should contain predominantly *para*-oriented aromatic units. This is the case of poly(*p*-phenylene terephthalamide) (PPTA) which is the polymer used to make the Kevlar fibre which has been available from du Pont since 1972. The form of the molecule is shown in table 2.1. PPTA is prepared from the low temperature polycondensation of *p*-phenylene diamide (PPD) and terephthaloylchloride (TCl) in an appropriate solvent. The reaction is given by

$$H_2N - \!\!\bigcirc\!\!- NH_2 \quad + \quad ClCO - \!\!\bigcirc\!\!- COCl$$

$$\text{(PPD)} \qquad\qquad\qquad \text{(TCl)}$$

$$\longrightarrow \quad \left[NH - \!\!\bigcirc\!\!- NH - CO - \!\!\bigcirc\!\!- CO \right] + \; 2HCl$$

$$\text{(PPTA)}$$

The PPTA polymer is produced at high molecular weights to give good fibre properties. The molecular weight of around 20 000 is similar to that found with polyamide 6.6. PPTA is insoluble in most solvents but a 10% solution of the polymer in concentrated (>98 wt%) sulphuric acid is one which results in the production of a mesophase or liquid crystal solution. The rod-like PPTA molecules are randomly oriented in the solution but, beyond a critical concentration, locally the molecules are attracted together and adopt an ordered arrangement in small domains to achieve better packing. The domains are still randomly arranged with respect to one another and the solution retains the flow properties of a liquid but in polarized light can be seen to be locally oriented. When the solution is passed through the holes of a spinneret the induced shear forces orient the molecular structure so that a highly oriented fibre can be produced without drawing.

The fibres can then be subsequently treated at high temperature and under tension to increase crystallinity and orientation. In this way a family of fibres can be produced. The fibres are spun through an air gap before entering a coagulating bath. The fibres are then washed and dried. The fibres can be then heat treated under tension to increase crystallinity and orientation.

Although the bonds are not completely straight they are nearly so and this results in a high tensile stiffness as well as thermal stability. The glass transition temperature of PPTA is over 375 °C and the polymer is thermally stable up to about 550 °C. The same type of fibre has been commercially produced under the name Twaron. It was developed by Akzo in Holland and is now produced there by the Japanese company Teijin-Twaron. These

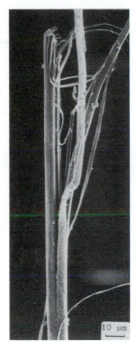

Figure 2.3. Aramid fibres are highly anisotropic and split on failure (source: PhD thesis, M-H Lafitte).

fibres were truly remarkable as they took organic fibres from a range of stiffness of less than 20 to 135 GPa.

Kevlar fibres are five times as strong as steel in air for the same weight and this difference rises to 30 times when they are used in water, because of their low density and the buoyancy provided by water. It means that ropes and cables for mooring large oil tankers or platforms can be made from Kevlar. The hydrogen transverse bonding holding the PPTA molecules together means, however, that the Kevlar absorbs water. The amount of water uptake varies with the conditions of humidity but it is reported that up to 6 wt% is possible. The highly anisotropic structure of aramid fibres also results in them being weak in all other directions than simply parallel to the fibre axis. As a consequence failure of these fibres is nearly always highly fibrillar, as can be seen from figure 2.3.

The structure of aramid fibres is very well ordered and aligned preferentially parallel to the fibre axis. The unit cell of the Kevlar fibre is shown in figure 2.4. Their highly anisotropic behaviour means that the fibres are not used in primary loading structures subjected to compressive forces.

The processes which limit the compressive properties of aramid fibres confer on them great tenacity, which means that they absorb much more

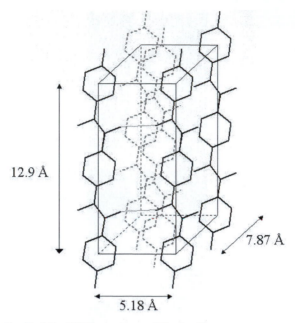

12.9 Å

7.87 Å

5.18 Å

Figure 2.4. Unit cell of the PPTA structure of Kevlar fibres.

energy than more brittle fibres when broken. This is because the fibre can deform plastically in compression. An elastic fibre can be bent to a minimum radius of curvature at which the tensile stresses in the convex surface attain the breaking stress of the material. When an aramid fibre is bent, slippage of the molecular structure occurs on and near the concave surface leading to the development of kink bands, as shown in figure 2.5.

The result of this plastic deformation is that the neutral stress axis is displaced from the fibre axis towards the convex surface. In this way the tensile stresses on the convex surface do not attain the failure stress of the fibre, which can be bent to a zero radius of curvature. This results in the fibre being difficult to cut as well as giving great tenacity and resistance to impact loading.

Another aramid fibre which has been successfully produced is the Technora fibre from Teijin, in Japan. This fibre is made from copolyamides containing PPTA and 3,4′-diphenyl ether. As it is produced from three basic products it is more expensive to make than the Kevlar fibre but has significantly different properties. The reaction mixture is filtered and spun into an aqueous coagulating bath. The fibres are then drawn to 6–10 times their original length and dried at 500 °C to form the final product. The fibre has a tensile stiffness of around 70 GPa due to its molecular structure not being straight, as can be seen from table 2.1, but it has superior fatigue properties than the PPTA fibres.

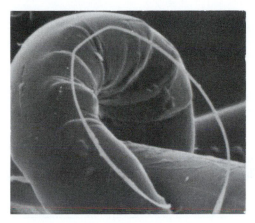

Figure 2.5. This Kevlar fibre is bent around another fibre. Both have diameters of 12 μm and it can be seen that the concave side of the fibre has developed slip bands by the plastic deformation in shear of the polymer. In this way the fibre accommodates the compression rather than breaking (source: PhD thesis, M-H Lafitte).

The organic fibre which has been commercially available since 1998 and which has the highest tensile stiffness is the Zylon fibre made by Toyobo in Japan. This fibre is 40% stiffer than steel for a fifth of the density. The fibre is made from poly-p-phenylenebenzobisoxazole (PBO). This is one of the polybenzazoles containing aromatic heterocyclic rings which were developed by the US Air Force as polymers which resist heat better than Kevlar. As the molecule is straight, the Zylon fibre, as shown in table 2.1, has a tensile stiffness twice as great as that of Kevlar. When Zylon fibres are first produced they absorb some water, around 2 wt%, but this is due to porosity in the fibre structure caused by solvent evaporation. Treatment of the fibre reduces this to around 0.2 wt%. Such a low water uptake is due to the transverse bonds being purely van der Waals.

Table 2.1 shows a fibre made from polypyridobisimidazole (PIPD) which was produced by the researchers from the Dutch company Akzo Nobel and now being developed by Magellan Systems International. It can be seen that the molecule is related to PBO and the Young's modulus claimed for the fibre of 330 GPa is even higher. The producers claim that there is a highly significant difference in the behaviour of the PIPD fibre as lateral cohesion is determined by hydrogen bonds. As mentioned above, the PBO fibre relies on van der Waals bonds (which are very weak) for lateral cohesion. Hydrogen bonds are ten times stronger which means that the M5 fibre has superior radial and compressive strengths. The presence of hydrogen bonds will probably mean that the fibre will absorb water, however. The M5 fibre seems to be very interesting from a technical view point and addresses the problem of anisotropy of aromatic fibres. It is attracting attention for use in body armour.

Table 2.3. Ultimate strengths of various polymers.

Polymer	Density (g/cm^3)	Ultimate strength (GPa)	Strength of commercial fibre (GPa)
Polyethylene	0.96	37	0.95
Polyamide 6	1.14	32	1.00
Polyvinylalcohol	1.28	29	1.00
Poly(*p*-phenylene terephthalamide)	1.43	30	2.65
Polyethylene terephthalate	1.37	28	1.00
Polypropylene	0.91	17	0.95
Regenerated cellulose	1.50	18	0.55

The more rigid molecules in aramid fibres confer not only higher mechanical properties but also better thermal resistance. This latter quality is a considerable advantage when these fibres are compared with other high modulus organic fibres made from simpler molecules such as polyethylene made by straightening out the molecular structure and exploiting the covalent bonds in the backbone of the molecule.

The production of aramid fibres involves complex and expensive chemistry, however. The high performance properties of the fibres is a result of the molecules being aligned parallel to the fibre's axis. Table 2.3 shows that although simpler polymers usually possess low mechanical properties they could, if all their molecules were aligned parallel to one another, give materials with the characteristics of high performance materials.

It is possible to calculate the theoretical maximum Young's modulus for a simple organic structure for the optimum case in which all the macromolecules are aligned in the direction of the applied load.

Consider the molecule shown in figure 2.6. The force f applied to the molecule deforms the A–B and B–C bonds and increases the valence angle

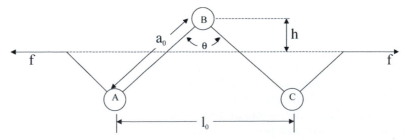

Figure 2.6. Schematic view of a section of a polymer chain stretched by a force, f. A, B and C are the carbon atoms in the backbone of the molecule, a_0 is the C–C bond, 1.53×10^{-10} m, and θ is the valence angle 112°.

ABC. Two force parameters can be defined relating the effort and the deformation produced such that

$$K_a = \frac{f}{\Delta a}\sin\theta/2, \qquad K_\theta = \frac{fh}{\Delta\theta}. \qquad (2.1)$$

These two parameters can be measured directly using infrared or Raman spectrometry and for an olefin molecule we find that

$$K_a = 460\,\mathrm{N\,m^{-1}}, \qquad K_\theta = 82 \times 20^{-20}\,\mathrm{N\,m}.$$

From the geometry of the molecule we see that

$$l_0 = 2a\sin\theta/2$$

so that the deformation produced by the force f is

$$dl = 2(a/2\cos\theta/2\,d\theta + \sin\theta/2\,da) \qquad (2.2)$$

and the strain

$$\varepsilon = \frac{d}{l_0} = \frac{a/2\cos\theta/2\,d\theta + \sin\theta/2\,da}{a\sin\theta/2}. \qquad (2.3)$$

The couple fh producing a change of angle of $d\theta$ results in an angular compliance which can be seen to be

$$\frac{d\theta}{f} = \frac{a/2\cos\theta/2}{K_\theta}. \qquad (2.4)$$

The deformation of the A–B and B–C bonds are produced by the component of force, $f\cos(90° - \theta/2)$, so that the deformation compliance is

$$\frac{da}{f} = \frac{\sin\theta/2}{2K_a}. \qquad (2.5)$$

The total compliance is obtained by replacing da and d in equation (2.3), using relations (2.4) and (2.5), and dividing by f so that

$$\frac{dl}{f} = a^2\frac{\cos^2\theta/2}{2K_\theta} + \frac{\sin^2\theta/2}{K_a}.$$

The Young's modulus E of the molecule can be calculated if its cross sectional area A can be estimated, as

$$E = \frac{f/A}{dl/l_0}.$$

It has been estimated that A is of the order of $18 \times 10^{-20}\,\mathrm{m^2}$ so that the theoretical maximum Young's modulus of a simple linear organic polymer is about 240 GPa. This is much greater than the moduli measured with most organic fibres because of their far from optimum molecular arrangements. However, ultrasonic measurements have revealed that, at a microscopic level, a simple polymer such as polyethylene possesses a modulus of 260 GPa. This observation has encouraged the optimization of the drawing processes of simple polymers with the aim of producing high modulus fibres and films, as is described below.

High modulus polyethylene fibres are produced using a dilute (<5%) sol-gel, in which the polymer is dissolved in a solvent and then spun. High modulus polyethylene fibres produced in Holland by DSM in collaboration with Toyobo in Japan under the name Dyneema and in the USA by Honeywell under the name Spectra have properties rivalling those of the aramid fibres with a lower specific gravity of 0.97. These fibres are limited in temperature to a maximum of 120 °C and suffer from creep.

An abrupt change in slope of the curve of strength as a function of temperature occurs around 5 °C due to a solid state phase change from orthorhombic to hexagonal. In the absence of an applied stress this phase change occurs around 140 °C, just below the melting point at 146 °C. However, the phase change can be induced at lower temperatures if a stress is applied. The level of stress which has to be applied to induce the phase change increases as the temperature is lowered until a critical point is reached around 5 °C, below which the fibre fails before the phase change can occur.

High modulus polyethylene fibres are finding markets as high strength mooring ropes for large ships and oil platforms as well as in flack jackets and similar antiballistic structures.

2.3   Glass fibres

Glass filaments have probably been formed since or before Roman times and more recently the production of fine filaments was demonstrated in the UK in the 19th century and used as a substitute for asbestos in Germany during the first World War. In 1931 two American firms, Owen Illinois Glass Co. and Corning Glass Works, developed a method of spinning glass filaments from the melt through spinnerets. The two firms combined in 1938 to form Owens Corning Fibreglas Corporation. Since that time extensive use of glass fibres has been made. Initially the glass fibres were destined for filters and textile uses, but the development of heat setting resins opened up the possibility of fibre reinforced composites and in the years following the Second World War the fibre took a dominant role in this type of material. Today, by far the greatest volume of composite materials is reinforced with glass fibres.

Fibres of glass are produced by extruding molten glass, at a temperature around 1200 °C, through holes in a spinneret with diameters of 1 or 2 mm and then drawing the filaments to produce fibres having diameters usually between 5 and 15 µm. The spinnerets usually contain several hundred holes

Table 2.4. Compositions (wt%) density and mechanical properties of various glasses used in fibre production. Type E is the most widely used glass in fibre production, types S and R are glasses with enhanced mechanical properties, type C resists corrosion in an acid environment and type D is used for its dielectric properties.

Glass type	E	S	R	C	D
SiO_2	54	65	60	65	74
Al_2O_3	15	25	25	4	
CaO	18		9	14	0.2
MgO	4	10	6	3	0.2
B_2O_3	8			5.5	23
F	0.3				
Fe_2O_3	0.3				
TiO_2					0.1
Na_2O				8	1.2
K_2O	0.4			0.5	1.3
Density	2.54	2.49	2.49	2.49	2.16
Strength (20 °C) (GPa)	3.5	4.65	4.65	2.8	2.45
Elastic modulus (20 °C) (GPa)	73.5	86.5	86.5	70	52.5
Failure strain (20 °C) (%)	4.5	5.3	5.3	4.0	4.5

so that a strand of glass fibres is produced. Several types of glass exist but all are based on silica (SiO_2) which is combined with other elements to create speciality glasses. The compositions and properties of the most common types of glass fibres are shown in table 2.4. The most widely used glass for fibre reinforced composites is called E-glass; glass fibres with superior mechanical properties are known as S and R-glass.

The strength of glass fibres depends on the size of flaws, most usually at the surface, and as the fibres would be easily damaged by abrasion, either with other fibres or by coming into contact with machinery in the manufacturing process, they are coated with a size. The purpose of this coating is both to protect the fibre and to hold the strands together. The size may be temporary, usually a starch–oil emulsion, to aid handling of the fibre, which is then removed and replaced with a finish to help fibre–matrix adhesion in the composite. There are two main types of coupling agent used in composites. These are organometallic compounds which are typically cobalt, nickel, lead, titanium or chromium, or, more commonly, organosilanes. They are applied as an aqueous solution. The organosilanes have the general structure X_3SiR, in which X is either an alkoxy or a halogen group and R is an organofunctional group which reacts with the glass provoking the hydrolysis of the alkoxy groups to form silanol groups. These groups then react with the silanol groups on the glass surface creating siloxane (Si–O–Si) bonds, as shown in figure 2.7.

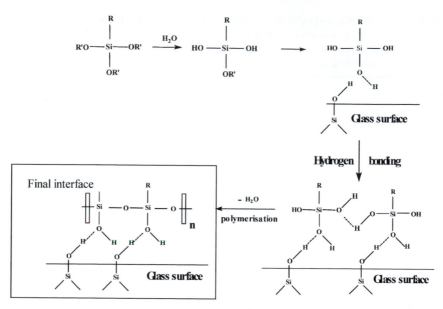

Figure 2.7. Interfacial bonding between the glass surface and an organosilane size.

The size may be of a type which has several additional functions which are to act as a coupling agent, lubricant and to eliminate electrostatic charges.

Continuous glass fibres may be woven, as are textile fibres, made into a non-woven mat in which the fibres are arranged in a random fashion, used in filament winding or chopped into short fibres. In this last case the fibres are chopped into lengths of up to 5 cm and lightly bonded together to form a mat, or chopped into shorter lengths of a few millimetres for inclusion in moulding resins. Examples of the use of different forms of glass fibres are given in chapter 4.

The structure is vitreous with no definite compounds being formed and no crystallization taking place. The structure of glass based on silica is shown schematically in figure 2.8. An open network results from the rapid cooling which takes place during fibre production with the glass cooling from about 1500 °C to 200 °C in 0.1–0.3 s. Despite this rapid rate of cooling there appear to be no appreciable residual stresses within the fibre and the structure is isotropic.

Glass is set to remain the most widely used reinforcement for general composites. Glass fibres are cheaper than most other relatively high modulus fibres and because of their flexibility do not require very specialized machines or techniques to handle them. Their elastic modulus is, however, low when compared with many other fibres; and the specific gravity of glass, which for E-type glass is 2.54, is relatively high. The poor specific

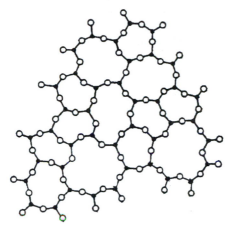

Figure 2.8. The structure of glass is based on a three-dimensional network of silica without any long range order.

value of the mechanical properties of glass fibres means that they are not ideal for structures requiring light weight as well as high properties, although considerations such as their relatively low cost compared with other high performance fibres and the ease of composite manufacture means that they are used in some such structures.

2.4 Chemical vapour deposition monofilaments

In the early 1960s, interest in the production of fibres which combined a high modulus with low density concentrated efforts on the lightest elements in the Periodic Table. This is because the ultimate strength and Young's modulus of a material are controlled by the outer electrons in the atoms and are not functions of atomic weight. The lightest element which possesses high mechanical properties is the fourth element in the Periodic Table, beryllium, which has a high Young's modulus of over 300 GPa and a density of 1.83. However, beryllium is highly toxic, especially as an oxide, and interest centred on the other light elements which were the fifth, sixth and fourteenth elements, boron, carbon and silicon.

The first fibre produced with a much increased Young's modulus compared with glass fibres was the boron fibre made by a chemical vapour deposition (CVD) technique onto a substrate which was a tungsten wire. Boron, the fifth element in the Periodic Table, is the lightest element with which it was found practical to make fibres. The first boron fibres were produced in the USA and then in France and in the USSR at the beginning of the 1960s. They had remarkable properties with a Young's modulus

Table 2.5. Properties of ceramic fibres produced by chemical vapour deposition.

Fibre type	SiC	—	B
Manufacturer	Specialty Materials Inc.	QinetiQ	Specialty Materials Inc.
Trade mark	SCS-6	Sigma	
Composition (wt%)	SiC on C core	SiC on W core	Boron on W core
Diameter (μm)	140	100	100–140
Density (g/cm^3)	2.7–3.3	3.4	2.57
Strength (GPa)	3.4–4.0	3.4–4.1	3–6
Failure strain (%)	0.8–1	0.8	1
Young's modulus (GPa)	427	400–410	380–400

exceeding 400 GPa, as shown in table 2.5, and quite extraordinary strength in compression. Silicon carbide fibres produced by a similar technique to that used to make boron fibres can be used as a reinforcement for metal matrix composites including titanium and intermetallic matrix composites. These fibres are based on materials which are readily available and the substrates can either be tungsten wire or a carbon filament, which is cheaper.

Chemical vapour deposition is a process in which one material is deposited onto a substrate to produce near-theoretical density and small grain sizes for the deposited material. CVD onto an appropriate fine filamentary substrate produces a thicker filament predominantly composed of materials which otherwise could not be made by drawing or other conventional processes. The fine and dense microstructure of the deposited material ensures maximum strength and Young's modulus.

The production of both boron and silicon carbide fibres by CVD takes place in a reactor which consists of a glass tube, which for commercial production is usually 1–2 m in length and most often vertical. Schematic representations of the arrangements for the production of both types of fibres are shown in figure 2.9. Several such reactors can be placed in series so as to allow staged deposition, but in practice this is avoided as the increase in the number of seals increases the risks of contaminants being introduced into the fibres. The wire substrate can run continuously through the reactor by the use of mercury seals at each end. These ensure that the reactive gases are contained in the reactor and also allow the substrate to be heated by electric current over the deposition length. The choice of the substrate is governed by the need to use a material which maintains its strength to around 1100 °C and is an electrical conductor. Heating the substrate, usually with a dc current, is a necessary part of the process which determines deposition rate, which must be constant or irregularities in the fibres' diameter occur with a subsequent fall in strength. Several potential substrate materials

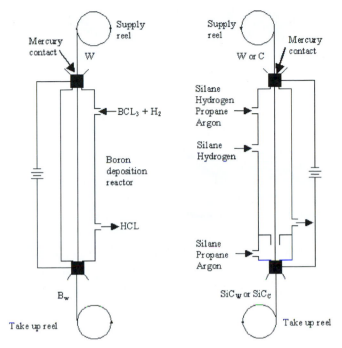

Figure 2.9. Schematic views of the reactors for producing boron and silicon carbide fibres by CVD.

were investigated including molybdenum and tantalum but tungsten was chosen as it was available in the form of fine wire (produced for electric light filaments) and showed greater compatibility with boron including having a similar coefficient of thermal expansion. Initial attempts to use carbon cores for the production of boron fibres were not successful due to differences in thermal expansion coefficients which led to the breakage of the carbon filament. During production, such breakages produce hot spots due to the locally raised electrical resistance in the substrate; in turn, these result in locally enhanced deposition rates and weak spots. This was overcome by the deposition of pyrolytic carbon onto the substrate which allowed the boron fibres to slide over the carbon fibres. Nevertheless, tungsten filaments are used for the commercial production of boron fibres, while both tungsten and carbon filaments are used as substrates for the production of SiC fibres.

2.4.1 Boron fibres

Boron fibres can be made by CVD onto a tungsten core of 12 μm diameter. Boron trichloride is mixed with hydrogen and boron is deposited according

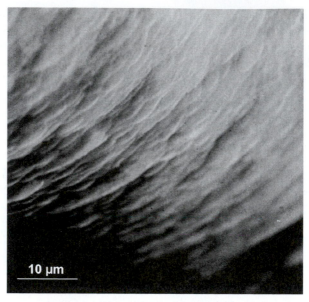

Figure 2.10. Boron is deposited on the tungsten wire substrate in the form of nodules which leads to the nodular surface shown.

to the reaction

$$2BCl_3(g) + 3H_2(g) \longrightarrow 2B(s) + 6HCl(g).$$

Passage through the reactor takes one or two minutes and results in a fibres with a diameter of 140 μm. During the deposition process the small boron atoms diffuse into the tungsten core to produce complete boridization and the production of WB_4 and W_2B_5. This leads to an increase in diameter to 16 μm which induces considerable residual stresses, putting the core into compression and the neighbouring boron mantle into tension. A final annealing process puts the fibres' surface into compression.

Boron fibres can be relatively easily used to reinforce light alloys but it is necessary to protect the fibres' surface during contact with the molten metal to avoid chemical reactions occurring with a subsequent fall in strength. For this reason boron fibres have been commercially produced with both silicon carbide (SiC) and boron carbide (B_4C) coatings.

Boron fibres coated with SiC were originally made with a diameter of 106 μm but were found to split longitudinally due to low transverse strength. This was avoided by increasing the diameter to 140 μm. However, splitting due to localized high stresses when fibres touch during composite manufacture was found to occur even with these latter fibres.

The tungsten wire substrates are produced by conventional wire drawing which results in their surface being striated. The boron is deposited preferentially onto the highest peaks of the surface of the wire and grows as conical nodules leading to a nodular surface to the fibres. Figure 2.9 shows the nodular appearance of boron fibres. A smoother substrate such as pitch-based carbon fibres leads to a smoother surface for the deposited fibres. The internal structure of the deposited boron is nanocrystalline with grain sizes of 2–3 nm consisting of α-rhombohedral and tetragonal polymorphs of boron.

In the absence of such major defects three types of failure initiation processes have been identified. The stress-concentration which is produced by the joining of two neighbouring nodules at the fibres' surface is the most frequent failure initiation mechanism and surface etching to reduce rugosity results in an increase in strength. Two types of defect are associated with the boron mantle–core interface. The growth of neighbouring nodules can cause radial defects which are due to the one nodule shadowing the deposition of the boron onto the tungsten wire. Elongated voids can be formed at the interface by preferential deposition on each side of striations on the tungsten.

Boron fibres are not only strong in tension but also lead to composites which are very strong in compression, probably at least in part due to the large diameter which inhibits buckling. The mechanical properties of boron fibres are given in table 2.5.

Boron fibres exhibit a linear axial stress–strain relationship at room temperature up to 650 °C. Strength is retained in short term tests to such high temperatures, but heating in air for periods of hours at temperatures above 300 °C leads to large losses in room temperature strength. This strength loss is attributed to oxidation of the nodular boron surface.

At higher temperatures boron fibres exhibit anelastic behaviour up to around 800 °C. A boron fibre formed into a loop and exposed briefly to an open flame which is then withdrawn retains its imposed curvature. On reheating the fibres returns to its original shape, demonstrating that the deformation contains no permanent plastic component. Boron fibres also creep at high temperatures.

The high cost of producing boron fibres has limited their use and many of the applications originally envisaged for them have employed small diameter carbon fibres which were produced several years afterwards.

2.4.2 Silicon carbide CVD fibres

Silicon carbide (SiC) shows high structural stability and strength retention, even at temperatures above 1000 °C, even though passive oxidation can take place. In bulk form the parabolic oxidation rates of the SiC means that it can be used, in air, up to 1500 °C. SiC is therefore an attractive material to have in fibre form for reinforcement in high temperature composite materials. Silicon carbide fibres made by CVD are produced on both tungsten and carbon cores in similar reactors to those used for the production of boron fibres, except that multiple injection points are used for the introduction of the reactant gases. Various carbon containing silanes have been used as reactants. In a typical process, with CH_3SiCl_3 as the reactant, SiC is deposited on the core as follows:

$$CH_3SiCl_3(g) \longrightarrow SiC(s) + 3HCl(g).$$

SiC fibres produced on a tungsten core of 12 μm diameter show a thin inter-facial layer between the SiC mantle and the tungsten core, caused by chemical interaction between the SiC and tungsten which forms reaction products of α-W_2C and W_5Si_3. Heating above 900 °C leads to growth of this reactive layer and degradation of the properties of the fibres. The carbon fibres cores, of 33 μm diameter, which are used for the CVD production of many SiC fibres are themselves coated with a 1–2 μm thick layer of pyrolytic carbon which covers surface defects on the core and produces a more uniform diameter. The fibres on a carbon core show a gradation of composition, which is richer in carbon near the interface and becomes stoichiometric SiC towards the fibres surface. The fibres on a tungsten core have a mantle of stoichiometric SiC. Both types of SiC fibre have smoother surfaces than do boron fibres, due the deposition of small columnar grains rather than conical nodules. The specific gravity of a 100 μm diameter SiC–W fibres is 3.35 whilst that of a 140 μm diameter SiC–C fibres is around 3.2.

The SiC fibres made on a carbon core and destined to reinforce light alloys are produced with a surface coating the composition of which is made to vary from being carbon-rich to silicon carbide at the outer surface. The fibres which are to be used to reinforce titanium have a protective layer which varies from being rich in carbon to being rich in silicon to a composition which is again rich in carbon at the surface as shown in figure 2.11. The outer sacrificial layer protects the fibres during contact with the molten and highly reactive

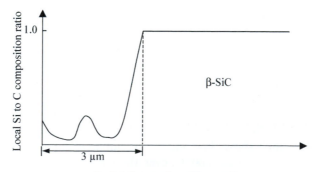

Radial distance from fibres surface

Figure 2.11. The SiC-6 fibres produced on a carbon core for the reinforcement of titanium possesses a protective layer which is 3 μm thick. This surface layer can be seen to be due to a first coating of carbon which is then made richer in silicon and then a second carbon-rich layer which plays a sacrificial role during composite manufacture.

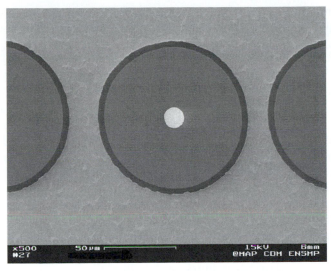

Figure 2.12. A cross section of a silicon carbide fibre, having a diameter of 103 μm, produced on a tungsten core and revealing a surface coating of carbon which has allowed it to be used to reinforce a titanium alloy (source: PhD thesis, S Hertz).

titanium during composite manufacture. This type of fibre, produced by Specialty Materials Inc. in the USA, is called SCS-6 as the coating increases the fibres diameter by 6 μm. The fibres shows no degradation after five hours at 900 °C when embedded in a Ti(6Al4V) matrix.

A cross section of a typical 100 μm diameter SiC fibre on a tungsten core is shown in figure 2.12. This fibre has a surface coating which is 3–4 μm thick and which varies from being carbon-rich to becoming a thin layer of less than 1 μm of TiB_2 at the surface.

Low failure stresses are due to surface flaws whereas higher strengths are controlled by defects at the core–mantle interface. The strengths of CVD SiC fibres has been shown to be anisotropic with the radial strength being significantly lower than longitudinal tensile strength. Strength decreases in air when the fibres are heated to above 800 °C for long periods, most probably due to oxidation of the carbon-rich outer layers. At temperatures above 900 °C interfacial reactions between the tungsten core and mantle cause degradation of properties as explained above. At still higher temperatures, grain growth in the SiC mantle may be the cause of further falls in strength.

2.5 Carbon fibres

Carbon is the sixth lightest element and the carbon–carbon covalent bond is the strongest in nature (4000 kJ/mole), however, the arrangements of the bonds and the distances between the carbon atoms can vary so giving many types of carbon, including graphite, diamond and amorphous forms. Graphite is a regular, highly anisotropic, two-dimensional array resulting from sp^2 hybridization of the electrons which means that there are two electrons occupying the outer energy shell of the carbon atom whereas sp^3 hybridization gives the three-dimensional structure of diamond. The latter degrades to the former type of hybridization under thermal ageing. The different energy levels of the valence electrons lead to varying mechanical properties in different directions. Graphite, therefore, is a well defined, highly anisotropic, state of carbon in which the carbon atoms occupy planes which are separated from one another by a distance of 0.34 nm. This should be borne in mind as many fibres which are described as graphite fibres contain no graphite; however, they have a close but less well ordered carbon phase and so it is more correct to speak of carbon fibres.

Many fibres can be converted into carbon fibres, the basic requirement being that the precursor fibre carbonizes rather than melts when heated. In 1901 Edison used bamboo fibres, which like all plants are made up of cellulose, and converted them into carbon filaments for his electric light bulbs. He therefore became the first person to patent a carbon fibre but they proved very brittle and after three years were replaced by tungsten. Interestingly, however, the carbon fibres developed in the USA in the 1950s and early 1960s used viscose fibres regenerated from cellulose. This proved a slow process as the carbon yield from cellulose is only 24% and mechanical properties were not high, although such fibres have low thermal conduction properties and are still used in carbon–carbon heat shields and brake pads.

Most carbon fibres, though, are made by a process developed first in the UK and then in Japan in the late 1960s based on the conversion of a modified form of polyacrylonitrile (PAN), which is often abbreviated to acrylic, textile fibre. Acrylic fibres are used in carpets and as artificial wool in some pullovers. PAN was developed in Germany in 1947 and has a carbon yield of 49%. The molecular structure of polyacrylonitrile is shown in figure 2.13.

The carbon backbone of the molecule should be noted. The properties of carbon fibres made from PAN precursors depend on the temperature of pyrolysis. The different stages in carbon fibre production are shown in figure 2.14.

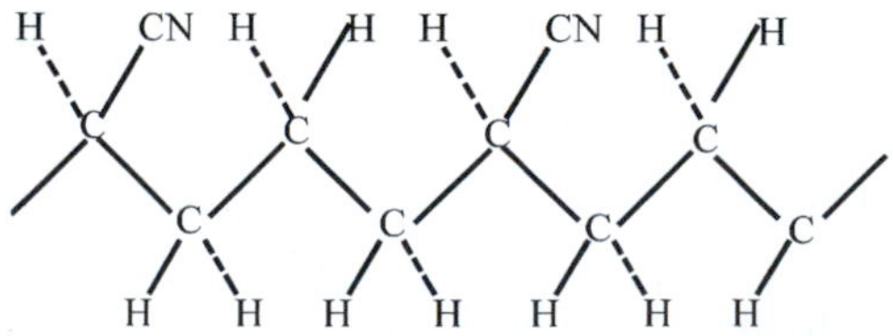

Figure 2.13. The molecule of polyacrylonitrile (PAN).

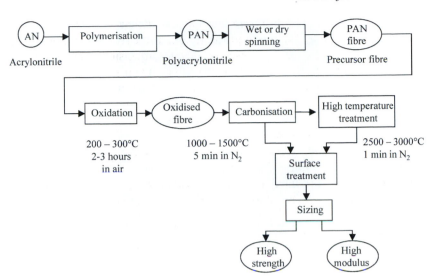

Figure 2.14. Carbon fibres made from PAN can be processed at different temperatures in order to change their properties.

The PAN fibres, held in tension, are first heated to about 250 °C in air which crosslinks the structure and makes it infusible. Figure 2.15 shows how the linear PAN molecule is converted to a ladder structure by the introduction of oxygen. This makes the molecule infusible and further heating under a nitrogen atmosphere leads to the loss of the oxygen and hydrogen so that at 1000 °C only carbon and about 7 wt% of nitrogen remain. Heating to above 1000 °C progressively eliminates the nitrogen and around 1500–1600 °C a wholly carbon fibre is achieved. At this temperature the strength of carbon fibres reaches a maximum. The fibre has to be held under tension as during pyrolysis half of its structure is lost so that shrinkage occurs. Without the tension the microstructural alignment given to the precursor fibre is completely lost whereas under load the alignment parallel

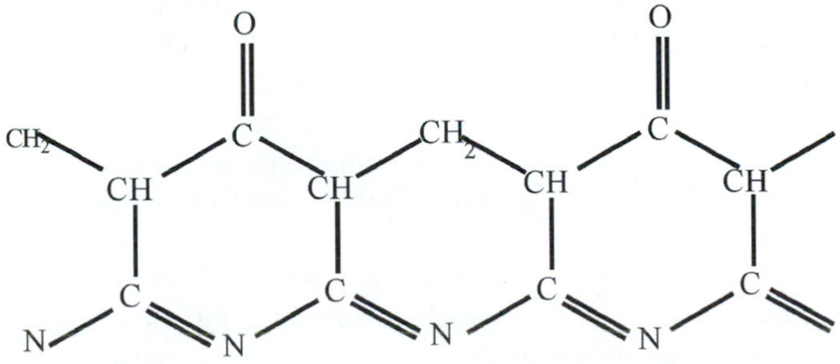

Figure 2.15. Conversion of the PAN molecule into an infusible ladder structure.

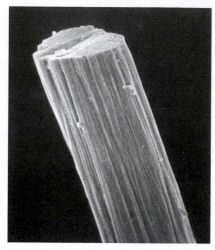

Figure 2.16. A high strength carbon fibre with a diameter of 7 μm made from a PAN precursor.

to the fibre axis is preserved and only a reduction in fibre diameter occurs. Although some stages of fibre production are time consuming the fibres are made in a continuous manner, with the lengthy processes occurring with the fibres winding back and forth in the heating furnaces. Fibre production rates are of the order of 100 m/min. Figure 2.16 shows a high strength carbon fibre, with a diameter of 7 μm, made from a PAN precursor. The original striations on the surface of the precursor fibre are maintained in the final carbon fibres despite the major changes which took place in the microstructure during manufacture.

Figure 2.17 shows a schematic view of the structure of a high strength carbon fibre made from PAN. The structure which results from the pyrolysis of PAN is highly anisotropic, with the basic structural units formed by the carbon atom groups aligned parallel to the fibre axis. They form imperfectly stacked layers called turbostatic layers with the interlayer spacing being greater than that of graphite. There is complete rotational disorder in the radial direction and the relatively poor stacking of the carbon atoms means that a graphite structure is never achieved. The structure contains pores which account for the density of the fibres being less than that of fully dense carbon. Table 2.6 shows some typical properties of carbon fibres. However, there exists a considerable range of properties for these fibres.

If the PAN-based fibres are heated to above 1500 °C the basic carbon structural units increase in size which results in a fall in strength whereas the Young's modulus increases continuously with increasing temperature up to around 3000 °C. The earliest and still the most widely used carbon

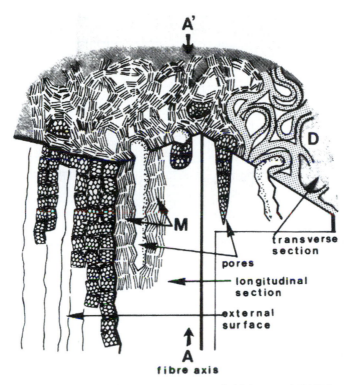

Figure 2.17. A schematic view of the microstructure of a high strength PAN-based carbon fibre (after Oberlin and Guigon).

Table 2.6. Typical characteristics of carbon fibres.

Fibre type	Diameter (μm)	Density (g/cm³)	Strength (GPa)	Failure strain (%)	Young's modulus (GPa)
Ex-PAN					
High strength (1st generation)	7	1.80	4.4	1.8	240
High strength (2nd generation)	5	1.82	7.1	2.4	294
High modulus (1st generation)	7	1.84	4.2	1.0	436
High modulus (2nd generation)	5	1.94	3.92	0.7	588
Ex-pitch					
Oil derived pitch	11	2.10	3.7	0.9	390
Oil derived pitch (high modulus)	11	2.16	3.5	0.5	780
Coal tar pitch	10	2.12	3.6	0.58	620
Coal tar pitch (high modulus)	10	2.16	3.9	0.48	830

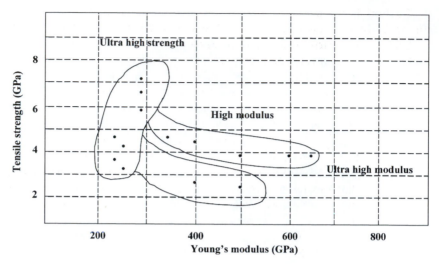

Figure 2.18. The elastic moduli of carbon fibres is increased by increasing the temperature of pyrolysis. The two bands of fibres have been produced by reducing the diameters from 7 to 5 μm which has produced fibres of greater strengths.

fibres made from PAN were made from similar precursors which gave fibres of 7 μm diameter. A range of high strength and high moduli fibres made from these precursors was developed by varying the pyrolysis temperatures. Figure 2.18 shows how by varying the pyrolysis temperature several types of carbon fibres were produced by the different producers by using the same precursor.

Improvements in precursor quality and drawing techniques have led to some increase in strength but as we can see there is a separate band placed at a higher strength level for both high strength and high modulus fibres. This later generation of higher strength fibres are produced by using finer precursors which give carbon fibres of 5 μm diameter and which allow higher strengths to be obtained, of up to 6.5 GPa. This increase in strength compared with earlier generation of fibres is accounted for by the probability of finding a defect in the fibre. As the volume of the fibres per unit length is reduced the chances of a significant defect existing in the length is considerably reduced and so the average strength of the fibres increases. The reduction from 7 to 5 μm may seem slight but the cross sectional areas of the fibres were reduced to a half of their former value and so the effect on fibre strength is very significant. Although carbon fibre tensile strength could theoretically be further enhanced, such improvements are not accompanied by increases in compressive strength. The higher Young's moduli are obtained by heating to around 3000 °C, which has considerable cost implications and so probably represents the upper limit of elastic moduli for carbon fibres made from PAN.

Carbon fibres made from pitch, which is the residue of petroleum refining or by coal tar distillation in the steel industry, were developed in the USA in the early 1970s and in Japan in the 1980s. The molecular weights of the polyaromatic compounds vary between 200 and 800. The high carbon yield of pitch, which approaches 90%, makes it an attractive and potentially cheap source for making carbon fibre precursors. Cost, however, is increased by the purification processes which is necessary for this naturally occurring material. In addition the production of high performance carbon fibres requires that the microstructure of the precursors be highly aligned. In order to achieve this it is imperative that the polycarbon layers in the precursor fibre be aligned with the fibre axis. Some carbon fibres are made from pitch with no attempt to align their structures and the result is fibres with a low Young's modulus of around 40 GPa. These fibres, which are made from anisotropic pitch, are used for their chemical inertness and low cost for the reinforcement of cement and concrete. The biggest producer of this type of isotropic pitch-based carbon fibre is Kureha in Japan. High performance carbon fibres made from pitch require that the pitch be converted into a mesophase or liquid crystal solution which is then spun, giving precursor fibres with aligned microstructures. This is possible as the pitch contains large planar polyaromatic molecules which show a tendency to align during the melt spinning process when passing through the spinneret due to local shear stresses. This is reminiscent of the production of aromatic organic high performance fibres. The softening point of pitches is not a single value as it depends on the exact composition of the pitch but is in the range 50–300 °C, which is much lower than other aromatic materials used to make high performance fibres. The mesophase pitch fibres are mechanically stretched during drawing to further improve alignment. The mesophase is formed by heating the pitch to a temperature of 400–450 °C and is formed of spherical liquid crystals of diameters of about 1 µm. From this point the fabrication route is the same as that for the PAN-based fibres although the final carbon fibre diameter is greater and usually around 10–11 µm. Unlike PAN-based fibres, there is no peak in strength at 1500 °C. Pitch-based precursors heated to around 2300 °C give fibres with Young's moduli as high as those obtained with PAN-based fibres at 2900 °C, and heating to these higher temperatures gives even greater stiffness of up to 4.5 times that of steel. It is therefore more economical to produce high modulus carbon fibres from pitch than it is from PAN. The properties are due to a less disordered microstructure of the pitch-based fibres which can become truly graphitized as the average stacking distances of the carbon atoms of such fibres made at temperatures above 1800 °C is less than 0.345 nm. The larger basic structural units in pitch-based carbon fibres when compared with PAN-based fibres coupled with a greater inherent anisotropy, however, leads to lower compressive strengths and increased difficulties and costs in producing high strength carbon fibres. The internal morphology of pitch-based carbon fibres can be varied so that various producers have made fibres with a random or radial structure, which can lead to radial cracks forming during the pyrolysis process, or a circumferential arrangement of the carbon layers, which is usually preferred.

Carbon fibres made by whatever route have surfaces in which the carbon atom cycles tend to be arranged parallel to the surface, which means that there are few available pendant atomic bonds with which a matrix resin can easily bond. The higher the Young's modulus of the carbon fibre the lower its tendency to interact with the matrix. For this reason carbon fibres are surface treated in order to increase interlaminar shear strength of the composites reinforced by the fibres. Surface treatment results in the oxidation of the fibre surface so as to create carboxylic COOH, hydroxylic OH and other oxygen-containing functional groups. There are several ways that the carbon fibre surface can be treated but the industrial process is by anodic oxidation.

In the absence of oxygen, carbon fibres are the ideal reinforcement and can be used to above 3000 °C. However, the fibres suffer from oxidation from around 400 °C so that they cannot be used for long term applications above this temperature.

2.6   Alumina-based fibres

Fibres based on alumina (Al_2O_3) were originally produced for refractory insulation such as furnace wall linings. They can be obtained from aluminium salt precursors which are transformed into aluminium hydroxides. Heating the precursor fibres induces the sequential development of transition phases of alumina, which if heated to a high enough temperature all convert to the most stable form which is α-alumina. However, this transformation is followed by a rapid growth of porous α-alumina grains giving rise to weak fibres.

If alumina is combined with silica (SiO_2) the transformation to α-alumina can be retarded and controlled. For a number of commercially produced fibres alumina is combined with silica and the fibres are produced from polymer precursors or sol-gels and then sintered. As silica softens at around 1000 °C the fibres are not suitable for higher temperature applications. Alumina-based fibres are inherently resistant to oxidation and have been successful in reinforcing light metal alloys. When heated to around 1200 °C, alumina combined with silica is partially converted to mullite which can have a range of compositions from $2Al_2O_3 \cdot SiO_2$ to $3Al_2O_3 \cdot 2SiO_2$. The interatomic bonds governing creep in alumina which are ionic and covalent lead to creep at temperatures above 1000 °C. Mullite itself is interesting as it has a complex crystallographic structure which means that it resists creep deformation. The development of fibres combining both alumina and another phase, such as mullite or zirconia, which can hinder creep processes is encouraging interest in the possibility of oxide fibres being used to reinforce ceramics.

There exists a wide range of alumina-based fibres which have different compositions, microstructures and properties.

2.6.1 Alumina silica fibres

The microstructures of these fibres depend on the highest temperature the fibres have seen during the ceramization. Very small grains of η-, γ- or δ-alumina in an amorphous silica continuum are obtained with temperatures below 1000–1100 °C.

The Young's moduli of these fibres are lower compared with that of pure alumina fibres, and such fibres are produced at a lower cost. This, added to easier handling due to their lower stiffness, makes them attractive for thermal insulation applications, in the absence of significant load, in the form of consolidated felts or bricks up to at least 1500 °C. Such fibres are also used to reinforce aluminium alloys in the temperature range 300–350 °C. Continuous fibres of this type can be woven due to their lower Young's moduli.

The Saffil fibre was developed by ICI in the UK and contains 4% of silica. It is produced by the blow extrusion of partially hydrolysed solutions of some aluminium salts with a small amount of silica, in which the liquid is extruded through apertures into a high velocity gas stream. The fibre contains mainly small δ-alumina grains of around 50 nm but also some α-alumina grains of 100 nm. The widest use of the Saffil-type fibre in composites is in the form of a mat which can be shaped to the form desired and then infiltrated with molten metal, usually aluminium alloy. It is the most successful fibre reinforcement for metal matrix composite.

The Altex fibre is a continuous fibre from Sumitomo Chemicals in Japan. Its properties are shown in table 2.7. The fibre is obtained by the chemical conversion of a polymeric precursor fibre, made from a poly-aluminoxane dissolved in an organic solvent to give a viscous product with an alkyl silicate added to provide silica. The precursor is then heated in air to 760 °C, a treatment which carbonizes the organic groups to give a ceramic fibre composed of 85% alumina and 15% amorphous silica. The fibre is then heated to 970 °C and its microstructure consists of small γ-alumina grains of a few tens of nanometres intimately dispersed in an amorphous silica phase. The presence of silica in the Altex fibres does not reduce their strength at lower temperatures compared with pure alumina fibres, but a lower activation energy is required for the creep of the fibre. At 1200 °C the continuum of silica allows Newtonian creep and the creep rates are higher than those of pure α-alumina fibres.

The 3M corporation in the USA produces a range of ceramic fibres under the general name of Nextel. The Nextel 312 and 440 series of fibres are produced by a sol-gel process. They are composed of 3 moles of

Table 2.7. Properties and compositions of alumina-based fibres.

Fibre type	Manufacturer	Trade mark	Composition (wt%)	Diameter (μm)	Density (g/cm³)	Strength (GPa)	Strain to failure (%)	Young's modulus (GPa)
α-Al₂O₃-based fibres	Du Pont de Nemours	FP	99.9% Al_2O_3	20	3.92	1.2	0.29	414
	Du Pont de Nemours	PRD 166	80% Al_2O_3 20% SiO_2	20	4.2	1.46	0.4	366
	Mitsui Mining	Almax	99.9% Al_2O_3	10	3.6	1.02	0.3	344
	3M	Nextel 610	99% Al_2O_3 0.2–0.3% SiO_2 0.4–0.7% Fe_2O_3	10–12	3.75	1.9	0.5	370
Alumina silica-based fibres	ICI	Saffil	95% Al_2O_3 5% SiO_2	1–5	3.2	2	0.67	300
	Sumitomo Chemicals	Altex	85% Al_2O_3 15% SiO_2	15	3.2	1.8	0.8	210
	3M	Nextel 312	62% Al_2O_3 24% SiO_2 14% B_2O_3	10–12 or 8–9	2.7	1.7	1.12	152
	3M	Nextel 440	70% Al_2O_3 28% SiO_2 2% B_2O_3	10–12	3.05	2.1	1.11	190
	3M	Nextel 720	85% Al_2O_3 15% SiO_2	12	3.4	2.1	0.81	260

alumina for 2 moles of silica with various amounts of boria to restrict crystal growth. The Nextel 312 fibre, which first appeared in 1974, is composed of 62 wt% Al_2O_3, 24 wt% SiO_2 and 14 wt% B_2O_3 and is mainly amorphous. It has the lowest production cost and is widely used, but has a mediocre thermal stability as boria compounds volatilize from 1000 °C, inducing some severe shrinkage above 1200 °C. To improve the high temperature stability in the Nextel 440 fibres the amount of boria has been reduced. These latter fibres have the following compositions: 70 wt% Al_2O_3, 28 wt% SiO_2 and 2 wt% B_2O_3 in weight and are formed of small γ-alumina in amorphous silica.

The Nextel 720, also from 3M, contains the same alumina to silica ratio as in the Altex fibre, that is around 85 wt% Al_2O_3 and 15 wt% SiO_2. The sol-gel route, higher processing temperatures together with the addition of seeds for α-alumina formation, has induced the growth of alumina-rich mullite and α-alumina. Unlike other alumina–silica fibres the Nextel 720 fibre is composed of mosaic grains of $\sim 0.5\,\mu m$ and elongated α-alumina grains. Figure 2.19 shows the mosaic microstructure of the Nextel 720 fibre.

Post heat treatment leads to an enrichment of α-alumina in the fibre as mullite rejects alumina to evolve towards a 3:2 equilibrium composition. Grain growth occurs from 1300 °C. This fibre shows a dramatic reduction, of two orders of magnitude, in creep strain rates when compared with pure alumina fibres. This improved creep resistance is attributed to the particular microstructure of the fibre which inhibits the flow of the material. The fibre exhibits the lowest creep rates of all commercial small diameter oxide fibres but have shown that the fibre surface is particularly reactive above 1100 °C, which leads to the development of grains at the surface. Slow crack growth, initiated by these defects, considerably reduces the time to failure during creep as well as the tensile strength of the fibre.

2.6.2 α-Alumina fibres

α-Alumina is the most stable and crystalline form of alumina to which all other phases are converted upon heating above around 1000 °C. As we have seen above, fibres based on alumina can contain silica as its presence allows the rapid growth of α-alumina grains to be controlled. However, the presence of silica reduces the Young's modulus of the fibre and reduces its creep strength. Pure α-alumina fibres with fine and dense microstructures should show better creep resistance and higher stiffness but the structure is difficult to obtain. The control of grain growth and porosity in the production of α-alumina fibres is obtained by using a slurry consisting of α-alumina particles, of strictly controlled granulometry, in an aqueous solution of aluminium salts. These particles act as seeds for the α-alumina growth. The rheology of the slurry is controlled through its water content. The particles can also be obtained by the spinning of a hydrated alumina

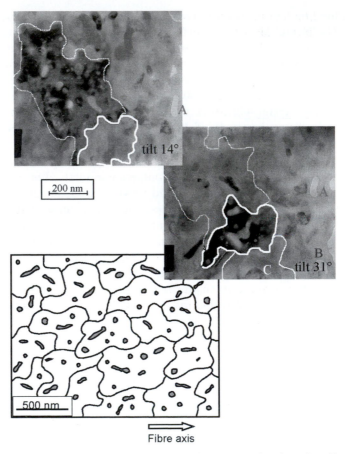

Figure 2.19. The Nextel 720 fibre is composed of a mosaic of grains of mullite, each of which made up of slightly differently oriented grains. α-Alumina grains, some of which are elongated, are embedded in the mullite grains (source: PhD thesis, F Deléglise).

sol containing additives to help α-alumina formation. The precursor filament, which is then produced by dry spinning, is pyrolysed to give an α-alumina fibre.

Fibre FP, manufactured by du Pont in 1979, was the first wholly α-alumina fibre to be produced. It was a continuous fibre composed of 99.9% α-alumina and had a density of 3.92 g/cm^3 and a polycrystalline microstructure with a grain size of 0.5 μm, a high Young's modulus 410 GPa but a strain to failure of only 0.4%. This brittleness combined with a relatively large diameter of around 20 μm made it unsuitable for weaving and, although showing initial success as a reinforcement for light alloys, production did not progress beyond the pilot plant stage and commercial production ceased. Nevertheless Fibre FP represents an example of an almost pure alumina in filament form and as such allows the fundamental mechanisms in this class of fibre to be investigated.

Up to 1000 °C, Fibre FP showed linear macroscopic elastic behaviour in tension. Above 1000 °C the fibre was seen to deform plasticity in tension and the mechanical characteristics decreased rapidly. At 1300 °C, strains in traction increased and could sometimes reach 15%. Creep was observed from 1000 °C. The strain rates from 1000 to 1300 °C were seen to be a function of the square of the applied stress and the activation energy were found to be in the range 550–590 kJ mol^{-1}. The creep mechanism of Fibre FP has been described as being based on grain boundary sliding achieved by an intergranular movement of dislocations and accommodated by several interfacial controlled diffusion mechanisms, involving boundary migration and grain growth. No modification of the granulometry was observed after heat treatment without load at 1300 °C for 24 hours but large deformations resulting from tensile and creep tests conducted at 1300 °C were observed to induce grain growth. There was no overall preferential direction for the grain growth but the development of cavities at some triple points was noticed due to the pile up of intergranular dislocations at triple points. The external surfaces of the FP fibres broken in creep at 1300 °C after large deformations showed numerous transverse microcracks, which were not observed for smaller strains. This fibre was seen to be chemically stable at high temperature in air, but its isotropic fine grained microstructure led to easy grain sliding and creep, excluding any application as a reinforcement for ceramic structures.

Other manufacturers have modified the production technique to reduce the diameter of the α-alumina fibres that they have produced. This reduction of diameter has an immediate advantage of increasing the flexibility and hence the weaveability of the fibres.

The Almax fibre is produced commercially by Mitsui Mining of Japan. The fibre has a lower density of 3.60 g/cm^3 compared with Fibre FP. It consists of one population of grains of around 0.5 μm. The fibre exhibits a large amount of intragranular porosity which accounts for its lower density and lower Young's modulus compared with a dense α-alumina fibre. The porosity indicates rapid grain growth of α-alumina grains during the fibre fabrication process without elimination of porosity and internal stresses. As a consequence, grain growth at 1300 °C is activated without an applied load. Creep occurs from 1000 °C and Almax fibres show a lower resistance to creep than dense α-alumina fibres.

A continuous α-alumina fibre, with a diameter of 10 μm, was introduced by the 3M Corporation in the early 1990s with the trade-name of Nextel 610 fibre. The fine grain fracture morphology of the fibre can be seen in figure 2.20. The manufacturer has indicated that the fibre is composed of around 99% α-alumina but their more detailed chemical analysis gives 1.15% total impurities including 0.67% Fe_2O_3 used as a nucleating agent and

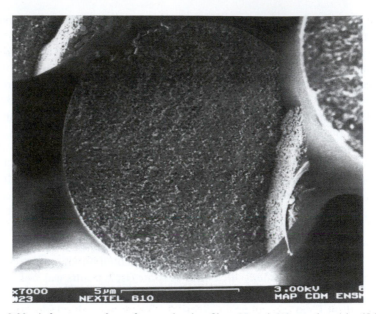

Figure 2.20. A fracture surface of an α-alumina fibre, Nextel 610, produced by 3M. The small diameter of 10 μm coupled with fine grains of 0.1 μm in size produce a strong fibre. The fracture has been initiated at a defect at the surface on the left of the photograph (source: PhD thesis, F Deléglise).

0.35% SiO_2 as grain growth inhibitor. The fibre is polycrystalline with a grain size of 0.1 μm, five times smaller than in Fibre FP. As shown in table 2.7, the strength announced by 3M is 2.4 GPa, which is twice the tensile strength measured on Fibre FP. However, the smaller grains and possibly the chemistry of its grain boundaries mean that when creep occurs from 900 °C the strain rates are 2–6 times larger than those of Fibre FP. A stress exponent of approximately 3 is found between 1000 °C and 1200 °C with an apparent activation energy of 660 kJ/mol.

2.6.3   Alumina and zirconia fibres

Both du Pont and 3M have produced fibres consisting of α-alumina and zirconia (ZrO_2) although neither were produced on a commercial scale. The first of these two fibres was the PRD-166 fibre from du Pont in which 20 wt% of partially stabilized zirconia was added to increase the elongation to failure of the fibre. The intention was to produce a fibre which, compared with Fibre FP, was easier to weave. The PRD-166 fibre had a diameter of 18 μm and a density of 4.2 g/cm^3. The dispersion of zirconia intergranular particles of 0.15 μm limits grain growth of the alumina grains, which had a mean diameter of 0.3 μm instead of 0.5 μm for Fibre FP for a similar initial alumina powder granulometry. These particles underwent a martensitic reaction in the vicinity of the crack tips resulting in the partial closure of cracks and an increase of the fibre strength. The resulting stiffness of the reinforced alumina was lower than that of Fibre FP, $E = 344\,GPa$, due to the lower Young's modulus of zirconia compared with that of alumina. The increase in strain to failure was not sufficient to allow weaving with the PRD-166 fibre and production of the PRD-166 fibre did not progress beyond the pilot stage.

3M produced a fibre called Nextel 650 which had a diameter of 11 μm and was composed of 90% α-alumina in the form of grains of 0.1 μm and stabilized zirconia grains of 5–10 nm. The role of the zirconia in this fibre was to enhance high temperature properties. For both fibres the introduction of dispersed particles of zirconia at grain boundaries limited the mobility of intergranular dislocations at high temperatures and the segregation of Y^{3+} and Zr^{4+} ions at grain boundaries lowered Al^{3+} intergranular diffusion rates. Elastic properties were maintained up to 1100 °C, that is 100 °C above pure alumina fibres, and the onset of creep delayed. Progressively increasing the temperature up to 1300 °C reduced the advantage of the zirconia-containing fibres over the single phase α-alumina fibres.

Existing polycrystalline fibres based on alumina have been seen to be resistant to oxidation but to lose their mechanical properties above 1200 °C due to microstructural modifications and enhanced facility of grain motion. Combining a second phase, such as mullite or zirconia with the α-alumina, is seen to improve high temperature creep behaviour although the former, Nextel 720 fibres, show sensitivity to alkaline contamination which leads to the growth of alumina grains and falls in strength.

2.7   SiC-based fibres

Better thermomechanical stability than that obtained with the above alumina-based systems can be achieved with materials having stronger bonds than the ionic/covalent bonds in alumina. For this reason SiC, which possesses covalent bonds, was the first candidate for the production of reinforcements for high temperature structural materials.

Fibres based on silicon carbide have been developed through the pyrolysis of organo-silicon precursor filaments in an analogous fashion to the technique of carbon fibres produced by carbonizing polyacrylonitrile precursors. These silicon carbide-based fibres first allowed ceramic matrix composites to be developed and are the most widely used reinforcement for this type of composite. For these applications the fibres are often made with a carbon-rich surface which is useful in controlling the interface in composites. An alternative surface coating could be boron nitride. Recently, complex surface coatings of SiC and pyrolytic carbon have been prepared to control interfacial debonding.

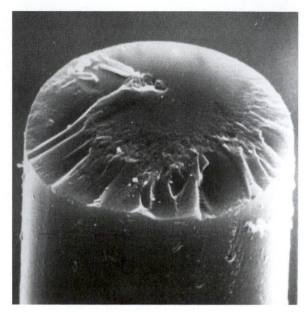

Figure 2.21. The Nicalon 200 series fibres have a glassy appearance when viewed in a scanning electron microscope. The diameter of the fibre is 15 μm (source: PhD thesis, G Simon).

The work of Yajima and his colleagues in Japan was first published in the mid-1970s and described the production of fine SiC ceramic fibres. Such fibres are produced in Japan by Nippon Carbon under the name of Nicalon. The manufacture of Nicalon fibres involves the production of polycarbosilane precursor fibres which consist of cycles of six atoms arranged in a similar manner to the diamond structure of β-SiC. The molecular weight of this polycarbosilane is low, around 1500, which makes drawing of the fibre difficult. In addition methyl groups ($-CH_3$) in the polymer are not included in the Si$-$C$-$Si chain so that during pyrolysis the hydrogen is driven off, leaving a residue of free carbon. The earliest (200 series) Nicalon fibres involved heating the precursor fibres in air to about 300 °C to produce cross-linking of the structure. This oxidation makes the fibre infusible but has the drawback of introducing oxygen into the structure which remains after pyrolysis. The ceramic fibre is obtained by a slow increase in temperature in an inert atmosphere up to 1200 °C and has a glassy appearance when observed in SEM, as can be seen in figure 2.21. The fibre contains a majority of β-SiC, of around 2 nm, but also significant amounts of free carbon of less than 1 nm and excess silicon combined with oxygen and carbon as an intergranular phase. The strengths and Young's moduli of Nicalon fibres tested in air or an inert atmosphere show little change up to 1000 °C. Above this temperature, both these properties show a slight decrease up to 1400 °C. Between 1400 and 1500 °C the intergranular phase begins to decompose, carbon and silicon monoxides are evacuated and a rapid grain growth of the silicon carbide grains is observed. The density of the fibre decreases rapidly and the tensile properties exhibit a dramatic fall. When a load is applied to the fibres, it is found that a creep threshold stress exists above which creep occurs. The fibre is seen to creep above 1000 °C and no stress-enhanced grain growth is observed after deformation. Creep is due to the presence of the oxygen-rich intergranular phase. The properties and composition of these fibres are shown in table 2.8.

A second Japanese company, Ube Industries, made SiC-based fibres using similar precursors to that used for the Nicalon fibres, however, with the addition of titanium to give polytitanocarbosilane. This route is no longer generally used by Ube Industries which found that the introduction of titanium by the use of titanium alkoxides meant that there was no way of easily reducing the oxygen in the fibres and this resulted, as with the Nicalon 200 fibre series, with creep from 1000 °C.

A later generation of Nicalon fibres has been produced by crosslinking the precursors by electron irradiation so avoiding the introduction, at this stage, of oxygen. These fibres are known as Hi-Nicalon, which contains 0.5 wt% oxygen. The decrease in oxygen content in the Hi-Nicalon compared with the NL-200 fibres has resulted in an increase in the size of the SiC grains to around 10 nm and a better organization of the free carbon. A significant part of the SiC is not perfectly crystallized and surrounds the ovoid β-SiC grains. Significant improvements in the creep resistance are found for the Hi-Nicalon fibre compared with the Nicalon-200 fibre which can further be enhanced by a heat treatment so as to increase its crystallinity.

Ube Industries produces a fibre using polyzirconocarbosilane (PZT) precursors which has allowed the titanium to be replaced by zirconium and the oxygen content to be reduced. The resulting fibres, known as Tyranno ZM and which contains 8 wt% oxygen, showed increased high temperature creep and chemical stability and resistance to corrosive environments. These fibres have found an application in diesel filters for buses and trucks due to their high temperature properties and stability.

Efforts to reduce the oxygen content by processing in inert atmospheres and crosslinking by radiation have produced fibres with very low oxygen contents. These fibres are not, however, stoichiometric as they contain significant amount of excess free carbon affecting oxidative stability and creep resistance.

Dow Corning produced stoichiometric SiC fibres from a PCS containing a small amount of titanium, as in the early Ube Industry fibres. The precursor fibres are cured by oxidation and doped with boron. In this way degradation of the oxicarbide phase at high temperature is controlled and catastrophic grain growth and associated porosity, as occurred with the previous oxygen-rich fibres, is avoided. The precursor fibre can then be sintered at high temperature (1600 °C) so that the excess carbon and oxygen are lost as volatile species to

Table 2.8. Properties and compositions of silicon carbide-based fibres.

Fibre type	Manufacturer	Trade mark	Composition (wt%)	Diameter (μm)	Density (g/cm^3)	Strength (GPa)	Strain to failure (%)	Young's modulus (GPa)
Si-C based	Nippon Carbide	Nicalon NL200	56.6% Si 31.7% C 11.7% O	14	2.55	3.0	1.05	220
	Nippon Carbide	Hi-Nicalon	62.4% Si 37.1% C 0.5% O	14	2.74	2.8	1.0	270
	Nippon Carbide	Hi-Nicalon Type-S	69% Si 31% C 0.2% O	12	3.05	2.5	0.63	400
	Ube Chemical	Tyranno Lox-M	54.0% Si 31.6% C 12.4% O 2.0% Ti	8.5	2.37	2.5	1.4	180
	Ube Chemical	Tyranno SA	68% Si 32% C 0.6% Al	11	3.02	2.8	1.45	375
	COI Ceramics	Sylramic	67% Si 29% C 2.3% B 2.1% Ti 0.8% O 0.4% N	10	3.05	3.2	0.8	400

Figure 2.22. The fracture surface of a near stoichiometric Hi-Nicalon Type-S fibre broken after creep at 1400 °C. A thin silica layer can be seen at the surface which has been produced by oxidation (source: PhD thesis, N Hochet).

yield a polycrystalline, near-stoichiometric, SiC fibre called Sylramic fibre. SiC grain sizes in the Sylramic fibre are 0.2 µm with smaller grains of TiB_2 also present. Characterization of the fibre in creep shows a considerably reduced deformation rate when compared with the Hi-Nicalon fibre. Sylramic fibres are now made by ATKCOI Ceramics.

Nippon Carbon has obtained a near-stoichiometric fibre, the Hi-Nicalon Type-S, from a PCS cured by electron irradiation and pyrolysed in a hydrogen-rich atmosphere. As a result the excess carbon is reduced from C/Si = 1.39 for the Hi-Nicalon to 1.05 for the Hi-Nicalon S. The SiC grain size is about 50 nm. Figure 2.22 shows a Hi-Nicalon Type-S fibre after a creeps failure at 1400 °C. This fibre maintains its strength and Young's modulus to 1400 °C and has the lowest creep rates of all available fine SiC fibres.

Ube Industries has developed a near stoichiometric fibre made from a polyaluminocarbosilane called Tyranno SA. The precursor fibre is cured by oxidation, pyrolysed up to 1700 °C to allow the outgassing of CO and sintered at a temperature above 1800 °C. The fibre SiC grain sizes of about 300 nm and shows a rougher fracture surface than the earlier fibres. A thin covering of SiC can be seen to have formed due to oxidation. Aluminium has been added as a sintering aid and, from the results of the manufacturer, gives a better corrosion resistance than TiB_2 which is found in the Sylramic fibre. The creep resistance has been considerably enhanced compared with other non-stoichiometric fibres.

The three near-stoichiometric SiC fibres show very good strength retention and creep properties when compared with earlier generations of the fine SiC-based fibres but they do not have the same characteristics because of differences in grain size and especially the presence of sintering aids. These are boron for the Sylramic fibre and aluminium for the SA fibre. The sintering aids lead to a slight fall in strength and faster creep rates from 1300 °C which are not seen in the Hi-Nicalon Type-S fibre which contains no sintering aids. All of the SiC-based fibres suffer from surface oxidation when heated in air above around 1200 °C.

2.8   Continuous monocrystalline filaments

Continuous monocrystalline filaments have been developed by the Saphikon company in the USA. These filaments are grown from molten alumina and as a consequence are produced at a slow rate and high cost and with large diameters usually in excess of 100 μm. The near stoichiometric composition of these fibres with the absence of grain boundaries ensures that they should be able to better withstand higher temperatures. Careful orientation of the seed crystal enables the crystalline orientation to be controlled so that creep does not occur up to 1600 °C. Published data on the strength of Saphikon fibres as a function of temperature reveals that strength variation is not a single function of temperature. The observed fall in strength around 300 °C, which is then followed by an increase in strength around 500 °C, could be due to stress corrosion followed by crack blunting. These fibres, are not without defects and characteristic bubbles can be seen in the fibres most probably due to convection during fibre growth at the meniscus point between the solid and the melt. The fracture morphology of the α-alumina Saphikon fibre shows clearly the cleavage planes of the crystalline structure.

The same manufacturing processes have been employed to produce a eutectic fibre consisting of interpenetrating phases of α-alumina and YAG (yttrium aluminium garnet, $Y_3Al_5O_{12}$). The structure depends on the conditions of manufacture, in particular the drawing speed, but can be lamellar and oriented parallel to the fibre axis. This fibre does not show the same fall in strength seen with the single phase alumina fibre. However, such fibres are seen to relax from 1100 °C but do not have as strong a dependence on temperature as the polycrystalline oxide fibres.

The growth process is extremely slow, typically 100 mm/h, but can easily be adapted to a wide range of ceramic systems for growing single crystal and directionally solidified eutectic filaments. Although the crystal structure is continuous and Saphikon produced lengths of up to 3000 m of fibre, usually the lengths of filaments produced in laboratories are short, being typically tens of centimetres.

An alternative approach, which has been pioneered in Russia, to making single crystal fibres and one which is potentially much cheaper is to infiltrate the molten ceramic in the channels formed by sandwiching molybdenum wires between molybdenum sheets. Seeds, which are used to control crystal growth and orientation are placed on the top surface of the molybdenum die. The crucible is filled with the raw material which is melted. As the raw material becomes molten the molybdenum die is lowered into it and the molten ceramic is drawn up the channels by capillary forces. The die is then withdrawn and the ceramic solidifies from the top down. Systems which have been made by this technique include Al_2O_3, $Al_2O_3-Al_5Y_3O_{12}$, $Al_2O_3-ZrO_2(Y_2O_3)$ and $Al_2O_3-AlGdO_3$. Mullite filaments have also been produced by this process. Many fibres can be produced simultaneously by this technique. Drawing rates are around 10 mm/min.

The molybdenum die material is finally removed by etching.

2.9   Whiskers

Whiskers are monocrystals in the form of filaments. The potential of SiC, alumina and other ceramic whiskers as reinforcements has been discussed for many years as their small diameters, usually between 0.5 and 1.5 μm, means that they contain very few defects and must possess extremely high strengths, perhaps up to the theoretical strength for matter, which is approximately one tenth of its Young's modulus. In addition their aspect ratios of length to diameter can be considerable as they can be produced with lengths between 20 μm and, it is claimed, several centimetres. A high aspect ratio is just what is required to achieve reinforcement. Considerable difficulties have to be overcome if whiskers are to be used as reinforcements, however. They are extremely small so that a plastic bag containing whiskers seems to contain dust. This means that alignment of the whiskers in a matrix is very difficult. There are potential uses for whiskers combined with more conventional fibres so as to provide some reinforcement of the matrix in the transverse direction. Their fineness is also another handicap in their exploitation as 1 μm is just the size to block up the alveolar structure of the lungs. For this reason above all whiskers remain an intriguing possibility as reinforcements but one which is little exploited.

2.10   Statistical analysis of fibre properties

Fibres are very fine filaments of matter and in any section of a composite structure there will be found thousands and most probably millions of fibres. Such large populations lend themselves to statistical analysis and the inherent scatter in fibre properties ensures that a statistical approach to describing their properties is necessary.

It is not possible to give a single value for the strength of a fibre because they show considerable scatter and the mean strength decreases with fibre

Hi-Nicalon SiC fibre

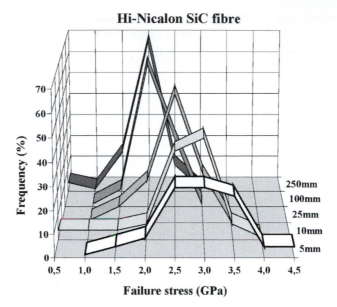

Figure 2.23. Strength distribution of a Hi-Nicalon fibre tested at five different gauge lengths.

length. Figure 2.23 shows the scatter of strength of Hi-Nicalon SiC-based fibres produced by Nippon Carbon. Both the scatter and length dependence is due to the distribution of defects in the fibres and the longer the fibre the greater chance there is of it containing a sizeable defect and so being considerably weakened. Failure of the fibre as a function of applied load is therefore controlled by the random distribution of defects and requires a statistical treatment for its analysis.

2.10.1   Weibull analysis

The effect of a random distribution of a single type of defect on the strength of a solid has been described by Weibull, who likened the failure of the solid to the breaking of a chain in which the weakest link controls failure, as seen in figure 2.24. This correctly reflects the behaviour of a brittle solid for which a crack, developed from the most critical defect, immediately induces the failure of the material. The assumptions of this model are: (1) the stress field in each link is considered as being uniform, (2) the failure stress of one link is independent of that of the other links and (3) all links have the same failure probability for a given applied stress. The first assumption implies that local modifications of the stress field around the defect are negligible so that each link can be considered as being subjected to a uniform stress field. The second assumption means that there is no

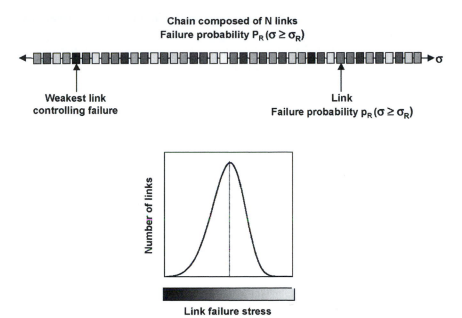

Figure 2.24. Schematic representation of a chain composed of N links presenting a distribution of failure stresses.

interaction between the defects. The third assumption implies that the defect distribution inside the material must be homogeneous and isotropic.

As in a chain under tensile loading, we will consider here that the stress field is uniform and uni-axial. The failure probability of each link under an applied stress σ, $P_R(\sigma)$, is the probability that this applied stress is greater than or equal to the failure stress of the link σ_R. The probability of survival of each link under an applied stress σ, $P_S(\sigma)$, is the probability that this applied stress is lower than the failure stress of the link, and is equal to $1 - P_R(\sigma)$. The survival of the chain for an applied stress σ requires the survival of each of its N links so that the survival probability of the chain is $P_S(\sigma) = P_S(\sigma)^N$. If $P_R(\sigma)$ is the probability of failure of the chain, we can write

$$1 - P_R(\sigma) = (1 - P_R(\sigma))^N \tag{2.6}$$

$$P_R(\sigma) = 1 - \exp\{N \ln[1 - P_R(\sigma)]\}. \tag{2.7}$$

In the case of a body of volume V, consider it divided up into small volumes V_0 containing one defect, analogous to links in a chain. In this case N is analogous to V/V_0:

$$P_R(\sigma) = 1 - \exp\{V/V_0 \ln[1 - P_R(\sigma)]\}. \tag{2.8}$$

$\ln[1 - P_R(\sigma)]$ can be replaced by any appropriate monotonic function $f(\sigma)$ decreasing from 0 when $\sigma = 0$, as $P_R = 0$, to $-\infty$ when $\sigma = +\infty$ as $P_R = 1$. Weibull took for $f(\sigma)$ a power law equal to $-[(\sigma - \sigma_u)/\sigma_0]^m$ for $\sigma \geq \sigma_u$ and zero for $\sigma < \sigma_u$, where σ is the applied stress, σ_u is a threshold stress below which there is a zero probability of failure and σ_0 and m are two material parameters, σ_0 is a scale parameter, and m, known as the Weibull shape parameter, measures the variability of the flaws in the materials. The greater the value of m the less is the scatter of results.

For a defect distribution in a material following a Weibull law, the failure probability, for an applied stress σ of a specimen, of volume V, is finally expressed as

$$P_R(\sigma, V) = 1 - \exp\left[-\frac{V}{V_0}\left(\frac{\sigma - \sigma_u}{\sigma_0}\right)^m\right] \qquad \text{for } \sigma \geq \sigma_u \qquad (2.9)$$

with $P_R(\sigma, V) = 0$ for $\sigma < \sigma_u$.

The two scale factors σ_0 and V_0 are often found grouped into one, also called σ_0 but which, in this case, no longer has the dimensions of stress, as it replaced $\sigma_0 V_0^{1/m}$ in equation (2.9). In this case the above expression can be rewritten more simply as

$$P_R(\sigma, V) = 1 - \exp\left[-\left(\frac{\sigma - \sigma_u}{\sigma_0}\right)^m V\right]. \qquad (2.10)$$

$P_R(\sigma)$ is the probability that the failure stress σ_R lies between 0 and the applied stress σ and is therefore a cumulative failure probability. The probability that σ_R lies between σ and $\sigma + \mathrm{d}\sigma$ is $g(\sigma)\,\mathrm{d}\sigma$ where $g(\sigma)$ is the density of failure probability giving

$$P_R(\sigma, V) = \int_0^\sigma g(\sigma, V)\,\mathrm{d}(\sigma) \qquad (2.11)$$

$$g(\sigma, V) = \frac{mV}{\sigma_0}\left(\frac{\sigma - \sigma_u}{\sigma_0}\right)^{m-1} \exp\left[-\left(\frac{\sigma - \sigma_u}{\sigma_0}\right)^m V\right]. \qquad (2.12)$$

Typical variation of P_R and g for $\sigma_u = 0$ are shown in figure 2.25.

Application to fibres with constant diameter

The first practical example of the statistical analysis of fibre fracture will be given for elastic fibres the diameter, D, of which can be considered not to vary. We will assume that σ_u can be put equal to zero. These two assumptions will be discussed at the end of this section. We obtain from equation (2.10) with the fibre length being written ℓ:

$$\ln[-\ln(1 - P_R(\sigma, \ell))] = m\ln(\sigma) + \ln(\ell) + \ln(\pi D^2/4) - m\ln(\sigma_0) \qquad (2.13)$$

where $\ln(\pi D^2/4) - m\ln(\sigma_0)$ is a constant.

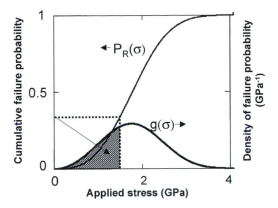

Figure 2.25. The cumulative failure probability for an applied stress σ is the area below the curve of the density of failure probability between zero and σ.

The statistical analysis of a series of tensile tests on fibres can be carried out in the following way. A number n of tensile tests are performed at a given gauge length ℓ so that P_R only depends on σ. The results are arranged in order of increasing failure stresses, $\sigma_1 < \cdots < \sigma_i < \cdots < \sigma_n$, as shown in figure 2.26(a). An experimental cumulative failure probability $P_R(\sigma_i)$ is calculated for each fibre of rank i having failed at a stress σ_i. Unless an infinite number of fibres are tested, which is clearly impossible, the most general expression must take into account the possibility of fibres breaking at lower stresses than that of the weakest specimen as well as breaking at higher stresses than the strongest fibre tested. Various expressions of this probability have been proposed, depending on the shape distribution, among which the most used is $P_R(\sigma_i) = i/(n+1)$. The variation of the experimental failure probability $P_R(\sigma)$ can then be plotted as a function of the strength of the fibres as shown in figure 2.26(b). The gradient of the curve $\ln[-\ln(1-P_R(\sigma))]$ as a function of $\ln(\sigma)$ gives a measure of the Weibull shape parameter, then σ_0 can be determined from the origin of the curve fibres as shown in figure 2.26(c).

In this way the distribution of fibre properties as described by Weibull statistics can be obtained from a series of tensile tests, usually not fewer than 30 depending on the scatter of results, with specimens of the same length.

As explained above, the average strength of fibres taken from a bundle decreases as a function of length ℓ, and this offers another approach to obtaining the Weibull shape parameter using the median stress. The median failure stress for a number of specimens of length ℓl, denoted by σ_{50}, is defined as the stress for which the failure probability is $P_R(\sigma_{50}) = 0.5$, as can be seen in figure 2.26(b). This is not necessarily the

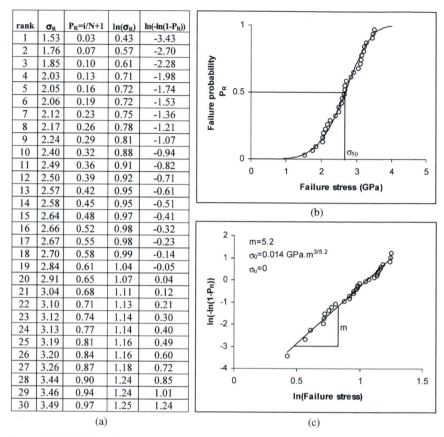

rank	σ_R	$P_R=i/N+1$	$\ln(\sigma_R)$	$\ln(-\ln(1-P_R))$
1	1.53	0.03	0.43	-3.43
2	1.76	0.07	0.57	-2.70
3	1.85	0.10	0.61	-2.28
4	2.03	0.13	0.71	-1.98
5	2.05	0.16	0.72	-1.74
6	2.06	0.19	0.72	-1.53
7	2.12	0.23	0.75	-1.36
8	2.17	0.26	0.78	-1.21
9	2.24	0.29	0.81	-1.07
10	2.40	0.32	0.88	-0.94
11	2.49	0.36	0.91	-0.82
12	2.50	0.39	0.92	-0.71
13	2.57	0.42	0.95	-0.61
14	2.58	0.45	0.95	-0.51
15	2.64	0.48	0.97	-0.41
16	2.66	0.52	0.98	-0.32
17	2.67	0.55	0.98	-0.23
18	2.70	0.58	0.99	-0.14
19	2.84	0.61	1.04	-0.05
20	2.91	0.65	1.07	0.04
21	3.04	0.68	1.11	0.12
22	3.10	0.71	1.13	0.21
23	3.12	0.74	1.14	0.30
24	3.13	0.77	1.14	0.40
25	3.19	0.81	1.16	0.49
26	3.20	0.84	1.16	0.60
27	3.26	0.87	1.18	0.72
28	3.44	0.90	1.24	0.85
29	3.46	0.94	1.24	1.01
30	3.49	0.97	1.25	1.24

(a)

(b)

(c)

Figure 2.26. (a) Experimental results of 30 tensile tests conducted with PAN-based carbon fibres of 7 μm in diameter at a gauge length of 25mm. The results are arranged in order of increasing failure stresses and the experimental failure probability is deduced. (b) Experimental (O) and theoretical (——) cumulative functions of failure probability. (c) The Weibull parameters are determined from the slope and origin of the straight line.

same as the mean stress. Equation (2.9) can be written as

$$\ln[-\ln(0.5)] = m\ln(\sigma_{50}) + \ln(\ell) + \ln(\pi D^2/4) - m\ln(\sigma_0). \quad (2.14)$$

As the diameter of the fibre, D, is supposed to be constant we obtain

$$\ln(\sigma_{50}) = -\frac{1}{m}\ln(\ell) + \text{constant}. \quad (2.15)$$

A plot of $\ln\sigma_{50}$ as a function of $\ln(\ell)$ for several gauge lengths gives a straight line curve with a gradient of $-1/m$ as shown in figure 2.27.

This analysis requires that the diameter of the fibre is almost constant, which is the case for carbon fibres but is not always true for fine diameter

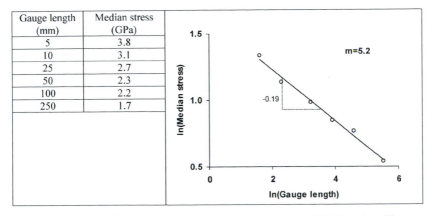

Gauge length (mm)	Median stress (GPa)
5	3.8
10	3.1
25	2.7
50	2.3
100	2.2
250	1.7

Figure 2.27. The logarithmic plot of the median failure stress of PAN carbon fibres as a function of the gauge length gives a straight line with a gradient of $-1/m$.

SiC or alumina fibres for which the iterative approach described later has to be followed.

Most of the statistical analyses of ceramic fibres can be carried out initially with the hypothesis that the threshold stress can be taken as zero. This means that there is a non-zero probability that one of the fibres encloses a defect inducing the failure for an applied stress which tends to zero. If this assumption does not allow the experimental strength distribution results to be correctly fitted to the theoretical curve, whereas the introduction of a threshold stress does, this means that the defect size distribution lies below a maximum size or that the fibre encloses internal stresses that must be overcome to induce the first fibre failures.

Significance of the Weibull parameter

The value of the Weibull shape parameter for most ceramic glass or carbon fibres is low, often around 5, which reflects the large dispersion found in their properties. This value of m is similar to that obtained with ceramics and can be compared with the values of m obtained for steel which give $m > 20$, as is illustrated in figure 2.28.

The curves presented in figure 2.29 illustrate the effect of the variation of one parameter in the Weibull expression of the failure probability with $\sigma_u = 0$.

Fibres with varying diameters

Fibre production processes can sometimes give rise to large variations in the diameters D from one fibre to another. In this case an effective failure stress σ^E has to be introduced to reflect the effect diameter variation on the failure probability, $\sigma^E = \sigma_R A^{1/m}$ with $A = \pi D^2/4$. The ranking is then carried out

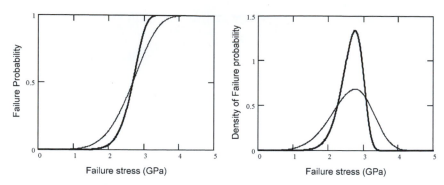

Figure 2.28. Probability curves for materials with the same volume and median failure stress (failure stress for $P_R(\sigma_{50})$ but different values of the Weibull modulus. The curves show that the materials have the same median stress (curves cross at $P = 1/2$), but for the material with $m = 5.2$ and $\sigma_0 = 0.014$ a broader distribution is seen than for the other material with $m = 10$, $\sigma_0 = 0.17425$ mm.

for each gauge length on these effective failure stresses and the probability of failure $P_R(\sigma_i^E) = i/(N+1)$ is assigned to the ith fibre $\sigma_{i-1}^E \le \sigma_i^E \le \sigma_{i-1}^E$. It can be seen that this ranking depends on the value of m which is sought. This value is estimated by an iterative calculation. Several values of m are systematically tested. For each m, the value of σ_0 is determined to give the best fit between the experimental and the theoretical distributions of failure probability and the pair (m, σ_0) allowing the best fit is chosen. The use of an effective failure stress has also the advantage of allowing results obtained at different gauge lengths to be plotted on the same curve, as shown in figure 2.30, by writing $\sigma^E = \sigma_R V^{1/m}$ with $V = Al$. As the precision in the determination of m and σ_0 is linked to the number of results treated, this method increases significantly the reliability of the values calculated.

Fibre tensile properties prediction

When it has been shown that the failure stress distribution can be described by a Weibull law, the median and mean failure stresses and the standard deviation and coefficient of variation of a population of fibres of any volume V can be calculated as follows:

Median failure stress σ_{50}, defined as $P_R(\sigma_{50}) = 0.5$:

$$\sigma_{50} = \sigma_u + \frac{\sigma_0}{V^{1/m}} (\ln 2)^{1/m}. \qquad (2.16)$$

Mean failure stress $\bar{\sigma}$, defined as $\bar{\sigma} = \int_0^\infty g(\sigma)\sigma \, d\sigma$:

$$\bar{\sigma} = \sigma_u + \frac{\sigma_0}{V^{1/m}} \Gamma\left(1 + \frac{1}{m}\right). \qquad (2.17)$$

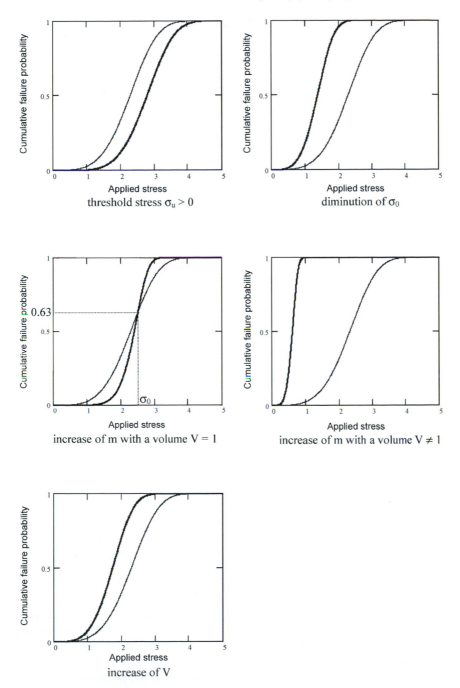

Figure 2.29. Effect of Weibull parameter variations on the stress distribution (bold curve) to be compared with a (fine) reference curve.

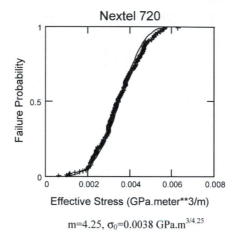

$$m=4.25,\ \sigma_0=0.0038\ \mathrm{GPa.m}^{3/4.25}$$

Figure 2.30. Determination of the Weibull parameters with results obtained at six different gauge lengths using effective failure stresses.

Standard deviation from the mean s, defined as $s = [\int_0^\infty g(\sigma)(\sigma - \bar\sigma)^2\,\mathrm{d}\sigma]^{1/2}$:

$$s = \frac{\sigma_0}{V^{1/m}}\left[\Gamma\left(1 + \frac{2}{m}\right) - \Gamma^2\left(1 + \frac{1}{m}\right)\right]^{1/2}. \tag{2.18}$$

The coefficient of variation $\mu = s/\bar\sigma$ can be calculated from equations (2.12) and (2.13). In the most usual case when $\sigma_u = 0$, μ is seen not to depend on V as

$$\mu = \left[\Gamma\left(1 + \frac{2}{m}\right) - \Gamma^2\left(1 + \frac{1}{m}\right)\right]^{1/2} \Big/ \Gamma\left(1 + \frac{1}{m}\right) \qquad \text{if } \sigma_u = 0. \tag{2.19}$$

The Γ function is defined as

$$\Gamma(x) = \int_0^\infty \exp(-u)u^{x-1}\,\mathrm{d}u \tag{2.20}$$

and is a generalization to the factorial as $\Gamma(n) = (n-1)!$ for integer values of $n \geq 1$. In the formulae (2.17) to (2.19) the values of $\Gamma(x)$ have to be known for $x \in [1,3]$ $(m \geq 1)$ and are shown in figure 2.31.

For a Weibull modulus $m \geq 1$, the values that can take $\Gamma[1 + (1/m)]$, $\{\Gamma[1 + (2/m)] - \Gamma^2[1 + (1/m)]\}^{1/2}$ and μ are shown in figure 2.32.

It can be seen that $\Gamma[1 + (1/m)]$ is close to 1 so that a first estimation of the mean failure stress can be given by $\bar\sigma \approx \sigma_u + \sigma_0 V^{-1/m}$. The mean and median failure stresses decrease as the standard deviation decreases and when the volume of the specimens increases, as can be seen in figure 2.29, but this decrease is less for higher m values. The curve shown as figure 2.32(c) shows that the coefficient of variation reduces when the Weibull modulus increases.

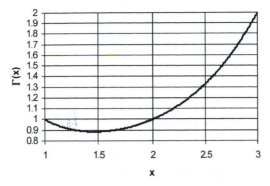

Figure 2.31. Variation of $\Gamma(x)$ for $x \in [1, 3]$.

Fibres with critical surface defects

Scanning electron microscope observations of fracture morphologies can often reveal that fracture is initiated from the fibre surface by defects generally created during the production or manipulation of the fibres, as is illustrated in figure 2.33, by a fracture surface of a Nextel 720 fibre.

In this case the failure probability must be written as

$$P_R(\sigma, S) = 1 - \exp\left[-\left(\frac{\sigma - \sigma_u}{\sigma_0} \right)^m S \right] \tag{2.21}$$

where $S = \pi D \ell$ is the fibre outer surface. The above methodology can then be followed in the same way.

Non-axial mechanical loading

Other mechanical loading configurations, such as loop tests, can be carried out which induce a variation of the stress in the fibre volume. The failure

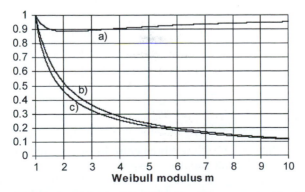

Figure 2.32. Variations of (a) $\Gamma[1 + (1/m)]$, (b) μ, and (c) $\{\Gamma[1 + (2/m)] - \Gamma^2[1 + (1/m)]\}^{1/2}$ for a Weibull modulus m between 1 and 10.

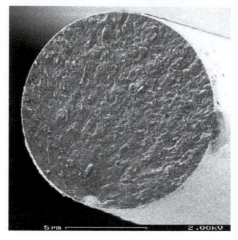

Figure 2.33. Fracture initiated by a defect at the surface of an alumina-mullite, Nextel 720 fibre.

probability is then expressed as

$$P_{\rm R}(\sigma, V) = 1 - \exp\left[-\int_V \left(\frac{\sigma - \sigma_u}{\sigma_0} \right)^m {\rm d}V \right]. \tag{2.22}$$

Experimental bias in results

The results obtained when testing fine fibres can be biased because of the difficulties in manipulating fine and brittle fibres. Defects may be created and, more significantly, weaker fibres can be broken during fabrication or mechanical test preparation before they can be tested. This difficulty inevitably increases with increasing specimen length so that merely handling the fibres provides an unavoidable selection process and modifies the failure distribution curve at the lower stress values. If it is assumed that any fibre with a strength lower than $\sigma_{\min}$ has been broken before the test, the failure probability is then expressed by equation (2.23) and its variation is illustrated in figure 2.34.

$$P_{\rm R}(\sigma, V) = 1 - \exp\left\{ -V\left[\left(\frac{\sigma}{\sigma_0} \right)^m - \left(\frac{\sigma_{\min}}{\sigma_0} \right)^m \right] \right\} \quad \text{for } \sigma \geq \sigma_{\min} \tag{2.23}$$

with $P_{\rm R}(\sigma, V) = 0$ for $\sigma < \sigma_{\min}$.

Fibres in which two populations of defects co-exist

As illustrated in figure 2.35, if the fibres contain two distinct populations of defects, a bi-modal Weibull law can be applicable so that

$$P_{\rm R}(\sigma, V) = 1 - \exp\left\{ -V\left[\left(\frac{\sigma}{\sigma_{01}} \right)^{m1} + \left(\frac{\sigma}{\sigma_{02}} \right)^{m2} \right] \right\}. \tag{2.24}$$

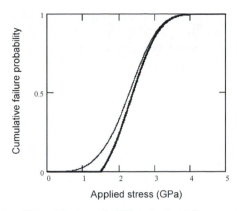

Figure 2.34. Variation of the failure probability obtained from tensile testing of a fibre population from which the weakest fibres have been eliminated ($\sigma_{min} = 1.5\,\text{GPa}$).

Ductile fibres

Rather than a single series configuration of the chain, a parallel configuration of chains is more appropriate if plasticity, around the tip of any crack, allows stress concentrations to be reduced and so hinders crack propagation, such as in ductile materials. After the failure of the weakest link, a redistribution of the load towards the other chains can permit the survival of the overall system.

All fibres show great dispersion in their properties and they do not all fail in the same manner. The mode of failure may involve slow crack propagation from surface or internal faults leading to eventual catastrophic failure.

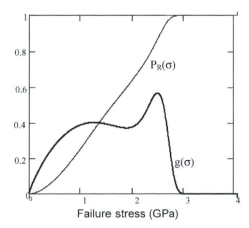

Figure 2.35. Cumulative failure probability P_R and density of failure probability of fibres containing two populations of defects.

Fibres such as carbon, glass, boron and ceramic fibres show no plastic deformation at room temperature and behave as purely brittle elastic bodies, their ultimate failure being determined by defect size as described by Griffith for elastic bodies. Organic fibres do have at least some plastic component to deformation at all temperatures. At high temperatures, often above 1000 °C, ceramic fibres are found to creep and they can deform plastically and even super-plastically.

2.11   Conclusion

A very wide range of possible reinforcements exist which can be used to reinforce all classes of materials. Fibres which are often thought of as being for textile uses find important applications in reinforcing rubber to allow tyres and industrial belting to be made. Other organic fibres with much higher performances can be used to reinforce both thermosetting and thermoplastic matrices which can also be reinforced with carbon and glass fibres. This latter class of materials represents the most important family of composite materials. Carbon and ceramic fibres can be used to reinforce light metal alloys. The most recent class of fibres to be produced is that of ceramic fibres, which can be either oxide fibres, usually based on alumina, or silicon carbide. This latter group of fibres offer the possibility of pushing the upper temperature limits of structural materials beyond those currently imposed by the most advanced metal alloys and may find use in the future in gas turbines, heat exchangers and other very high temperature devices.

The fineness of fibres means that a very great number are used in any application so that a statistical approach to analysing their properties is necessary.

2.12   Revision exercises

1. Fibres can be classified according to three groups. Give examples of each group. In which decade were synthetic fibres first commercially produced?
2. Give a definition of what constitutes a fibre and an approximate value of the diameter of a typical fibre used in composite materials.
3. If continuous circular fibres of 10 μm diameter are combined into a parallel tow so that they all touch so as to make up a hexagonally packed parallel tow, calculate the packing efficiency. If the diameter of the tow is 10 mm calculate how many fibres make up the tow. (Answer 907 000 fibres)
4. Calculate the tex of a bundle of 1000 carbon fibres which are 7 μm in diameter and have a density of $1.8 \, \text{g/cm}^3$.

5. Explain why even though glass fibres are much stronger than bulk glass the Young's moduli of both are similar.

6. What happens to the molecular structure of an organic fibre during drawing? What consequences does this have on its mechanical properties?

7. Explain the significance of the numbers in 'polyamide 66'. Give the chemical formula for this polymer. Under what name is this fibre commonly known?

8. What is the most widely produced organic fibre? Give its molecular formula. Why is this fibre inherently stiffer than a polyamide fibre?

9. The fibres mentioned in problems 6 and 7 are drawn from the melt. Explain what this means. Explain how the macromolecules are arranged in the fibres and why there is a difference in residual stresses between the surface and core of the fibres.

10. In what type of composite structure are thermoplastic fibres used as reinforcements?

11. Explain how it is necessary to use liquid crystal technology to produce high modulus organic fibres. Give examples. What consequences on the mechanical properties does this have and how does this affect the fracture behaviour of the fibres?

12. Explain why aramid fibres are not used in primary load bearing structures but for the same reason are used for improving the impact resistance of composite structures.

13. On which mineral are glass fibres based and approximately what percentage of it makes up the composition? Why are other minerals added to their composition? What is the most commonly used glass fibre? What modification is made to the composition to produce higher performance varieties of glass fibre?

14. Compare the Young's moduli of glass fibres with those of other materials such as aluminium, titanium and steel as well as those of carbon and aramid fibres.

15. Why are glass fibres immediately coated with a size? What are the four principal roles for the size put on glass fibres? What type of structure do many common sizes have?

16. The first high performance fibres to be made were produced by CVD. Explain what this means and give examples of such fibres including the chemical reactions involved. How do these fibres differ from high performance fibres which were produced later? Are they still produced and if so for what type of composite materials? Explain how some types of these fibres can be used to reinforce metal matrices.

17. Why are carbon fibres so important? Discuss properties of carbon compared with other light elements.

18. The first carbon fibres were produced from cellulose but most are now produced from PAN. Explain what this means and how the PAN is

converted into a carbon fibre. What chemical process is necessary to stop the precursor fibres decomposing during carbon fibre manufacture? Why is it so important to keep the fibres under tension during their production?

19. What are the typical diameters of carbon fibres made from PAN? Discuss the atomic structure of high strength carbon fibres and how it determines their properties.
20. How can a range of carbon fibres be produced from the same precursors? How are high strength and high modulus fibres produced?
21. Earlier generations of carbon fibres were generally weaker than those which have been produced more recently. Explain why this is the case and why the improvement has been for both high strength and high modulus fibres.
22. Explain why pitch-based carbon fibres are of interest and in which types of composite structures they find use.
23. Ceramic fibres are produced for high temperature applications. Give examples of such applications and identify the two main families of fibres which have been developed.
24. Explain how the control of the microstructure of alumina-based fibres is necessary if high performance fibres are to be produced. Explain how alumina can exist in many forms. Why are some alumina-based fibres produced containing other minerals such as silica, mullite and zirconia?
25. Discuss the implications of the different microstructures of alumina-based fibres on the behaviour of the fibres, both at room temperature and high temperature and also on their sensitivity to contamination.
26. Explain why the first generation of small diameter SiC fibres had only half the Young's modulus of bulk SiC and crept at a much lower temperature. How was the manufacturing technique changed to partially overcome this difference?
27. The fracture surfaces of the latest near stoichiometric small diameter SiC fibres are much more irregular than those of the first generation. Why is this? Explain the changes in manufacturing processes which have brought this about.
28. How do Weibull statistics explain why a bundle of fibres loaded parallel to the fibre direction will be much stronger than the same volume of the material in bulk form?

References

Morton W E and Hearle J W S 1975 *Physics of Textile Fibres*, 2nd edition (London: Heinemann)
Moncrieff R W 1979 *Man-made Fibres*, 6th edition (Guildford: Butterworth)
Bunsell A R 1988 *Fibre Reinforcements for Composite Materials* (Amsterdam: Elsevier)

Kostikov V I 1995 *Fibre Science and Technology* (London: Chapman & Hall)

Chawla K K 1998 *Fibrous Materials* (Cambridge: Cambridge University Press)

Bunsell A R and Berger M-H 1999 *Fine Ceramic Fibres* (New York: Marcel Dekker)

Wallenberger F T 2000 *Advanced Inorganic Fibres: Processes, Structures, Properties, Applications* (Dordrecht: Kluwer Academic Publishers)

Hearle J W S 2001 *High Performance Fibres* (Cambridge: Woodhead)

Chapter 3

Organic matrices

3.1 Introduction

Composite materials are most often thought of as consisting of fibres embedded in a polymer matrix. This is because of the market which has developed for fibre reinforced plastics but it should not be forgotten that fibres can be used to reinforce all classes of materials. Polymers, however, have established themselves as matrix materials for composites in many applications and well established markets exist for them whereas most composites based on other classes of matrix materials are still primarily at the development stage.

   The attraction of polymers for the role of matrix is that they are light in weight, often with a density little more than that of water, and they can be used, either in solution or molten, to impregnate the fibres at pressures and temperatures which are much lower than those which would be necessary for other materials, for example metals. This results in low density of the composite material, low costs of composite manufacture and of forming tools and relative ease of fibre impregnation. The polymers are often highly resistant to corrosive environments which results in useful properties for the composites produced. The low elastic moduli of most polymers, linked to their ease of deformation, allows load transfer between fibres by shear of the matrix materials and so an effective use of the fibre properties. Certain polymers can produce composites which are particularly resistant to impact and fatigue damage.

   A disadvantage of polymers as matrix materials is that they do not provide high performance mechanical properties at right angles to the fibre directions, which makes unidirectional composite materials inherently anisotropic. They also usually absorb water, which can modify the composite properties and compromise the adhesion between the fibres and the matrix, so weakening the composite. This can also lead to out-gassing of water vapour under low pressure conditions such as those found in space structures; water condensation on electronic circuitry on space mirrors could present difficulties. The absorption of water can lead to hydrolysis and matrix degradation as well as major changes in matrix properties. This is discussed in chapter 11. Degradation of polymers by irradiation, including ultraviolet light, as well as attack by atomic oxygen in low altitude space applications can be problems for some composite applications.

The polymers used as matrix materials can be divided into two primary families, the natures of which depend on their molecular structures. These are thermosetting and thermoplastic resins, for example, respectively, epoxy resin and polyamide (Nylon). Thermosetting resins have been used the longest as matrices for composites and are materials which undergo a transformation of their molecular structure during the manufacture of the composite material. This is because the long macromolecules which make up thermosetting resins possess reactive bonds which can be opened by a hardener to form strong covalent lateral links with other molecules. At room temperature, the resin starts as a viscous liquid, which in the presence of the hardener to initiate the crosslinking reaction and usually heating, changes to a rigid solid possessing a three-dimensional molecular network. Cooling to its original temperature reveals a completely modified material which is solid. The reactions are therefore irreversible. Most people will be familiar with epoxy glues which are activated by mixing two viscous components, the resin and hardener, which then become hard. The time for such a glue to set can be modified by changing the temperature as the process is thermoactivated. Such glues are examples of thermosetting resins without reinforcements. These resins lend themselves to composite manufacture as fibre impregnation is facilitated by the viscous liquid phase of the resin which can then be made solid by the crosslinking of the molecular structure. As the crosslinking of the thermosetting resins is thermoactivated it is necessary to store the resins at low temperatures, usually $-18\,°C$, until they are required for composite manufacture. Crosslinking of many resin systems takes place at relatively high temperatures, for example 120–180 °C for epoxy resins, and requires heating for perhaps an hour or more. This time-dependent process can present difficulties for large volume production, as can the storage of resins at low temperature, the impossibility of modifying the final form after manufacture and the difficulties of recycling thermosetting materials.

Thermoplastics are polymers which undergo dramatic changes to their mechanical properties when they are heated above their glass transition temperature (T_g) but these changes are reversed on lowering the temperature and no structural modification at the molecular level occurs. Unlike the thermosetting resins the macromolecules making up the thermoplastics possess no reactive lateral bonds which can link them strongly with other molecules. Lateral bonding is through secondary forces such as hydrogen and van der Waals bonds and entanglement of the molecules. The effect of heating above the T_g is to supply enough energy to the polymer to liberate the molecules from these secondary bonds and allow freer movement of the linear molecular structure. This is exploited in the forming of such thermoplastics, which can be rapid as it requires only heating the material to a high enough temperature. There is no need to store the polymers at low temperatures. Everyday examples of thermoplastics can be seen in many plastic bottles, which are usually made from thermoplastic polyester, Nylon and polyester fibres and PMMA, sold under a variety of names such as Perspex and Plexiglas, which although fairly rigid at room temperature soften on heating, for example in steam; and can be deformed easily under these conditions. Cooling the thermoplastic freezes in the new structure which can be further modified at will by further heating. Thermoplastic materials such as polyamide, thermoplastic polyester and polypropylene have become widely used as matrix materials for short fibre reinforced composites made by injection moulding and more recently are increasingly being considered for long or continuous fibre reinforced materials. Thermoplastics in principle can be recycled and the final form of a structure can be modified by reheating and further forming. Processing times depend on the time to melt the thermoplastic and can be extremely rapid. Fibre impregnation is not as easy as with thermosets and is usually the responsibility of specialist material suppliers as the thermoplastic has to be molten and fibre–matrix adhesion controlled.

A group of resins, which can be both thermosetting or thermoplastic, are known as thermostable resins. Such resins conserve their properties indefinitely at 200 °C and can be used for limited lengths of time up to and sometimes above 400 °C. Thermostable resins find use in composite applications which require high temperature performance such as in the aerospace industry. Elastomers, such as natural rubber, consist of long macromolecules which are often lightly linked together. Rubber, reinforced with a variety of fibres and used in such applications as tyres and industrial belting, is just as much a composite as other fibre reinforced materials sharing many of their characteristics and represents as big a market. The fibres are placed in specific regions of the tyre and are in the form of twisted tows rather than aligned or randomly arranged, and represent a small overall fibre volume fraction. Rubber can be seen as a lightly crosslinked thermosetting type material operating above its glass transition temperature. It represents a rather special class of materials, the details of which are more tied up with rubber chemistry than with composites and will not be treated in this book.

3.1.1 Common polymer matrix systems

Figure 3.1 shows examples of the most important families of thermosetting resin matrix systems used in composites. There can be wide variations within the thermosetting resins as they can be combined with other functional groups and crosslinking agents to give resins with different mechanical and chemical resistance properties. Thermosetting resins are primarily

Unsaturated polyester (UP) with styrene as the crosslinking agent

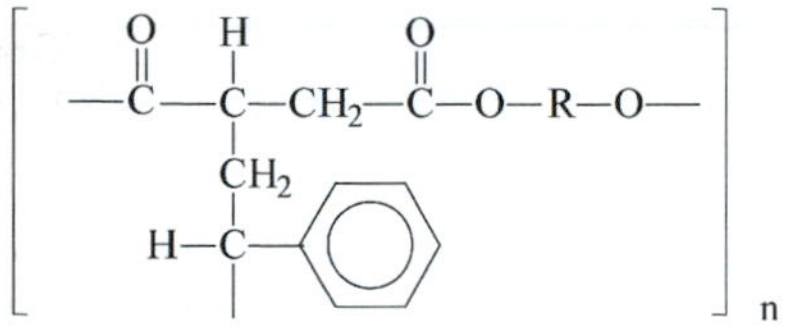

Epoxy resin with an amine crosslinking agent

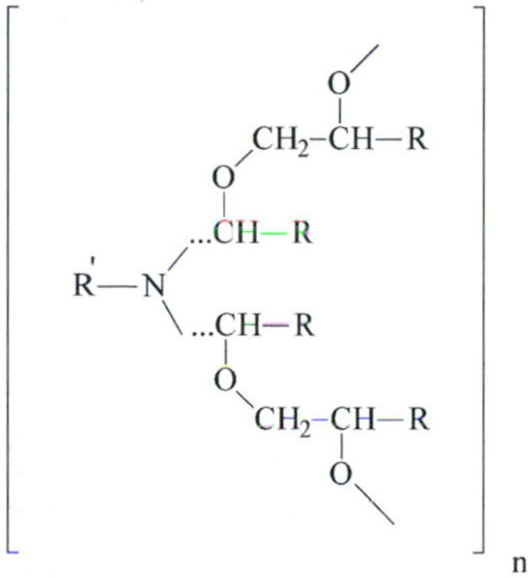

Polyimide (PI)

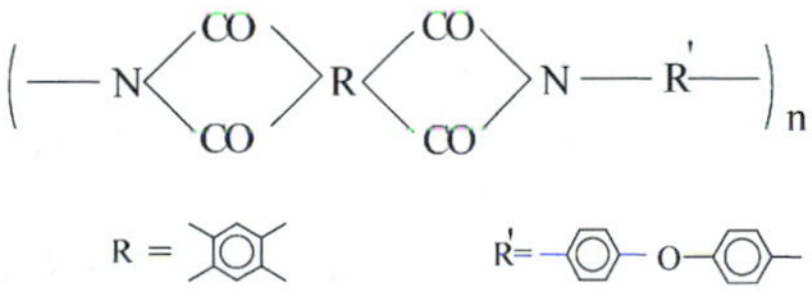

Figure 3.1. Common thermosetting resins.

produced from low molecular weight monomers whereas thermoplastics have high molecular weights. The first resin shown, in figure 3.1 is that of an unsaturated polyester resin which has been crosslinked with styrene. Unsaturated means that there are bonds available for further chemical reactions. This type of polyester can therefore be crosslinked to give a three-dimensional molecular structure. The saturated polyester which will be mentioned later is, by contrast, a thermoplastic polymer which does not possess additional bonds for crosslinking and so remains a network of linear macromolecules. The thermosetting polyester crosslinked with styrene, shown in figure 3.1, is the most widely used thermosetting matrix used in general composites. This type of resin accounts for more than 80% of the matrix systems used in all composites. It is cheaper than other resins which will be mentioned but it is less tough, shrinks considerably on cross-linking and gives off dangerous smoke in the event of a fire. Epoxy resins

Table 3.1. A comparison of typical properties of the two most widely used families of thermosetting resins.

Property	Polyester	Epoxy
Heat deflection temperature	85–125 °C	155 °C
Tensile strength	35–80 MPa	90 MPa
Tensile modulus	2.8–3.9 GPa	3–4 GPa
Elongation	1.4–2.5%	4–10%
Flexural strength	80–140 MPa	80–140 MPa
Flexural modulus	3.5 GPa	3 GPa

represent around 10% of the matrix systems used in composites. They are used for higher performance applications as they are more expensive but tougher than polyester resins, shrink less and are generally considered to stand up to wet environments better. The epoxy shown in figure 3.1 will be discussed in more detail later but it can be noted that it is a DGEBA, the most widely employed in composite materials, crosslinked with an amine hardener. A comparison of typical mechanical properties of polyester and epoxy resins is given in table 3.1.

Polyimides can be thermoplastic but are more usually used as thermosetting resins. They are known as high performance matrices capable of being used to higher temperatures than the two resin systems previously mentioned. The have good mechanical properties, good chemical resistance and low smoke output in the event of fire.

Figure 3.2 shows the most widely used thermoplastic systems used for composite materials. These composite materials can be made by injection moulding, which is an extension of the moulding of plastic parts by the addition of short fibres. However, there is an expanding market for long or continuous fibre reinforced thermoplastics as this family of matrix materials offers rapid manufacturing times and the possibility of being recycled. Polypropylene is the cheapest of these polymers and for that reason, coupled with its chemical inertness, is attractive as a matrix material for mass produced products. It is, however, difficult to produce high quality bonds between polypropylene and reinforcing fibres. The development of appropriate bonding agents either placed on the fibres' surface, in a size, or grafted onto the polymer molecular structure is, however, overcoming many of these difficulties and polypropylene seems set to become a major matrix material.

Figure 3.2 shows three of the family of polyamides, which are also known as Nylons. Polyamide 6 is shown together with polyamide 66 and polyamide 11 so as to illustrate the significance of the numbers defining the type of polymer. The numbers refer to the number of carbon atoms

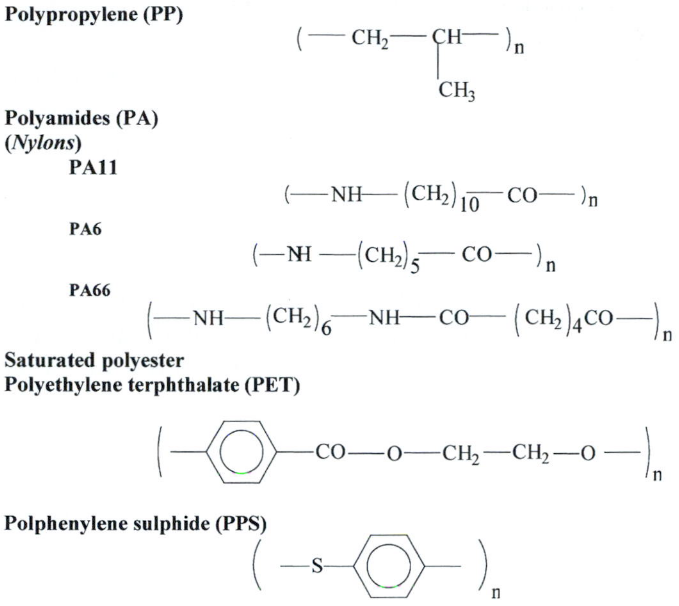

Polypropylene (PP)

$$(-CH_2-\underset{\underset{CH_3}{|}}{CH}-)_n$$

Polyamides (PA)
(*Nylons*)

PA11

$$(-NH-(CH_2)_{10}-CO-)_n$$

PA6

$$(-NH-(CH_2)_5-CO-)_n$$

PA66

$$\left(-NH-(CH_2)_6-NH-CO-(CH_2)_4CO-\right)_n$$

Saturated polyester
Polyethylene terphthalate (PET)

$$\left(-\bigcirc-CO-O-CH_2-CH_2-O-\right)_n$$

Polphenylene sulphide (PPS)

$$\left(-S-\bigcirc-\right)_n$$

Figure 3.2. Common thermoplastic matrix systems.

which is repeated to make up the macromolecule. Polyamide 11 is interesting as it absorbs less moisture than PA66, which is the most widely used polyamide for composite materials.

Figure 3.3 shows two thermostable matrix materials which maintain their properties at temperatures above 200 °C.

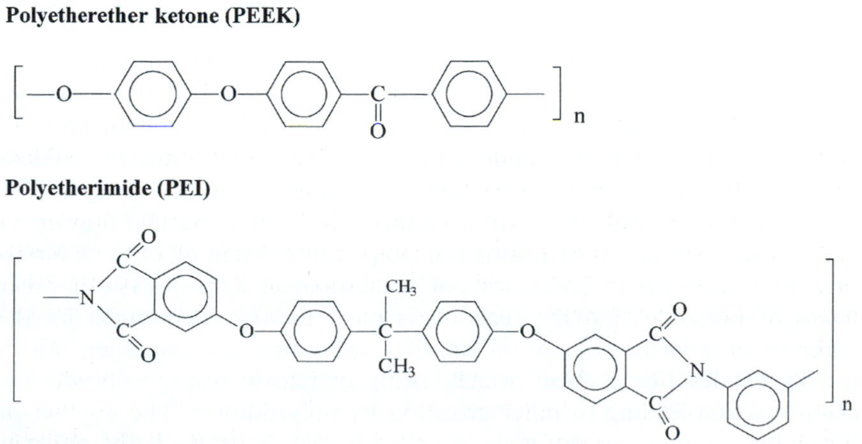

Polyetherether ketone (PEEK)

$$\left[-O-\bigcirc-O-\bigcirc-\underset{\underset{O}{\|}}{C}-\bigcirc-\right]_n$$

Polyetherimide (PEI)

Figure 3.3. Common thermostable matrix systems.

3.2   Resin structure

If synthetic composites were not possible without the development of synthetic fibres, which have mainly been developed in the second part of the 20th century, the same is certainly true for the development of synthetic resins which are the polymers used in most composite materials. This is perhaps underlined by noting that the German chemist, H Staudinger, who first described the macromolecular structure of polymers in the 1920s, received a Nobel Prize for this work in 1952.

Polymers are composed of long molecules, known as macromolecules, consisting of thousands and sometimes millions of atoms resulting in very long chain lengths. Usually the macromolecules consist of a backbone of carbon atoms as the C−C bonds are particularly stable and this is the case for the resins, which are of interest for composite materials. The simplest polymer consisting of long linear macromolecules is polyethylene, which has the structure

$$-CH_2-CH_2-CH_2-CH_2-CH_2-.$$

Polymers based on other atoms do exist such as the silicones (polysiloxanes):

$$\begin{array}{ccc} CH_3 & CH_3 & CH_3 \\ | & | & | \\ -Si-O-Si-O-Si- \\ | & | & | \\ CH_3 & CH_3 & CH_3 \end{array}$$

The macromolecules are rarely straight or planar so that one image is that of a confused mass of intertwined long lengths of molecules, much like spaghetti on a plate. However, the molecular morphology of polymers, including those used as matrix materials for composites, can be much more complex than this description.

Polymers are made by the linking together of their basic units called monomers, which are groups of atoms which can exist in their own right but which can be encouraged to open up so providing the opportunity to create two, three or more bonds, so permitting linkage, through covalent bonds, with other similar monomers or other chemical groups. The opening of these bonds can occur spontaneously due to molecular movement and this is further aided by raising the temperature. Opening of these bonds can also occur due to bombardment by ultraviolet light, x-rays or other forms of radiation, but the reaction is most usually encouraged by the addition of a small amount of another substance, the hardener, which usually creates free radicals which, being extremely reactive, produce a chain reaction leading to polymerization by polyaddition. The number of bonds that a monomer can make is called its functionality. If the polymer is a combination of a number of different molecules (n_A, n_B, n_C etc.) the functionality $\bar{F}$ is given by

$$\bar{F} = \frac{n_A F_A + n_B F_B + n_C F_C + \text{etc.}}{n_A + n_B + n_C + \text{etc.}}.$$

In the case of bifunctional systems, for which $\bar{F}$ is 2, the resulting macro-molecules are linear. Entanglements of the linear molecular chains together with secondary bonding, such as hydrogen or van der Waals bonding, give these polymers bulk cohesion which in some cases, for example cellulose, make their structure stable even to the point of resisting most solvents. Alternatively linear polymers with weak intermolecular bonding give elastomers, which in the classic case of natural rubber requires the light cross-linking of the molecules through a process known as vulcanization by the addition of 6–8% of sulphur. Even vulcanized elastomers are characterized by great deformability, up to 1000%, due to the low density of crosslinking. Tri- or multifunctional monomers allow three-dimensional molecular net-works to be created with a much greater density of crosslinking occurring between the molecular chains.

The polyaddition process occurs when the addition of another monomer leads to the repeated linking of molecular chains to create macromolecules. When only one type of monomer is involved a homopolymer is produced giving

$$-A-A-A-A-A-A-A-A-$$

for the monomer A. The simultaneous polymerization of two or more different monomers, A and B for example, can lead to copolymerization which can give several arrangements, one of the simplest of which might be an alternate sequence such as

$$-A-A-B-B-A-A-B-B-A-A-B-B-.$$

Other possible combinations are a random arrangement such as

$$-A-A-B-A-B-A-B-A-B-B-B-A-B-$$

or arranged in sequential blocks such that

$$-AA-(A-BB)-(BB-A)-AA-$$

Alternatively the copolymers can be grafted together such that

$$
\begin{array}{c}
-A-A-A-A-A-A-A-A-A-A-A-A- \\
\quad\;\; | \qquad\qquad\qquad\qquad\qquad | \\
\quad\;\; B \qquad\qquad\qquad\qquad\qquad B \\
\quad\;\; | \qquad\qquad\qquad\qquad\qquad | \\
\quad\;\; B \qquad\qquad\qquad\qquad\qquad B \\
\quad\;\; | \qquad\qquad\qquad\qquad\qquad | \\
\quad\;\; B \qquad\qquad\qquad\qquad\qquad B \\
\quad\;\; | \qquad\qquad\qquad\qquad\qquad |
\end{array}
$$

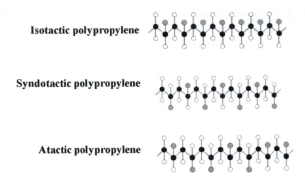

Figure 3.4. Polypropylene molecules can show three forms of tacticity.

Such reactions can lead to several possible molecular arrangements. If the molecules are long and linear with regular repetition of the basic units without any change in their arrangement in space they are known as isotactic. If the arrangement is regular but the groups are regularly alternatively arranged in space they are known as syndotactic. An irregular repetition of similar units is known as an atactic polymer. Polypropylene can exist in the three forms, as shown in figure 3.4. The black dots represent carbon atoms, the open circles hydrogen and the grey dots methyl groups.

Cyclic arrangements, in which one end of the molecule links with the other end, can also occur and are generally to be avoided. In addition three-dimensional networks and copolymerization in which the macro-molecules of the second polymer are grafted just at their ends onto the first polymer chain can also occur. If the polymer is to crystallize, the macro-molecules have to be linear and regular in their arrangement, that is isotactic or syndotactic, so that they can fit alongside one another in the crystalline phase.

Polycondensation is another reaction used for producing macro-molecules and differs from that described above by the loss of reaction products which are often, but not always, water molecules. To initiate such a reaction catalysers can be used but are not always necessary. Polycondensation begins with small polyfunctional molecules which lose water either through heating or to a solvent and form, by combination, long macromolecules. If the molecules are bifunctional then linear molecules are produced, whereas functionality greater than two will result in three-dimensional molecular networks.

The molecular structure of three-dimensional network polymers can therefore be seen to be modelled by three main components, linear or planar, often rigid units such as benzene rings, rotational units and trifunctional bonding units whereas the structures of two-dimensional network polymers lack the trifunctional units. Three-dimensional crosslinked polymers cannot generally be sufficiently organized to crystallize, but exceptions exist which do not readily spring to mind as being polymers. Examples are diamond, quartz and graphite. The two-dimensional organization of carbon atoms gives great strength in the plane of the atoms due to the strong C–C bond and this is exploited in polymers and for example carbon fibres. In graphite, the layers of carbon atoms are held together in a regular lattice by secondary, van der Waals bonds with a distance between the planes of 0.338 nm. This leads to the great anisotropy of graphite which in the plane of the C–C bonds is extremely strong and rigid in tension whereas shearing of the planes is very easy, allowing graphite to be used as a lubricant.

Both thermosetting resins and thermoplastic matrices used for composite materials are made in the ways described above. Thermosetting resins can be used, before three-dimensional crosslinking, to impregnate the fibres which are placed in a mould or otherwise positioned in the arrangement finally desired for the composite structure and then heated, usually under pressure, to finish the curing of the resin. The result is a solid matrix with a three-dimensional molecular structure bonded together by strong covalent bonds enveloping the fibres. The process is irreversible. As the crosslinking of thermosetting resins is thermoactivated it is necessary to store the unprocessed resins at low temperature, usually around $-18\,^\circ$C. Other complications which can be associated with thermosetting resins are degassing during curing leading to porosity and bubbles, imperfect mixing of reacting agents leading to a heterogeneous molecular structure, excesses of uncombined hardener which can lead to degradation in the presence, for example, of water, difficulties in the control of viscosity during composite manufacture and shrinkage during crosslinking. In addition the crosslinking process is exothermic, which can lead to difficulties in controlling the kinetics of the chemical reactions during composite manufacture. Thermosetting resins normally decompose rather than melt on heating to high temperatures and recycling presents major difficulties. Thermosetting resins were, however, the first class of polymers used as matrices for composites and represent at present approximately two thirds of the organic matrix materials used.

Thermoplastic materials have found wide use for the production of fibres, such as polyamide and polyester fibres and in injection moulding processes as their ability to melt whilst retaining their macromolecular structure can be exploited in these processes. Thermoplastic materials are used as the matrix materials for both short and continuous fibre reinforced composites. The processes involved in producing short and continuous thermoplastic composites are radically different. Short fibre thermoplastic composites are usually made by injection moulding, which involves a mixture of the fibres and molten polymer being injected by a system of screw threads into a mould. The injection process shears the molecules, so making them more reactive to bonding with the fibres than otherwise. This shearing process is not a feature of continuous fibre reinforced composite manufacture, so the polymers must either be made more reactive by the addition of functional groups or specially adapted sizes have to be placed on the fibres to ensure good bonding. The two-dimensional macromolecules, of which thermoplastics are composed, are made up of atoms linked together by strong covalent bonds; however, interactions, other than mechanical entanglement, between macromolecules are through weak or secondary bonds. The reactions and strengths of the bonds between the macromolecules depend on the temperature so that below a critical temperature known as the glass transition temperature the bonding is sufficiently strong to result in a rigid material and the degree of bonding determines the interaction of the polymer with solvents. Above the glass transition temperature, T_g, the molecules have much more freedom of movement and the material becomes more rubbery in behaviour and can be deformed easily. Reducing the temperature to below the T_g returns the material to its former rigid state. This behaviour is therefore reversible and is typical of thermoplastics. If the thermoplastic is atactic the polymer is amorphous and generally transparent. If the polymer is isotactic or syndotactic the macromolecules can become aligned through the influence of the secondary bonds and locally can form into a regular arrangement. These localized crystalline zones form spherulites made of folded macromolecules which can occupy a considerable percentage of the volume of the semi-crystalline thermoplastic, typically 30 or 40%. The crystallinity of a semi-crystalline thermoplastic can be altered by melting the polymer and controlling the rate of cooling. The faster the cooling rate the lower the degree of crystallinity. When thermoplastics are drawn into fibres the crystal zones are drawn out to form microfibrils made up of both crystalline and amorphous zones. Thermoplastics melt on heating to sufficiently high temperatures and this is used to impregnate the reinforcing fibres during the manufacture of thermoplastic matrix composites. Recycling is possible with thermoplastics although practical considerations often limit its effectiveness.

The T_g of amorphous thermoplastics is usually well defined whereas the mixture of phases in a semi-crystalline thermoplastic makes the identification of the T_g less easy. Thermosetting resins can also be said to possess a T_g above which the resin loses rigidity although this is generally less obvious than in thermoplastics.

As polymers are composed of long molecular chains made by reactions which are controlled by the local chemistry at each point in the polymer, the lengths of the macromolecules are not all the same and show considerable dispersion. This dispersion can have important consequences on the behaviour of the polymer, for example the shorter molecules can be more reactive to solvents or more easily pyrolysed than the rest of the polymer. In defining polymers it is useful therefore to define their average molecular weights and the molecular weight distributions. The average molecular weight can be defined and determined by the mass of all the molecular chains divided by the number of chains such that

$$\overline{Mn} = \frac{\sum n_i \cdot M_i}{\sum n_i}.$$

The value of Mn can be measured experimentally by osmometry and other techniques.

Alternatively the average molecular weight can be defined as the ratio of the sum of all the masses of molecules with the same mass divided by the sum of all the individual masses so that

$$M^- \overline{Mw} = \frac{\sum m_i \cdot M_i}{\sum m} \quad \overline{Mw} = \frac{\sum m_i \cdot M_i}{\sum m_i} = \frac{\sum n_i \cdot M_i^2}{\sum n_i \cdot M_i}$$

where m_i is the fraction of the weight of the polymer which contains all the macromolecules of molecular weight M_i. The value of M_i can be determined by the measurement of opacity of the resin.

It is easy to imagine the consequences of dissolving a polymer in a solvent. The movement of the macromolecules and their disentanglement in the solution will be all the easier the shorter they are, so that the rheology of the solution will be affected. As a consequence a polymer can also be characterized by its viscosity in solution. If the viscosity of the solution with a certain concentration C is η, and η_0 is the viscosity of the solvent, an expression for the intrinsic viscosity of the polymer can be written such that

$$[\eta] = \lim_{C \to 0} \frac{(\eta - \eta_0)}{\eta_0 C}.$$

This value is related to another expression for the average molecular weight $\overline{Mv}$ through

$$[\eta] = k\overline{Mv}^\alpha \qquad \text{with } 0.5 < \alpha < 1$$

in which k and α are two constants which are characteristic for each type of polymer and are found in published tables:

$$\overline{Mv} = \left[\frac{\sum n_i \cdot M_i}{\sum n_i \cdot M_i} \right]^{1/\alpha} \qquad \text{and is equal to } \overline{Mw} \text{ if } \alpha = 1.$$

The values of Mn, Mw and Mv are clearly related, but values may differ as a result of the methods used to evaluate them.

The molecular weight distribution can be determined by several techniques. In fractionation the solution is progressively diluted which provokes the precipitation of those parts of the polymer with progressively lower molecular weight. It is by measuring the molecular weight of each of these fractions of the polymer, M_i, and plotting the weight of each fraction,

m_i, as a function of M_i that the molecular weight distribution can be obtained. This technique has the disadvantage of being lengthy to carry out and to require a considerable amount of matter to analyse.

An alternative method is by gel chromatography, in which a diluted solution of the polymer passes through a chromatography column containing a porous gel consisting of a crosslinked polymer containing pores of various dimensions. The longer molecules can only traverse the gel through the largest of pores, linking each side of the gel by the shortest routes so that these molecules cross the gel the fastest. The rate at which the molecules cross the gel is therefore in a decreasing order of molecular weight. The detection of the molecules can be made by the interaction of light or ultraviolet or infrared spectrophotometry, and calibration of these methods is achieved by comparison with the results obtained from fractionated polymers, either the same as the one being studied or one very similar, so as to obtain molecules of a constant length.

3.3   Matrix mechanical behaviour

3.3.1   The curing of thermosetting resins

The rate of cure, or crosslinking, of a thermosetting resin system can be monitored by simply controlling the viscosity of the resin. The crosslinking process proceeds asymptotically towards the condition in which all of the resin would be combined with the hardener, as is illustrated schematically in figure 3.5. This means that the process is never quite finished and optimizing curing cycles can be problematic so that post curing is often performed. This involves a curing cycle which is subsequent to the initial curing process. It can also mean that the properties of the resin evolve with time and temperature. The crosslinking process is illustrated schematically in figure 3.6. The resin, before curing, begins as a liquid which, initially, under the

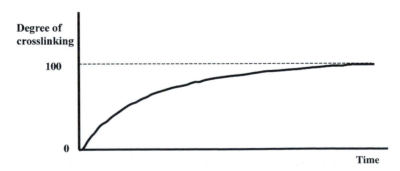

Figure 3.5. The crosslinking process progresses asymptotically towards full crosslinking but never quite reaches this stage.

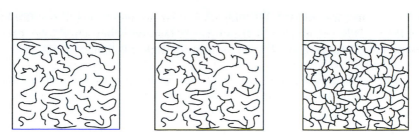

Figure 3.6. The molecular structure is irreversibly modified by the crosslinking process as is shown from left to right. The resin is initially not crosslinked, then lightly crosslinked (at which point it gels), and finally nearly fully crosslinked.

effects of heat, becomes less viscous; however, as the crosslinking process progresses it becomes increasingly viscous and gels. Finally the resin becomes solid as the process approaches a fully crosslinked resin. The crosslinking process is exothermic and can generate considerable heat, so that in some manufacturing processes little or no heat is required. This can cause problems in the manufacture of thick specimens as it becomes difficult to maintain a constant temperature throughout the thickness.

The viscosity of the resin increases during crosslinking as shown in figure 3.7. Such a curve can be obtained by stirring the resin with a glass rod connected to a motor and a transducer to monitor the couple necessary to keep the rod rotating. The resin is said to have gelled when the viscosity reaches 500 poise and this represents the beginning of the formation of a

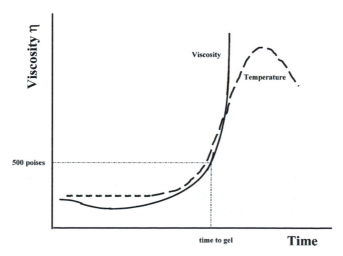

Figure 3.7. The viscosity of the resin first decreases as a function of the temperature at which the composite is cured and then increases as crosslinking proceeds. The process is exothermic so that even if no heat is supplied the temperature of the resin increases during curing.

three-dimensional network through crosslinking. The pot life of the resin is the period between the mixing of a resin with a hardener and the gelling of the resin. This represents the useful life of the resin if it is to be used to impregnate fibres.

The overall reaction of the resin can be monitored by a differential scanning calorimeter, or DSC, in which the temperature, θ, of a known mass of the resin positioned in a container is compared with the temperature of an exactly similar but empty container during a continuous and slow temperature rise. As the weight M_r and specific heat C of the resin and the temperature difference $\Delta\theta$ are known, the heat flow φ, either to or from the resin, can be calculated as

$$\varphi = \frac{M_r \cdot C \cdot \Delta\theta}{\Delta t}.$$

Figure 3.8 shows schematically the results obtained by differential thermal analysis or differential scanning calorimeter of the exothermic heat flow during the crosslinking of a thermosetting resin. Endothermic reactions such as melting, crystallization or the vaporization of solvents are also revealed by this technique.

Shrinkage during the curing of thermosetting resins can present serious difficulties which have to be countered, usually by the use of additives which physically inhibit the shrinkage. If ρ_0 is the specific gravity of the resin hardener mixture before crosslinking, say at $20\,^\circ\text{C}$, ρ_r and ρ_h, the specific gravities respectively of the resin and hardener separately, M_r the weight of the resin, M_h the weight of the hardener, we can see that the volume of the resin used is given by M_r/ρ_r and similarly the volume of the hardener is given by M_h/ρ_h. The total volume of the mixture, Vol_{mix}, is therefore

$$\text{Vol}_{\text{mix}} = \frac{M_r}{\rho_r} + \frac{M_h}{\rho_h}.$$

We can now see that the specific gravity of the mixture, ρ_0, is given by the combined weights of the resin and hardener divided by the total volume, giving

$$\rho_0 = \frac{(M_r + M_h) \times (\rho_r \times \rho_h)}{(M_r \times \rho_h) + (M_h \times \rho_r)}.$$

If ρ_1 is the specific gravity, at room temperature, of the final crosslinked resin the total volumetric shrinkage is given by

$$\frac{\rho_1 - \rho_0}{\rho_1}.$$

Shrinkage of polyester resins can attain values approaching 10%.

The monitoring of the cure of resins during the manufacture of composite structures can be simulated by impregnating a mat of fibres with the

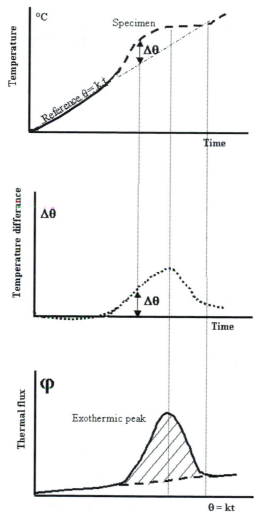

Figure 3.8. The reactivity of a thermosetting resin is exothermic so that the temperature of the resin specimen increases faster than that of the reference cell. The opposite occurs in the case of solvent evaporation.

resin suitably mixed with the hardener and monitoring the increasing stiffness of the composite specimen. The specimen is subjected to torsion by the motor which turns it repeatedly through a constant angular amplitude, and as a consequence a couple is transmitted to the transducer by the composite. As the resin hardens the couple registered increases. The time to gel can be measured by tracing the tangent of the curve to the point of intersection with the abscissa. The curing cycle for a thermosetting resin

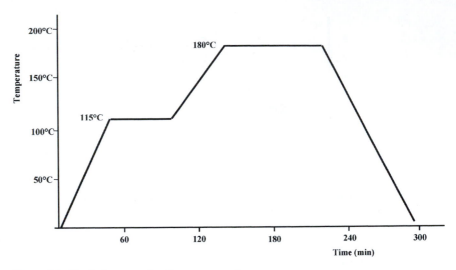

Figure 3.9. Typical cure cycle of an epoxy resin.

has to be carefully controlled so as to allow elimination of solvents and control the rate of crosslinking to obtain optimum matrix properties. The rate of heating as well as dwell times at different temperatures have to be controlled, as can be seen from figure 3.9 which is typical of an epoxy resin.

Other techniques used to monitor the curing of thermosetting resins during composite manufacture are often based on the measurement of electromagnetic impedance over a range of frequencies from Hz to MHz; alternatively the impedance of ultrasonic signals can be used.

3.3.2   Thermoplastic and cured thermosetting resin mechanical behaviour

Thermoplastic and thermosetting resins, once cured, can be evaluated by the same procedures although their responses to mechanical loading will depend on their different molecular structures. Conceptually the simplest test is a straight tensile test, often using waisted specimens so as to attempt to control the position of the failure point. Such a test can give the failure strength of matrix materials, and with the use of suitable strain gauges the elastic or Young's modulus of the polymer can be obtained. However, in practice, tensile tests can present difficulties, as gripping sufficiently hard to avoid slippage can damage specimens and cause failure in the grips. Also, many polymers are far from elastic in behaviour so that plastic deformation, creep and striction of specimens during the test complicates interpretation.

One of the most popular ways of characterizing polymers and composite materials is by three-point bend tests in which the specimen rests on two rounded supports and is loaded, on the upper surface, centrally with respect to the other two supports, as shown in figure 3.10. The type of

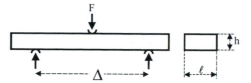

Figure 3.10. Three-point bending test.

curve obtained is shown in figure 3.11. In order to determine the apparent tensile strength (σ_u) of the specimen the ratio of the distance between supports to the specimen thickness must be greater than 16. The failure load obtained with a smaller ratio is modified as shear stresses in the specimen come to dominate failure. This is exploited in testing composite specimens, which can break by inter-laminar shear in the centre of the specimen, for which a ratio of 4 is used.

The apparent tensile strength (σ_u) and Young's modulus, E, in bending, both in MPa, can be obtained using the expressions

$$\sigma_u = \frac{3}{2} \times \frac{F_u \times \Delta}{l \times h^2} \qquad E = \frac{\Delta^3}{4lh^3} \times \tan\alpha$$

in which the failure force, F_u, is in N and the dimensions are in mm.

Typical values for resins used for matrices are $\sigma_u = 10\,\text{MPa}$ and $E = 3\,\text{GPa}$.

It is clear that these types of measurements could be made at different temperatures and the variation of different mechanical characteristics determined as a function of temperature. This is effectively what happens in viscoelastic techniques although only small displacements are used so as to facilitate analysis. These techniques subject the specimens to cyclic loads which induce deformation of the material. The induced strains and applied stresses are only in phase if the material is perfectly elastic, which is rarely the case for a matrix material for a composite, as is shown schematically in figure 3.12. The response of the material depends on the nature and time

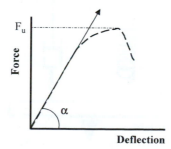

Figure 3.11. Force-deflection curve of a specimen loaded in three-point bending.

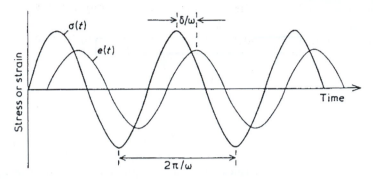

Figure 3.12. The stress and induced strain are not in phase in the case of a viscoelastic material subjected to a cyclic load.

of reaction of its microstructure, which in case of polymers varies considerably with temperature, and their viscoelastic behaviour is a sensitive means of identifying the molecular mechanisms governing their characteristics.

If the material is considered to be viscoelastic it can be pictured as an elastic spring in parallel with a dashpot, as shown in figure 3.13. This means that its response to an applied load will be time-dependent. Such behaviour can be seen to be the sum of an elastic component obeying Hooke's law, characterized by an elastic modulus E', such that

$$\frac{\sigma}{\varepsilon} = E'$$

and another viscous component obeying Newton's law, characterized by a loss modulus E'' such that

$$\sigma = \eta \times \frac{\mathrm{d}\varepsilon}{\mathrm{d}t}$$

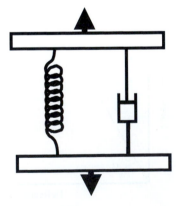

Figure 3.13. A viscoelastic material can be pictured as a spring in parallel with a dashpot.

where η is the viscosity of the damping mechanism and $\mathrm{d}\varepsilon/\mathrm{d}t$ is the strain rate.

The modulus of the material can therefore be seen to be composed of a real and imaginary component so that we can write

$$E^* = E' + \mathrm{i}E''.$$

The viscoelastic characterization of the material involves the application of a sinusoidal load to the specimen. The difference in phase, δ, shown in figure 3.12 is a direct measurement of the damping characteristics of the material and is given by

$$\tan \delta = E''/E' \quad \text{or} \quad G''/G'.$$

E' and G' are known respectively as the storage modulus and shear storage modulus. E'' and G'' are known respectively as the loss modulus and shear loss modulus.

The results of a viscoelastic analysis are thermomechanical spectra consisting of the storage and loss functions plotted as a function of temperature. A $\tan \delta$ spectrum is characterized by a series of peaks, each of which can be attributed to the onset of motion of elements of the molecular structure. Such results can be complicated to interpret as the molecular structures of many polymers, including those used for matrix materials, are complex, providing several mechanisms which can influence the viscoelastic behaviour. Figure 3.14 shows the variation of the shear storage modulus (G') and the shear

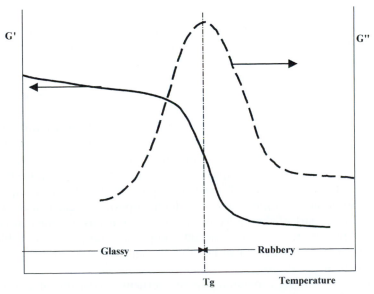

Figure 3.14. The shear storage modulus (G') and the shear loss modulus (G'') of a simple linear polymer.

loss modulus (G'') of a simple linear polymer. It can be seen that as the behaviour of the polymer passes from a glassy state to a rubbery state at the glass transition temperature the shear storage modulus falls and the shear loss modulus passes through a maximum corresponding to a peak of energy dissipation due to the liberation of the molecular structure. A similar curve could be drawn for variations of the elastic moduli E' and E''.

The effect of crosslinking a polymer is to increase its glass transition temperature. However, this also depends on the nature of the crosslinking mechanisms. The T_g of an epoxy resin crosslinked with a rigid aromatic structure will be higher than that of a similar epoxy crosslinked with a linear aliphatic structure which is more flexible. The two crosslinking processes give very different T_g with the more rigid arrangement typically having a T_g of 177 °C whereas the aliphatic crosslinking gives a T_g of 93 °C.

3.4   Thermosetting matrix systems

3.4.1   Unsaturated polyester resins

This is the commercially most important resin system used with composite materials. It is not considered to be a high performance matrix system and so is not found in advanced composites, but it is very widely used for general purpose composite applications. The basis of the unsaturated polyester resins are recurring ester group $-(-CO-O)-$ combined as linear molecules with aliphatic groups, which do not contain aromatic groups and which provide sites which can react with other unsaturated monomers such as styrene. Variation in the basic components of the polyester chain and in the ratio of the saturated and unsaturated components allows a wide range of resins to be produced to meet different performance requirements. Reactions generally occur quickly, but in a controllable manner to give a crosslinked thermoset structure.

The most widely used polyester resins, accounting for around 80% of the market, are described as orthophthalic (ortho) resins for which orthophthalic acid is used as the saturated acid part of the backbone of the polymer. They are the cheapest form of polyester resin and are used in contact moulding, typically of large structures. A slightly more expensive resin is isophthalic (iso)polyester for which isophthalic acid is used as the saturated acid part of the molecule. Isopolyester resins are used in closed moulding processes, corrosion resistant composites and gel coats. Table 3.2 shows the ingredients which can be used to make the unsaturated polyester resin backbone.

Figure 3.15 shows a common polyester structure consisting of a saturated acid, an unsaturated acid and one or more glycols, this latter group being a broad range of related structures containing two hydroxyl (alcohol) groups.

Table 3.2. The ingredients which can be used to make the unsaturated polyester resin backbone.

Ingredient		Function
Unsaturated acids and/or anhydrides	Maleic anhydride	Provides cure site
	Fumaric acid	Provides best cure site (maleic isomerizes to fumaric)
Saturated acids and/or anhydrides	Phthalic anhydride	Low cost and hard, balance of properties
	Isophthalic acid	Improved strength and chemical resistance
	Adipic acid and homologues	Flexibility and toughness
	Halogenated acids/anhydrides	Flame retardance
Glycols	Propylene glycol	Balance of properties at lowest cost
	Diethylene glycol	Flexibility and toughness
	Dipreopylene glycol	
	Bisphenol A/PG adduct	Chemical resistance and high heat deflection temperature
	Neopentyl glycol	Chemical resistance and toughness

The molecular weight of the backbone polymer is controlled by the manufacturer but is generally in the range 1000 to 4000. The site for cross-linking is provided by the unsaturated acid or anhydride group. Commonly maleic anhydride provides cure sites for crosslinking by isomerization to a maleate group which in turn transforms into a fumarate group, as shown in figure 3.16.

The isomerization of the maleate group depends on the type of glycol used. These groups are copolymerized with other monomers which contain a reactive function. General purpose polyester resins are most often based on a saturated phthalic acid, used as phthalic anhydride and unsaturated maleic acid, used as maleic anhydride, respectively used for the saturated and unsaturated acids with propylene glycol as the glycol. On average, one

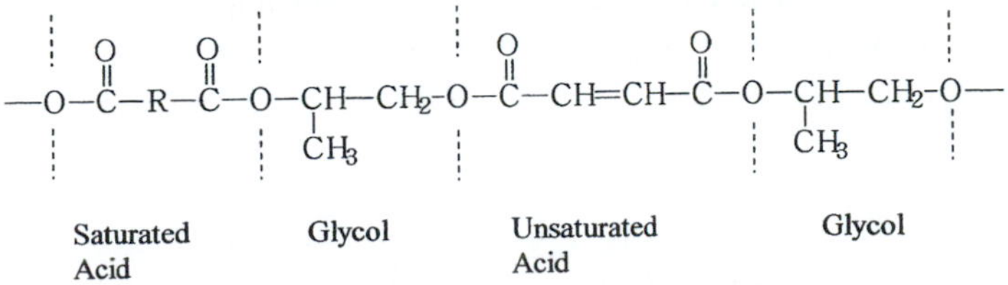

Figure 3.15. A typical composition of a polyester resin before crosslinking.

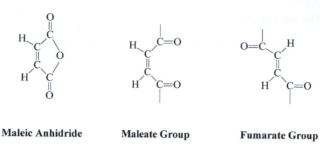

| Maleic Anhidride | Maleate Group | Fumarate Group |

Figure 3.16. The isomerization of maleic anhidride to fumerate.

or two monomer groups link the polyester macromolecules. Polyesters with improved environmental and chemical resistance and mechanical properties are often based on maleic anhydride, isophthalic or terphthalic acids and glycols, such as neopentyl glycol. Flame resistance is achieved through the use of saturated halogenated acids or anhydrides. Such unsaturated polyester resins are liquids with low viscosity which can readily be used to impregnate fibres.

The monomers which are used to crosslink the polyester must be miscible in the polyester and capable of providing two reactive sites. The result is a relatively low viscous solution which lends itself to the easy impregnation of fibres. The general form of a crosslinking monomer is shown in figure 3.17.

The most widely used crosslinking agent is styrene, shown in figure 3.18 and accounts for the distinctive smell in polyester work places. Figure 3.19 shows the preparation of the polyester resin formulation and the ongoing crosslinking process is shown in figure 3.20.

Sometimes methyl methacrylate, shown in figure 3.21, mixed with styrene is used to crosslink the polyester in order to provide greater transparency and external weather resistance. A range of crosslinking

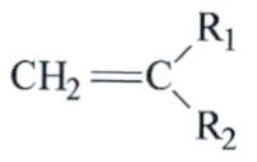

$$CH_2 = C \overset{R_1}{\underset{R_2}{\diagdown}}$$

Figure 3.17. The general form of the crosslinking monomer in which R_1 and R_2 are reactive sites allowing bonds to be made to the polyester.

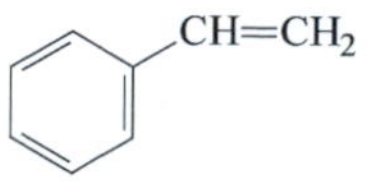

$$CH=CH_2$$

Figure 3.18. The styrene monomer.

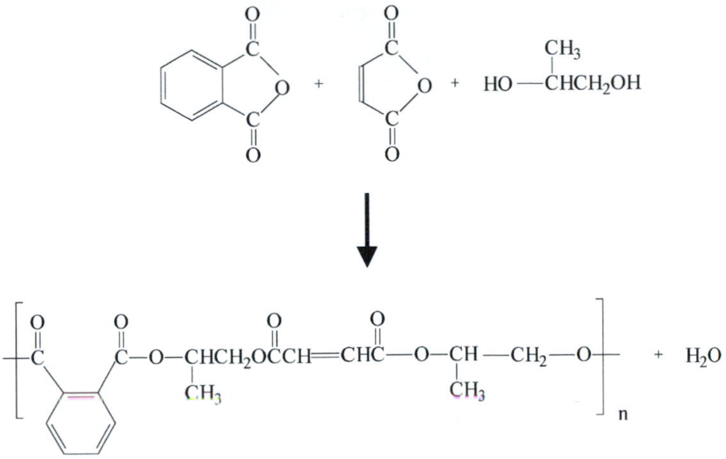

Figure 3.19. The preparation of a polyester resin.

monomers can be used to give resins with different characteristics, as shown in table 3.3.

The crosslinking reaction has to be started and this is done by an initiator or promoter, also known, inappropriately, as a catalyst, which is often an organic peroxide which breaks down at elevated temperatures to provide free radicals.

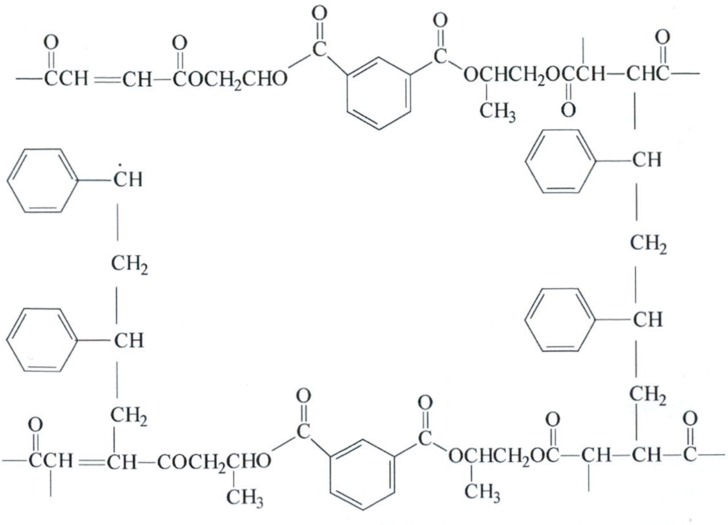

Figure 3.20. The crosslinking (or curing) of a polyester resin. The process is shown as being ongoing as the carbon atom, denoted as C with a dot over, has an unused bond which is available for further crosslinking.

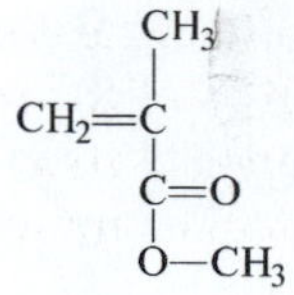

$$CH_2{=}\underset{\underset{\underset{O{-}CH_3}{|}}{\underset{C{=}O}{|}}}{\overset{\overset{CH_3}{|}}{C}}$$

Figure 3.21. Methyl methacrylate used to crosslink polyester.

A hydroperoxide R$-$O$-$O$-$H is used which decomposes when heated to provide free radicals: R$-$O$-$O$-$H $\rightarrow$ RO$^{\bullet}$ + OH$^{\bullet}$. The free radicals begin the crosslinking process

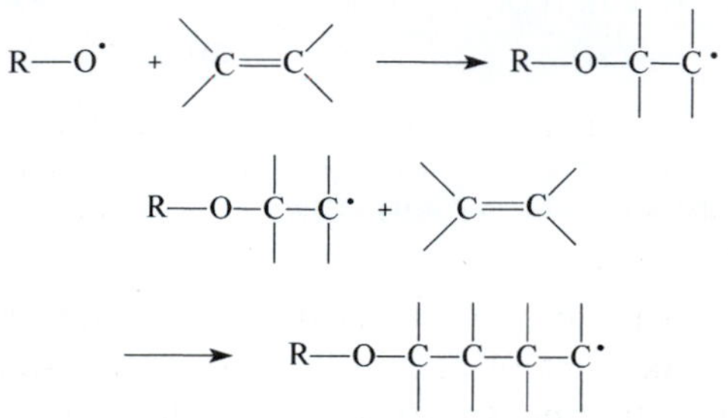

$$R{-}O^{\bullet} + \ \underset{/}{\overset{\backslash}{C}}{=}\underset{\backslash}{\overset{/}{C}} \ \longrightarrow \ R{-}O{-}\overset{|}{\underset{|}{C}}{-}\overset{|}{\underset{|}{C}}{}^{\bullet}$$

$$R{-}O{-}\overset{|}{\underset{|}{C}}{-}\overset{|}{\underset{|}{C}}{}^{\bullet} + \ \underset{/}{\overset{\backslash}{C}}{=}\underset{\backslash}{\overset{/}{C}}$$

$$\longrightarrow \ R{-}O{-}\overset{|}{\underset{|}{C}}{-}\overset{|}{\underset{|}{C}}{-}\overset{|}{\underset{|}{C}}{-}\overset{|}{\underset{|}{C}}{}^{\bullet}$$

A common peroxide which is used is methyl ethyl ketone peroxide (MEKP) which is thermally stable up to fairly high temperatures, having a half life at 100 °C of about 12 hours. Curing can occur at room temperature, however, if an accelerator is added which encourages the breakdown of the peroxide. It should be noted that the initiator and accelerator must not be mixed without the resin as an explosion will result. MEKP will readily decompose at room temperature in the presence of small amounts of

Table 3.3. Monomers used for the crosslinking of polyesters.

Monomer	Boiling point	Characteristics
Styrene	145 °C	Low cost, good reactivity, high shrinkage
Chlorostyrene	188 °C	Faster curing, high exotherms, lower shrinkage
Vinyl toluene	172 °C	Lower shrinkage
α-Methyl styrene	165 °C	Less exothermic reaction slower curing
Methyl methacrylate	100 °C	Improved weatherability
Diallyl phthalate	160 °C	Low volatility; useful in prepregs
Triallyl cyanurate (TAC)	Melting point 27 °C	Good high tempertature performance up to 200 °C
T-butyl styrene	219 °C	Low volatility

transition metal ions. Cobalt naphthenate is often used as such an accelerator with a concentration of only 0.01% of the resin weight being required.

The manufacturer of a composite structure needs to be able to control the rate of curing of the resin system to allow all the necessary manufacturing steps to take place. The rate of reaction of a polyester resin system depends on the amount of initiator, the accelerator and the temperature. Often the resins are supplied containing the accelerator, so that manufacturer can control gel time through the initiator and temperature. Inhibition of the crosslinking process through oxidation and evaporation of the styrene can cause difficulties during manufacture. This is sometimes minimized by the addition of small amounts of wax dissolved in the resin and which, during curing, exudes to the surface to form a film which keeps oxygen out and the styrene in.

Typically the failure stresses of the polyester resins is in the range 40–100 MPa, their elastic moduli between 2 and 4 GPa and the T_g of most polyester falls in the range 80 to 120 °C. The heat deflection temperature is often a property given for polyester resins and is a short-term indication of the ability of the resin structure to resist the thermally induced molecular movements. The properties of some typical polyester systems in the form of cast unreinforced specimens are shown in table 3.4.

Flammability and smoke production are factors which can determine the choice of a resin system for a particular application. Fire redundancy is achieved using halogenated resin systems or with the use of an additive, which is often alumina trihydrate. Smoke density depends in part on the crosslinking monomer. Styrene, the most widely used product, gives the highest smoke output and methyl methacrylate the lowest. The glycol which gives the highest smoke output is the widely used propylene whereas increasing the unsaturated acid content for isophthalic-maleic resins decreases smoke production. Certain fire retardant additives such as antimony trioxide increase smoke production, but the presence of alumina trihydrate has the opposite effect.

Polyester resins are generally resistant to short-term exposure to common acids, bases, solvents and water. However, long-term exposure, especially at elevated temperatures, can lead to degradation. Hydrolysis can occur with long-term exposure to water and this is accelerated in the presence of H^+ or OH^- ions.

The unsaturated polyesters represent a large family of resins and the environmental resistance of each system clearly depends on the exact chemistry used and the conditions in which the resin is to be used. Orthophthalic polyester resin does not resist boiling water well. However, it is widely and satisfactorily used in composite boat hulls. The quality of the fibre–matrix interface is often an important factor in determining resin composite degradation, as a poor interface allows rapid ingress of the environment into the composite.

Table 3.4. Typical characteristics of common unreinforced thermosetting resins used in composite materials.

Resin	Physical properties						Chemical resistance properties			
	Flexural strength (MPa)	Tensile strength (MPa)	Tensile modulus (GPa)	Failure strain (%)	Heat deflection temperature (°C)	Normal maximum temperature limit (°C)	Water	Solvent	Acid	Alkali
Unsaturated polyester										
Ortho-phthalate	100–135	50–75	3.2–4.0	1.2–4.0	55–100	80–100	fair	poor	fair	poor
Iso-phthalate	110–140	55–90	3.0–4.0	0.8–2.8	100–125	100–130	good	fair	good	poor/fair
Modified bisphenol type	125–135	65–75	3.2–3.8	0.9–2.6	130–180	130–180	very good	fair	good	fair/good
Epoxy (bisphenol)										
Aliphatic polyamide cure	85–125	50–70	3.5	1.0–3.5	60–90	100	good	fair/good	fair/good	fair/good
Boron trifluoride complex	110	85	3.0–4.0	1.0–2.5	120–190	90–150	good	fair/good	good	good
Aromatic amine cure	80–130	60–75	3.0–3.5	1.5–3.5	85–170	120–180	excellent	good	fair/good	good
Aromatic anhydride cure	90–130	80–105	2.65–3.5	2.0–2.5	130–200	150–220	poor/fair	poor/fair	good	poor
Vinyl ester										
Polyimide	110–130	70–85	3.3	1.0–4.0	90–125	90–125	good	fair/good	good	good
Friedel–Crafts	75–130	50–120	3.1–4.7	2.0–3.5	250–360	250–360	low	—	good	low
Phenolic furane	110–120	95–110	4.1	1.5–3.0	160–240	150–300	excellent	good	good	fair/good
Silicone	100–120	60–75	2.5–3.5	0.5–1.0	180–220	250–300	good	excellent	good	poor
						250–300	good	poor	fair/good	poor/fair

3.4.2 Vinyl ester resins

Vinyl ester resins represent a link between the polyester group of resins described above and the epoxy resins described further on in the chapter. As with polyester, they react with styrene in order to form a crosslinked structure through addition polymerization which is promoted by peroxide catalysts. There are usually two ester groups per molecule and two unsaturated groups at the end of an epoxy polymer chain. The unsaturated groups are monocarboxylic (COOH) acids, usually methacrylic acid but also acrylic acid. The resins can be cured at temperatures above 10 °C or at elevated temperatures using peroxide initiators, and the results are resins with properties which are intermediate between those of polyester and epoxy resins.

Figure 3.22 shows a typical chemical structure of a general purpose vinyl ester based on a bisphenol A epoxy backbone, which results in a combination of corrosion resistance, high modulus and strength, toughness, uniformity of the cured structure and reduced internal stresses.

Typical properties of such a resin are shown in table 3.5. Increasing the molecular weight of the epoxy backbone increases toughness and chemical resistance although this is accompanied with some loss in heat deflection temperature.

Most vinyl ester resins used in composite materials employ methacrylates as the unsaturated end groups as they are more resistant to base hydrolysis. As the reactive sites occur only at the ends of the molecules, the total number of ester linkages is lower than in a typical unsaturated polyester resin, which results in comparatively improved resistance to caustic degradation. In addition the reaction of the acid with epoxy is complete in vinyl esters so that no excess glycol or similar reactive groups remain available to promote processes such as osmosis which cause blistering of gel coats, used as protective coatings of composite structures, and of unsaturated polyester composite laminates.

The hydroxyl groups in the vinyl ester chain promotes good wetting and adhesion to polar surfaces such as those of glass fibres.

Specialized vinyl ester resins are produced for particular purposes. Acrylic vinyl ester resins can be cured by exposure to ultraviolet light. Vinyl esters based on multifunctional epoxy backbones, which results in a

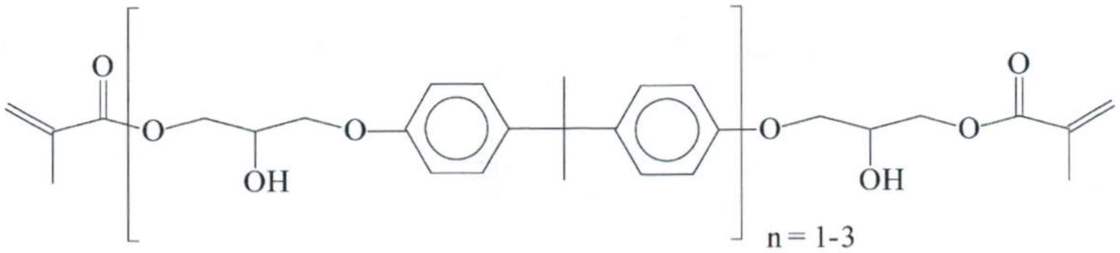

Figure 3.22. Molecular structure of vinyl ester.

Table 3.5. Typical properties of a vinyl ester resin based on bisphenol A epoxy.

Liquid properties		Cast properties	
Specific gravity	1.4	Tensile strength	80 MPa
Gel time at 25 °C	28 min	Failure strain	6%
Gel time at 80 °C	12 min	Flexural modulus	3.1 GPa
		Heat deflection temperature	100 °C

high density of crosslinking, can be made which enhances resistance to solvents and increases T_g. Fire retardant properties can be achieved by the use of a backbone structure of tetrabromobisphenol A which also improves fatigue resistance. Impact resistance is improved through rubber toughening by incorporating carboxy terminated acrylonitrile butadiene.

The uses of vinyl esters are many and include storage tanks and piping because of their resistance to corrosive environments and parts for reasonably high temperature applications. The resin system is suitable for most manufacturing processes including hand lay-up, projection techniques, filament winding, resin transfer moulding (RTM) and pultrusion.

3.4.3 Epoxy resins

Epoxy resins are generally considered to possess more attractive properties than polyester resins for high performance composite materials as they provide a resin which has low shrinkage, high adhesive strength, excellent mechanical strength and, provided that the correct system is used, chemical resistance.

The molecular structure of epoxy resins is based on the epoxy, or oxirane, group, shown in figure 3.23, consisting of two carbon atoms and an oxygen atom within a molecular chain so that R and R' represent the continuations of the chain on each side of the group.

The epoxy cycle is strained, which accounts for its high reactivity.

A type of epoxy used in composites is known as cycloaliphatic epoxy resins, in which the oxygen atom is attached to a six-member carbon ring, as shown in figure 3.24. This type of resin is recommended for use in

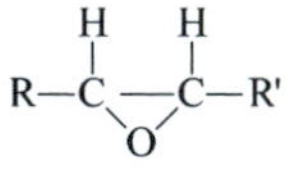

Figure 3.23. The epoxide group.

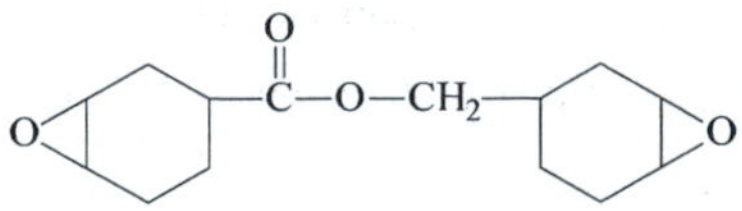

Figure 3.24. Cycloaliphatic resin.

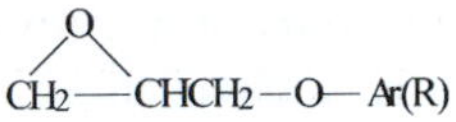

Figure 3.25. Glycidyl ether.

combination with bisphenol A epoxy resin for filament winding to increase their heat distortion temperature and to reduce viscosity of the resin.

However, the epoxide group which is most often used is glycidyl ether, which is shown in figure 3.25 in which Ar(R) means an aromatic ring.

The commercially most important epoxy resin known as diglycidyl ethers of bisphenol A (DGEBA) in which the epoxy group is attached to a hydrogen atom (R) and to glycidated polyhydroxyphenols (R'), accounts for 90% of epoxy resins produced. They are synthesized from epichlorohydrin and bisphenol A in the presence of sodium hydroxide. The synthesis of DGEBA epoxy resin is shown in figure 3.26.

The value of n in figure 3.26 indicates the number of times the structure within the brackets is repeated and this can be varied so that the density of crosslinking can be altered. In commercially available resins, n falls in the range 0–25. As n increases so does the number of hydroxyl groups. The epoxy resins with low values of n usually require curing agents that react with the epoxy group, whereas resins with higher n values are cured through the hydroxyl groups. In any given resin there will be a range of

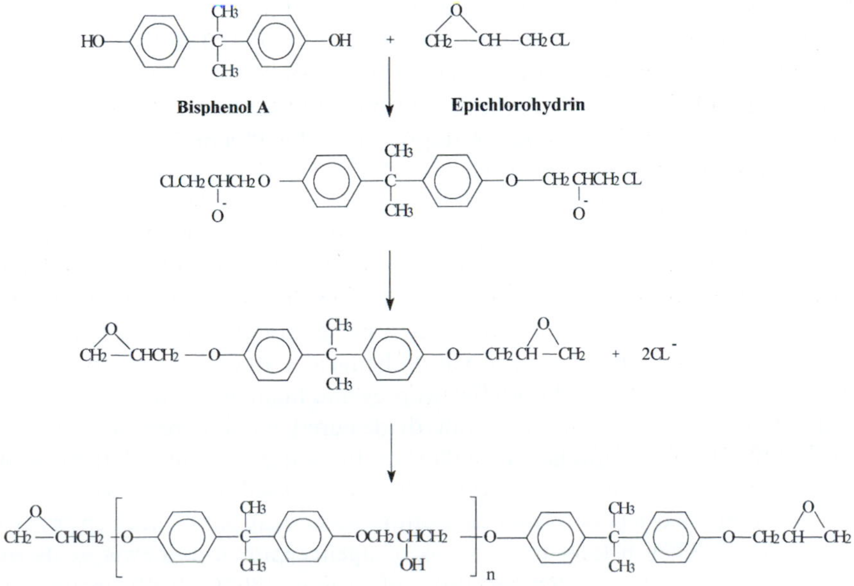

Figure 3.26. Synthesis of diglycidyl ethers of bisphenol A (DGEBA) epoxy resin.

chemical structures so that the value of n is an average value and so can be less than 1. Liquid epoxy resins, such as those used in composite materials, have values of 0–1 with values greater than 2 being solid.

Many applications including encapsulation, casting and filament winding, which require fluid resins epoxies with n values less than 1, are used as they offer the best balance of handling, reactivity and performance properties. Epoxy resins with n values higher than 1 can be used for composite manufacture if they are heated, so as to bring down their viscosity. This is the case for some resins cured with aromatic amines; however, increasing the temperature also increases the rate of reaction so that the pot life or usable lifetime for the resin is reduced. Solvents which do take part in the crosslinking process can also be used to reduce the viscosity of some resins, and this is applicable in the manufacture of some laminates in which the solvent can be removed by evaporation before curing. If the solvent is not completely removed during composite manufacture its presence can seriously degrade the composite properties, so such systems are not used in processes such as filament winding in which thick layers of fibre may be produced, thus inhibiting solvent removal before resin crosslinking.

Curing of epoxies can be carried out using a variety of agents. However, the most widely used systems consist of three families of material. These are amines, anhydrides and catalytic agents. Amines are the most widely used and include both aliphatic and aromatic amines. The aliphatic amines are used if room temperature curing is required as they are more reactive than anhydride amines. For this reason they are used for contact moulding processes for which their short curing time is useful, but they are not used for prepegging for which a slower reaction is required. The high volatility of the aliphatic amines can lead to handling problems including skin irritation and for this reason they can be combined with other agents to reduce their reactivity. Aromatic amines require raised temperatures to produce crosslinking of epoxides and so can be used for processes such as filament winding and prepegging. Figure 3.27 shows some aliphatic and aromatic curing agents.

Anhydrides are widely used to cure epoxies as they give long pot lives and low exotherms. They require high curing temperatures, typically 180 °C, for periods of several hours to achieve crosslinking and the cured resins have good high temperature and chemical stability, except to caustic soda, NaOH. They are used both for prepreg and filament winding processes. It should be noted, however, that anhydride cured epoxies are susceptible to degradation by water uptake, particularly if the water is warmer than around 40 °C. Figure 3.28 shows some anhydride systems used to cure epoxy resins.

Catalytic hardeners promote homopolymerization of the epoxy to achieve cure. They belong to a group of agents known as Lewis acids or bases and require curing temperatures of around 180 °C. Such agents are shown in figure 3.29.

ALIPHATIC AMINES

$$H_2NCH_2CH_2H\!-\!\!-\!CH_2CH_2NH_2$$

Diethylene Triamine (DETA)
liquid

$$H_2NCH_2CH_2\overset{\overset{\displaystyle H}{|}}{N}CH_2\overset{\overset{\displaystyle H}{|}}{N}\!-\!CH_2CH_2NH_2$$

Triethylene Tetramine (TETA)

AROMATIC AMINES

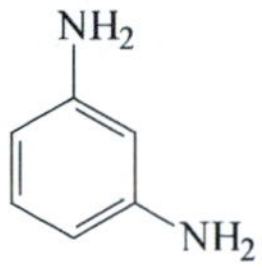

MPDA (meta-phenylenediamine)
white solid

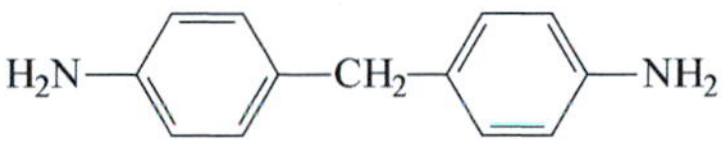

MDA (methylene dianiline)

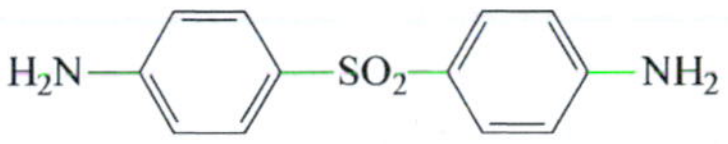

DDS (diaminodiphenyl sulphone)

Figure 3.27. Aliphatic and aromatic amine curing agents used with epoxides.

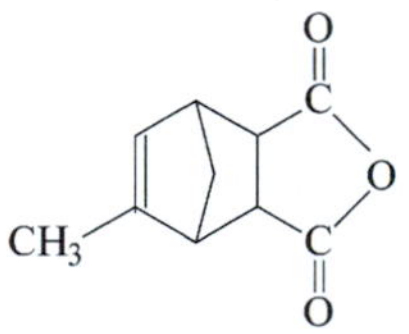

NMA (nadic methyl anhydride)
liquid

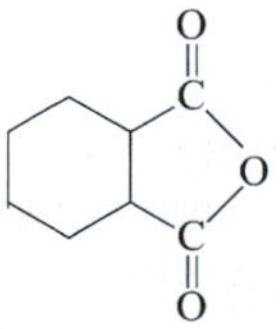

HHPA (hexahydrophthalic anhydride)
solid

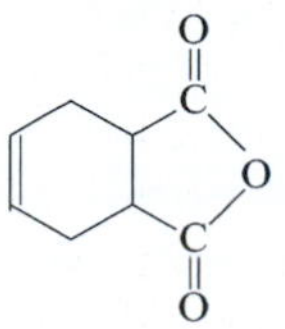

PA (phthalic anhydride)
solid

Figure 3.28. Anhydride curing agents for epoxides.

BF₃ NH₂CH₂CH₃ — Boron trifluoride monoethyl amine (BF₃ MEA). White solid; melting point 210 °C

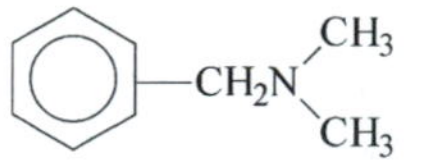

Benzyl dimethyl amine (BDMA). Liquid; boiling point 180 °C

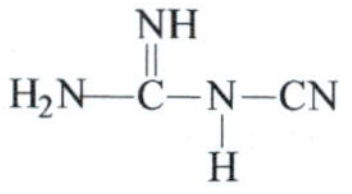

Dicyandiamide. White solid

Figure 3.29. Catalytic curing agents.

The process of crosslinking is very dependent on the type of curing agent used. One of the most used systems is shown in figure 3.30 in which an anhydride hardener is used.

An amine cured epoxy system is illustrated in figure 3.31. The reaction involves the epoxide group opening to combine with the hydrogen atom in the amine. In the case of aliphatic amines curing can occur at room temperature but higher temperatures around 150–180 °C are necessary in the case of aromatic amines. The latter system is known as a B-staged resin.

The above discussion is primarily concerned with the most widely available and used epoxy system based on bisphenol A, which is restricted in use at temperatures below 150 °C. Higher temperature stability can be obtained using epoxy systems which have functionalities greater than 2. A commonly used system of this type is the tetrafunctional resin tetraglycidylamine, TGMDA, shown in figure 3.32. This type of resin is gradually being overtaken by more complex systems with higher T_g and greater resistance to water. Such epoxy systems extend the long-term temperature range of epoxies to 175 °C.

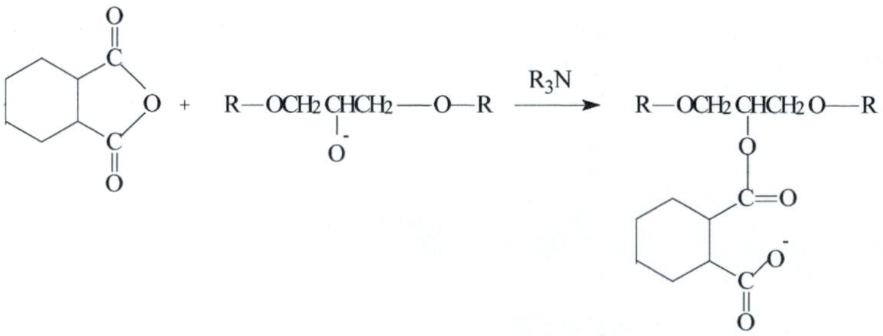

Figure 3.30. Crosslinking reactions with an anhydride hardener.

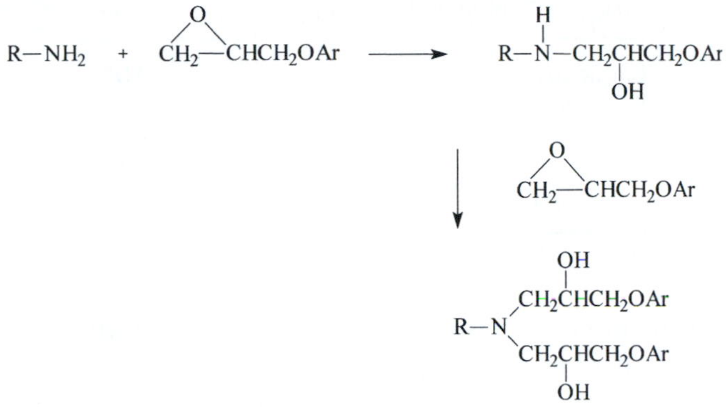

$$R-NH_2 \quad + \quad CH_2{-}CHCH_2OAr \quad \longrightarrow \quad R-\underset{|}{\overset{H}{N}}-CH_2\underset{|}{CHCH_2OAr}$$

$$\downarrow \quad CH_2{-}CHCH_2OAr$$

$$R-N\begin{array}{c} CH_2CHCH_2OAr \\ \\ CH_2CHCH_2OAr \end{array}$$

Figure 3.31. Crosslinking reactions with an amine hardener.

3.4.4 Polyurethane resins

These resins represent a large family of resin systems, most of which are made into foams, elastomers and surface coatings. They are not usually associated with composite materials; however, there are some exceptions which are becoming important and finding use in reinforced reaction injection moulding (RRIM) and continuous manufacturing processes such as pultrusion. These manufacturing techniques are described in chapter 4.

Polyurethane are polymers which are made by a reaction of an organic isocyanate with a polyol, such as an alcohol, which is any compound containing multiple hydroxyl groups. The general formula for an isocyanate is [$R-N=C=O$] in which R is a reactive, usually large aromatic group. The most widely used isocyanates are toluene diisocyanate (TDI) [$CH_3C_6H_3(NCO)_2$] and methylene diphenyl isocyanate (MDI) [$OCNC_6H_4CH_2C_6H_4NCO$]. Isocyanates contain the extremely reactive group [$-N=C=O$]. It will react with any hydroxyl [$-OH$] and can be combined with many different polymers so that a large range of possibilities exist. It should be noted that, because of the high reactivity of this group, it poses a considerable health risk as inhalation will cause a reaction with the [$-OH$] groups in the lung tissue, causing a respiratory illness called isocyanate asthma. Damage can be very debilitating and can be fatal. One variation of isocyanate chemistry was the development of phosgene poison gas which was used in the First World War. However, once made, the

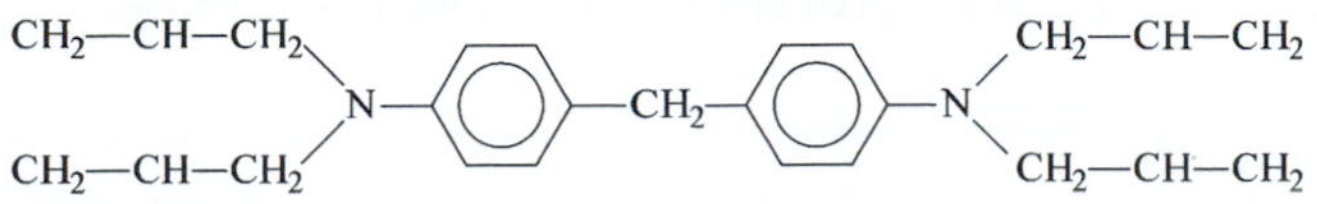

Figure 3.32. TGMDA (tetraglycidyl-4,4′-diaminodiphenylmethane) high performance epoxy resin.

Table 3.6. Typical properties of phenolic resole resins.

Tensile strength	61 MPa
Tensile modulus	3.9 GPa
Strain to failure	1.7%
Shear modulus	1.65 GPa
Specific gravity	1.26

resins which are produced have the isocyanate combined with other reactive agents so that after curing they pose no danger. For example they can be combined with epoxy systems to give resin systems which have continuous in-service temperature capabilities of 200 °C.

These resins are of interest because they allow rapid crosslinking, usually in times of a few minutes, and at temperatures which are low, from 25 to 60 °C. The resins are generally tough, due to their high impact strength and strain to failure which is greater than most other systems. They adhere well to fibres. The composites which are produced with these resins absorb much less water than, say, epoxy resins and therefore of interest for applications for which this could present a problem.

3.4.5   Phenolic resins

Phenol formaldehyde, usually known as phenolic resins, are the oldest completely synthetic polymers and were developed at the beginning of the 20th century. Phenolics are divided into two classes which are known as resoles and novolacs, also called, respectively, one-step and two-step resins. Resoles are the most used as a matrix system and prepared under alkaline conditions with formaldehyde/phenol ratios of more than one. The reaction is stopped by cooling to give a reactive and soluble polymer which can be stored at low temperatures and subsequently crosslinked by heating, in the temperature range 130–200 °C, as a continuation of the initial condensation reaction. This is the origin of the 'one-step' nomenclature. Resoles are used to impregnate papers or fabrics for both electrical and structural applications. Novolacs are not often used as a matrix for composite structures. They are produced under acid conditions with formaldehyde/phenol ratios of less than one. The reaction is carried to completion to give an unreactive thermoplastic which in a second step can be cured by the addition of hexamethylene-tetramine. See table 3.6 for properties of phenolic resole resins.

3.4.6   Polyimides

High temperature stability is achieved with the use of polyimide resins, which can be used as matrix materials up to temperatures around 300 °C. For

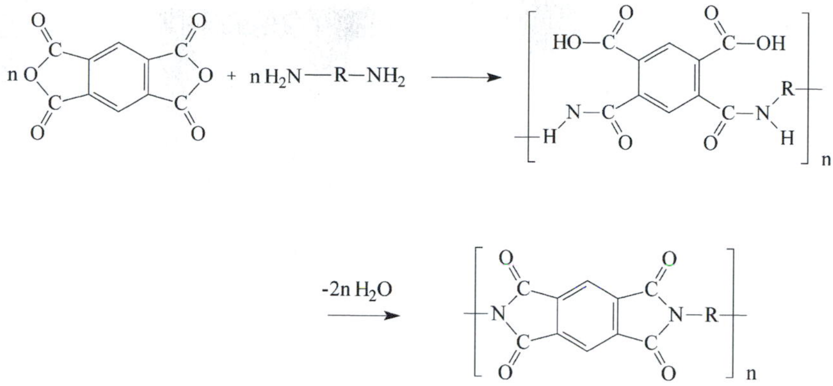

Figure 3.33. The formation of polyimide.

this reason polyimides are attractive not only for structural composite applications but also for electrical insulation. Polyimides can be made both in thermosetting and thermoplastic forms. These resins are generally made in two stages, consisting of the polycondensation of a dianhydride and an aromatic diamine which gives an acid which is soluble in polar solvents. Heating, usually above 250 °C to remove water results in the formation of the polyimide cycle, shown in figure 3.33. Typical properties of polyimide resins are shown in table 3.7.

A variety of routes exist for making polyimides, so polymerization by addition is also possible. Generally it is difficult to control the chemistry of

Table 3.7. Typical properties of a polyimide resin.

Tensile strength	
23 °C	100 MPa
93 °C	80 MPa
Tensile modulus	
23 °C	2.5 GPa
93 °C	2.2 GPa
Strain to failure	
23 °C	6.54%
93 °C	5.0%
T_g	
Dry	210 °C
Wet	190 °C
Specific gravity	1.37
Moisture absorption	0.9%

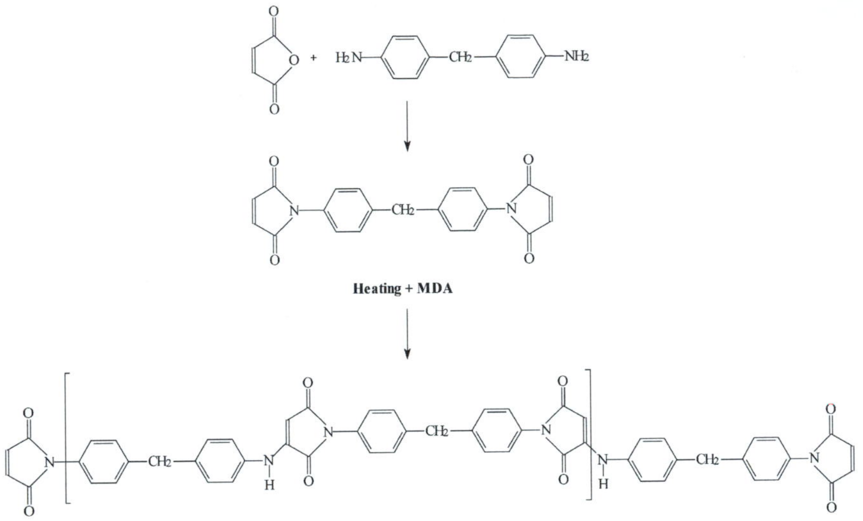

Figure 3.34. The chemistry of bismaleimide.

the reactions taking place to create polyimides, so problems associated with the stability of the reactants, their volatility, the molecular weight distribution and porosity can occur. The latter can be a particular problem in thick structures and can lead to bubbles and delamination of the structure. The manufacture of composites using polyimide resin systems usually requires long curing cycles.

Bismaleimides are thermosetting resins which are based on the polyimide chemistry but made by addition curing, and can be more easily processed than most resins in the family. The chemistry of their manufacture is shown in figure 3.34, resulting from the reaction of a maleic anhydride with methylene dianiline. However, the MDA is increasingly replaced by other aromatic diamines as it presents a health risk. The reaction on heating is with a linear function with a carbon backbone. The resulting resin is very brittle, but this can be countered by the addition of an aromatic diamine which is copolymerized with the bismaleimide. Processing occurs at high temperatures, above 175 °C, and leads to crosslinking of the resin.

Bismaleimides have an upper temperature use of around 225 °C and can be processed almost as easily as epoxies. In addition they are inexpensive and require only low pressures during composite manufacture. However, in general, bismaleimides are more brittle than epoxies, showing poor fracture toughness and as a consequence suffer from microcracking in laminates.

In situ polymerization of monomeric reactants (PMR) gives the most widely used resins with the highest in-service temperature capabilities.

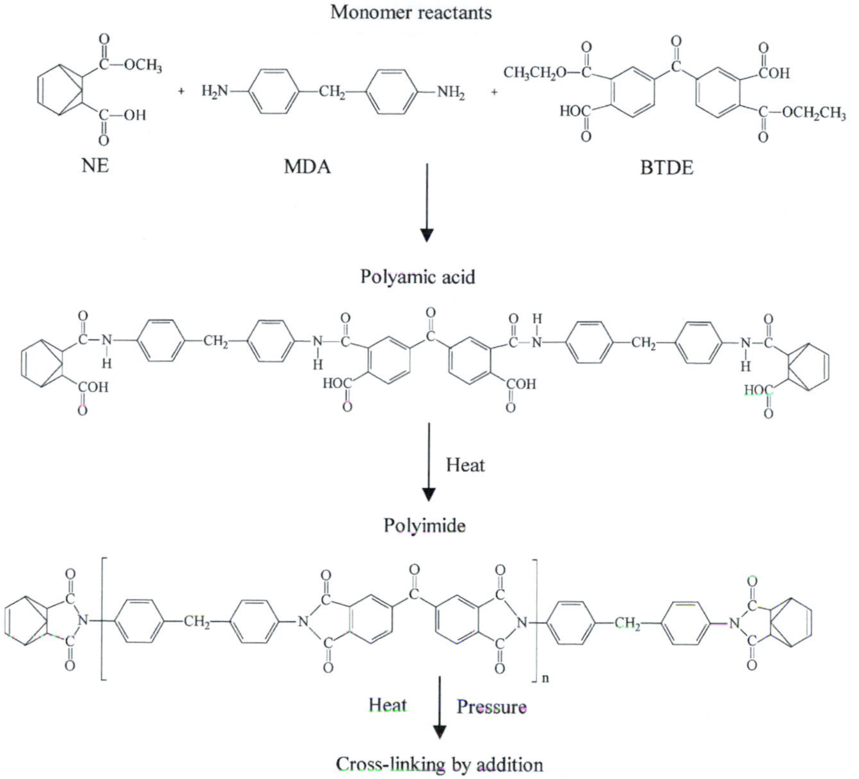

Figure 3.35. The chemistry of PMR-15 production *in situ*.

PMR-15 is a thermosetting polyimide with temperature capabilities up to 316 °C for at least 1500 hours in air. It is made by impregnating the fibres with a low molecular weight polymer consisting of nadic ester (NE), 4,4′-diaminodiphenyl methane (MDA) and benzophenone tetracarboxylic acid diethyl ester (BTDE). The MDA component is carcinogenic, which has led to much interest in producing resins with similar high temperature capabilities but without this component. The thermal crosslinking addition process for producing PMR-15 is shown in figure 3.35.

PRD-15 has a molecular weight of 1500, hence its name. It is brittle and is of primary interest to the aerospace industry for high temperature uses. It is destined most probably to be displaced with modified polyimides which should be tougher and less toxic during processing. Properties of PMR-15 are shown in table 3.8.

Most resins which are said to retain their properties to above 300 °C suffer, however, from a fall in T_g to around 225 °C when subjected to temperature cycles in a humid atmosphere.

Table 3.8. Properties of PMR-15 resin.

Tensile modulus	3.2 GPa
Tensile strength	55 MPa
Compressive strength	190 MPa
Specific gravity	1.3
T_g	350 °C

3.5   Thermoplastic matrix materials

The plastics industry produces far more thermoplastics than it does thermo-setting plastics, approximately in the ratio of 4 : 1; however, this ratio is not maintained in the area of composite materials which represent about 3% of the total plastics industry. Approximately twice as much thermosetting matrix material is used for composites than is thermoplastic matrix material. The rate of growth of thermoplastic matrix composites is, however, consid-erably higher than that of the thermosetting, composites. This, in part, is because of the widening use of composite parts and the need for faster pro-duction rates with thermoplastics than are possible with most thermosetting resins. Thermoplastics can simply be melted and formed with no delay, as occurs with thermosets while they crosslink. In addition thermoplastics do not need to be stored at low temperatures, which reduces costs and handling problems, and they are inherently tougher than thermosetting resin. Costs are further reduced as there are fewer scrapped parts because thermoplastics can be reprocessed and recycled. The natures of the two types of composites are, in the main, different, thermoplastics being primarily used for short fibre injection moulded composites and only relatively recently finding their way into long or continuous fibre composites. This difference is accounted for partly by historical reasons, with thermosets being employed for impregnat-ing fibre structures well before the use of thermoplastics, but above all for reasons of processing. The thermosetting resins can be easily used to impreg-nate fibre tows as they are either liquid or easily put into solution when in their non-crosslinked state. The thermoplastics are solids, so they have to be put into intimate contact with the reinforcements when in a molten state. This clearly is the case for injection moulded composites; and in the case of continuous fibres, analogous to most thermosetting systems, the thermoplastics are placed in contact with the fibres either in the form of a powder or in the form of filaments. This intimate combination is then heated so as to melt the thermoplastic matrix material. Adhesion with the fibres can be a difficulty with some thermoplastic matrices, such as poly-propylene, which have to be modified for good fibre–matrix bonding to occur. The high viscosity of molten thermoplastics and their lack of available free bonds is the cause of processing difficulties and the difficulty of creating

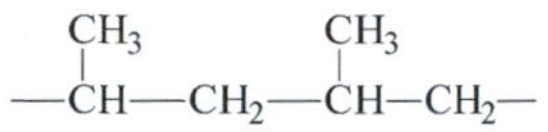

$$\begin{array}{ccc} CH_3 & & CH_3 \\ | & & | \\ -CH-CH_2- & CH-CH_2- \end{array}$$

Figure 3.36. The structure of polypropylene.

good adhesion with fibres. This is more of a difficulty with continuous fibre composites than with injection moulded parts as the shearing which occurs in the injection process can create bonding sites in the matrix material.

Thermoplastic composites can be based on readily available materials, such as polypropylene, polyamide (Nylon) and saturated polyethylene terephthalate (polyester) but also other more exotic and high performance thermoplastics are used such as polyetheretherketone (PEEK).

3.5.1 Polypropylene

Polypropylene, together with polyethylene and polybutadiene, make up the important polyolefine group of polymers. Polypropylene is a commodity plastic and together with polyvinyl chloride (PVC) and polysulphone, which are also commodity plastics, they represent 70% of the unreinforced thermoplastic market. These three plastics share the attribute of being of relatively low cost, but polypropylene is interesting as a composite matrix material as it is more rigid and stronger than PVC, more easily worked than polysulphone and has the lower density ($900\,kg\,m^{-3}$). Polypropylene is semi-crystalline when in the isotactic, most usual form and has a crystallinity of about 68%. The polymer can therefore be seen as consisting of an amorphous phase, which is in the rubbery phase at room temperature, in which are embedded the spherulitic crystals. Polypropylene is a linear polymer based on a backbone of C–C bonds with lateral methyl groups, as shown in figure 3.36.

The physical and mechanical properties of unreinforced polypropylene are shown in table 3.9. The low melting point should be noted. The polymer shows several transition temperatures at which changes to the structure occur. The most important of these transitions is the glass transition

Table 3.9. Mechanical and physical properties of undrawn bulk polypropylene.

Density	$905\,kg\,m^{-3}$
Melting point	$165–170\,°C$
Elastic modulus	$1.0–1.4\,GPa$
Yield stress	$25–38\,MPa$
Strain to failure	300%
Coefficient of linear expansion	$175 \times 10^{-6}\,°C$
Processing temperature	$190–285\,°C$

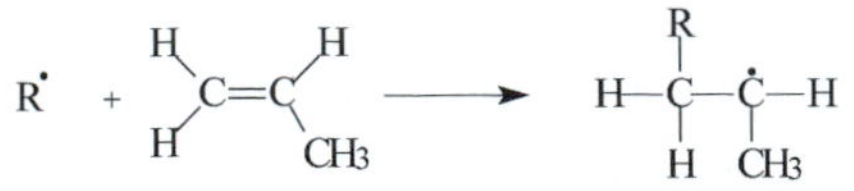

Figure 3.37. Reaction of an initiator radical R' with polypropylene.

temperature or T_g. As polypropylene is semi-crystalline there is a T_g for the amorphous regions which is around 5 °C and one for the crystalline regions, around 90 °C. Above these temperatures the molecular structure acquires sufficient energy to allow much freer movement. There is another, secondary transition temperature around −80 °C corresponding to localized movements in the amorphous phase. Such transitions can be revealed by viscoelastic tests, as are described above, in which the energy absorbed during cyclic loading is revealed by the phase difference between the applied strain and the induced load in the material. At such transitions the energy absorbed is a maximum.

Polypropylene is not very reactive because of the non-polar nature of the molecule and this leads to problems in forming bonds between the matrix and the fibres in a composite. Two approaches are used to improve bonding. The more usual technique is to incorporate, into the size on the fibres, a coupling agent which can bond both to the glass and to the polymer. Such coupling agents are often based on silanes which contain the $-Si(OH)_3$ group and which can bond to the silanol $-Si-O-$ groups on the glass fibre surface, as described in chapter 2. The silane can then bond to the polymer matrix so achieving adhesion. The lack of functionality of pure polypropylene means that to achieve bonding the polymer is often modified by the addition of reactive groups to the polymer chain, as shown in figure 3.37. Bonding of the polypropylene matrix to the glass fibres takes place in two steps, during which the macroalkyl radicals are formed by shearing of the polymer chains making up the matrix in the injection moulding process.

An adduct is then produced by grafting a modifier or functionalized group to the activated chain. The modifier must be bifunctional with one end group reacting with the size on the fibres and the other with the polymer. The addition of such modifying agents as acrylic acid or maleic anhydride, as shown in figure 3.38, leads to improved bonding as the matrix is able to form strong bonds with the size or directly with the glass.

3.5.2 Polyamide 6.6

The polyamides are a group of polymers often referred to as Nylons, which is a commercial name. They are engineering plastics which mean that they possess some useful properties, such as strength, stiffness, toughness and resistance to chemical attack when in the unreinforced state. The most

$$CH_2\!=\!CHCOOH$$

Acrylic Acid

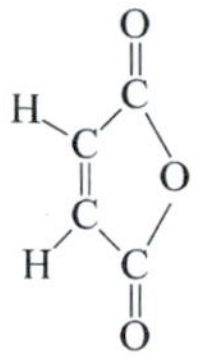

Maleic Anhydride

Figure 3.38. Commercial modifying agents used to improve the bonding between polypropylene and glass fibres.

$$H_2N - (CH_2)_6 - NH_2 \; + \; HOOC - (CH_2)_4 - COOH \longrightarrow$$

Hexamethylene diamine   adipic acid

$$-[\;-\,NH - (CH_2)_6 - NH - CO - (CH_2)_4 - CO\,-]_n -$$

Polyamide 6.6

Figure 3.39. The production of polyamide 6.6 by the copolymerization of hexamethylene diamine and adipic acid.

widely used polyamide, both in fibre and bulk form, is polyamide 6.6 which, as its name indicates, consists of linear molecules based on repeated units of two groups of six carbon atoms, as shown in figure 3.39.

Polyamide 6.6 is made by the copolymerization of hexamethylene diamine and adipic acid, both of which contain six atoms of carbon. It is semi-crystalline, ranging in crystallinity from 25 to 50%. It has one of the highest melting points of the polyamide family, 264 °C, and is typically processed around this temperature, but in common with other polyamides has relatively low viscosity in the molten form which aids processing. It adheres well to reinforcing fibres and shows good resistance to most chemical environments, although is sensitive to strong acids. In its unreinforced form it is used in many light engineering structures, from fans and door handles in cars and curtain fittings, which make use of the low coefficient of friction of polyamides. As a short glass fibre reinforced composite it finds uses around car engines due to its excellent resistance to oil. Some typical properties of bulk polyamide 6.6 are shown in table 3.10.

Table 3.10. Mechanical properties of polyamide 6.6.

Density	$1140 \, \text{kg} \, \text{m}^{-3}$
Melting point	264 °C
Elastic modulus	1.5–2.5 GPa
Yield stress	60–75 MPa
Strain to failure	40–80%
Coefficient of linear expansion	$90 \times 10^{-6} \, °\text{C}^{-1}$
Processing temperature	260–325 °C

Polyamide 6.6 absorbs significant amounts of water, which causes large falls in glass transition temperature. For example, a completely dry specimen will show a T_g of around 70 °C but as it absorbs water the T_g will fall as the water acts as a lubricant to the molecular structure. This can increase toughness as it reduces stress concentrations; however, water absorption, at saturation, can reach 6% of the dry weight of an immersed specimen at which point the T_g is 0 °C. This effect is largely reversible if the polymer is dried. Polyamide 11 and polyamide 12 have extra methylene groups between the amide groups which make these forms of Nylon more flexible and tougher, and also reduces the amount of water they absorb, so for some marine applications these polymers could be considered as matrix materials.

3.5.3 Polyester thermoplastic matrix

All polyesters contain the ester group, shown in figure 3.40. This group can be hydrolysed at temperatures around 150 °C by the process shown in figure 3.41.

For this reason it is most important to ensure that the polymer is dry when processing it and making composites. The most widely used polyester is polyethylene terephthalate (PET), which is shown in figure 3.42.

It is a semi-crystalline homopolymer made by the polycondensation of terephthalic acid and ethylene glycol with the crystalline phase typically making up 35% of the material. Its wide use is due to it being the lowest cost and highest volume polyester produced. It is melt blown for such products as drink bottles and spun into fibres. It has a low viscosity which helps the processing of composites. The presence of the aromatic ring means that the PET matrix is stiffer, stronger and can operate at higher temperatures than say polypropylene. It is used in many of the same types of application as polyamide 6.6. Compared with this material it absorbs

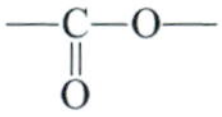

$$-\underset{\underset{O}{\|}}{C}-O-$$

Figure 3.40. The ester group.

$$R-\underset{\underset{O}{\|}}{C}-OR' + HOH \longrightarrow R-\underset{\underset{O}{\|}}{C}-OH + R'OH$$

Figure 3.41. The hydrolysis of a polyester.

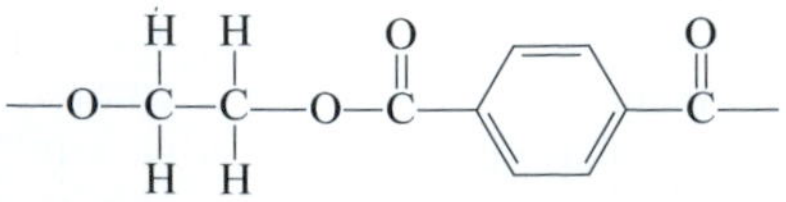

Figure 3.42. Polyethylene terephthalate (PET).

Table 3.11. Typical properties of PET.

Specific gravity	1.35
Melting point	264 °C
Elastic modulus	2.8 GPa
Yield stress	80 MPa
Strain to failure	70%
Coefficient of linear expansion	$70 \times 10^{-6}\,°C^{-1}$
Processing temperature	225–350 °C
Water uptake at 100% relative humidity (20 °C)	0.5%

less water and its T_g, which is around 75 °C, does not decrease with water uptake. The effect of water at temperatures above the T_g is to broaden the peak revealed by DSC measurements as hydrolysis sets in and cuts the long PET molecules. This enables the molecules to reassemble and increase crystallinity up to as much as 55% and broaden the range of crystallite sizes in the matrix. The effect of increased crystallinity is to reduce toughness, which if taken to extreme values can induce cracking in the matrix. Typical properties of unreinforced PET are shown in table 3.11.

3.5.4. Polyimides

This class of matrix material is interesting because of its high transition temperatures. However, partly as a result of this, processing is difficult. Linear polyimides are thermoplastics but they degrade before reaching their melting point so that they are made from solution or converted into a polymer in situ. Most polyimides are viscous and this complicates processing. Figure 3.43 shows polyether imide (PEI) which is marketed by GE Plastics under the name of ULTEM. This matrix material is derived from phthalic anhydride, bisphenol A and meta-phenylenediamine (MPD). The inclusion of an ether bond in the molecular backbone allows it to be processed from the melt. The molecular structure is irregular in shape, as can be seen from figure 3.43, and for this reason it is amorphous.

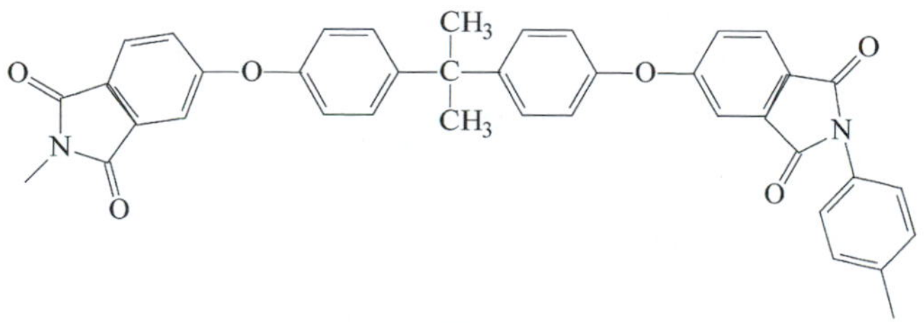

Figure 3.43. The molecular structure of PEI is irregular which results in the resin being amorphous.

Table 3.12. Typical properties of polyimides.

Specific gravity	1.26
Elastic modulus	2.7–4.0 GPa
Yield stress	105 MPa
Strain to yield	5–8%
Strain to failure	5–90%
Coefficient of linear expansion	$50 \times 10^{-6}\,^{\circ}\mathrm{C}^{-1}$
Melting point of PEI	349 °C
Upper service temperature of PEI	200 °C
Processing temperature of PEI	335–370 °C
Water uptake at 100% relative humidity (20 °C)	1.25%

PEI is a general purpose moulding resin with excellent processing characteristics, excellent resistance to chemical attack and hydrolysis. It has high mechanical strength, high heat deflection temperature and excellent flame resistance. PEI is used in composite applications because of its high in-service temperature, toughness, good interfacial adhesion and nonflammability. Typical properties of polyimide resins are shown in table 3.12.

3.5.5 Sulphur-containing polymer matrices

Sulphur (S) and the sulphone group of compounds containing the radical SO_2 united with two hydrocarbon radicals can be placed in the backbone of aromatic polymers to give useful high temperature matrix systems. The chemical formulae of several such polymers are given in figure 3.44. The presence of sulphur in the backbone increases flexibility whereas sulphone in more rigid. Polyphenylene sulphide (PPS) is a homopolymer and is

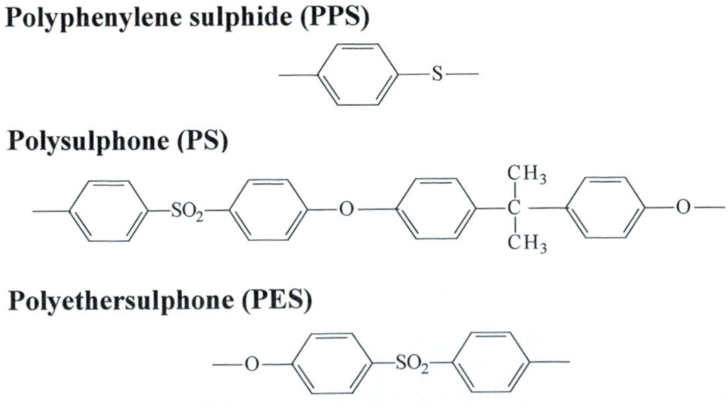

Polyphenylene sulphide (PPS)

Polysulphone (PS)

Polyethersulphone (PES)

Figure 3.44. Sulphur-containing matrix systems.

Table 3.13. Typical properties of sulphur-containing matrix systems.

Polymer	Specific gravity	Elastic modulus (GPa)	Failure stress (MPa)	Failure strain (%)	Coefficient of expansion ($^\circ C^{-1}$)	Processing temperature ($^\circ C$)
PPS	1.35	3.35	65	2.5	35	330 °C
PS	1.24	2.6	67	75	55	360 °C
PES	1.37	2.6	85	60	50	350 °C

crystalline. It is used as a composite matrix because of its chemical resistance, good mechanical properties and low moisture uptake. It has a high melting point and is non-flammable. Typical properties are shown in table 3.13.

Both polysulphone and polyethersulphone are amorphous. PS has a T_g in the range 180–265 °C and resists thermal oxidation well. Both are therefore used in applications such as food trays which require repeated sterilization. These amorphous polymers are very resistant to hydrolysis as well as to aqueous acids and bases, but they are susceptible to chlorinated solvents and most highly polar organic solvents. They can be used continuously in the 150–220 °C range.

3.5.6 Polyethers

Polyethers contain oxygen bonded to carbon in their backbone. The ether group introduces flexibility into the molecule which makes melt processing easier. There are many polyether thermoplastic resins available and some are used for composite materials. However, the best known is polyetheretherketone (PEEK) which is semi-crystalline, possesses a melting point around 380 °C and can be used for high temperature applications. Its molecular formula can be seen in figure 3.45. It possesses high mechanical properties,

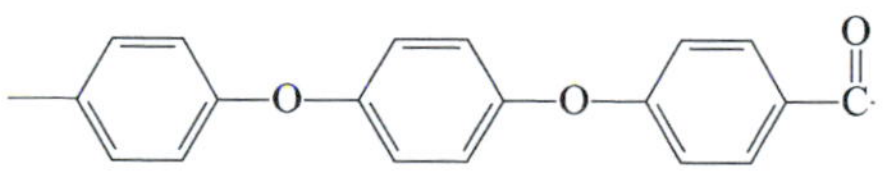

Figure 3.45. Molecular structure of polyetheretherketone (PEEK).

Table 3.14. Typical properties of PEEK.

Specific gravity	1.3
Elastic modulus	3.6 GPa
Yield stress	90 MPa
Strain to failure	20%
Coefficient of linear expansion	$45 \times 10^{-6} \,^\circ C^{-1}$
Melting point	340 °C

as can be seen in table 3.14, as well as being highly resistant to chemical attack and being nonflammable. It shows excellent adhesion to carbon fibres and gives very good interlaminar shear strengths for the composites for which it is used as a matrix. It shows transcrystalline growth from the irregularities on the carbon fibres surfaces. In order to control the microstructure of PEEK it is recommended to consolidate structures made from it at around the melting point and then to transfer the specimen to a cooler press at 150 °C.

3.6   Properties of fibre reinforced composites

When resins are reinforced the composite which is formed has properties which depend on both the resin properties and those of the fibres and their arrangements, as is discussed in chapter 7. Table 3.15 shows values of short glass fibre reinforced injection moulded composites with different matrices. In this type of composite the fibres are short, with lengths of just a few millimetres. They are generally randomly arranged although the flow of the matrix material can cause alignment of the fibres. The use of glass fibres, which have a relatively low modulus, together with the random arrangement means that the overall properties are not high but this type of composite is relatively cheap to produce and is widely used.

Table 3.16 shows typical properties of unidirectional composites with an epoxy matrix and reinforced by various fibres. The values given for the carbon fibre composites considered are produced from PAN and the high modulus type values are for fibres with a modulus of 500 GPa. Unidirectional composites are highly anisotropic, as discussed in chapter 7. Loading these composites in the direction of the fibres exploits their properties to a maximum and the role of the matrix is to ensure the integrity of the composite, provide shear strength and ensure load transfer to the fibres. The balanced glass fibre specimens were loaded parallel to the direction of half of the fibres.

Table 3.17 shows typical properties of some of the thermoplastic resins discussed above, reinforced with high strength carbon fibres which are all aligned in the same direction. Values for a reinforced epoxy resin are also

Table 3.15. Typical values of short glass fibre reinforced injection moulded composites.

	PP	PA66	PET	PPS	PEEK	PEI	PES
Weight%	40	40	40	40	30	40	30
Tensile strength (MPa)	139	138	110	138	157	214	196
Flexural modulus (GPa)	5.6	9.5	12.0	14.1	10.3	11.0	11.0
Heat deflection temperature (dry 182 MPa)	148	245	215	260	315	210	285

Table 3.16. Typical values of fibre reinforced epoxy resin.

	E-glass	S-glass	Balanced woven E-glass	High strength carbon	High modulus carbon	Kevlar
Fibre volume fraction (%)	0.55	0.50	0.45	0.63	0.57	0.60
Specific gravity	2.10	2.00	2.20	1.58	1.57	1.38
Longitudinal modulus E_1 (GPa)	39	43	30	142–180	294	87
Transverse modulus E_2 (GPa)	8.6	8.9	29.7	10.5	6.4	5.5
Shear modulus G_{12} (GPa)	3.8	4.5	5.3	7.4	4.9	2.2
Poisson's ratio ν_{12}	0.28	0.27	0.17	0.27	0.23	0.34
Poisson's ratio ν_{21}	0.06	0.06	0.17	0.02	0.01	0.02
Longitudinal tensile strength (GPa)	1.08	1.28	0.37	2.28–2.86	0.59	1.28
Transverse tensile strength (MPa)	39	49	367	49–57	29	30
In-plane shear strength (MPa)	89	69	97	71–83	49	49
Longitudinal tensile failure strain (%)	2.8	2.9	2.5	1.4–1.7	0.3	1.5
Transversal tensile failure strain (%)	0.5	0.6	0.25	0.6	0.5	0.5
Longitudinal compressive strength (GPa)	0.62	0.69	0.55	1.44–1.9	0.49	0.34
Transverse compressive strength (GPa)	0.13	0.16	0.55	2.3–2.5	0.98	0.16
Longitudinal coefficient of thermal expansion ($\times 10^{-6}\,°C$)	7.0	5.0	10.0	−0.9 to −0.3	−0.1	−2.0
Transverse coefficient of thermal expansion ($\times 10^{-6}\,°C$)	21.0	26.0	10.0	27–30	26	60.0

Table 3.17. Typical properties of unidirectional carbon fibre thermoplastic matrix composites and an epoxy resin matrix composite.

	Epoxy	PPS	PEEK (APC2)	Polyimide	Polyamide
Carbon fibre volume fraction (V_f)	0.6	0.6	0.6	0.6	0.6
Tensile strength (GPa)	1.9	2.3	2.5	1.9	2.4
Flexural strength (GPa)	1.7	1.2	1.5	1.5	1.2
Initial elastic modulus (GPa)	139	139	139	139	139
Flexural modulus (GPa)	130	130	130	130	125
Short beam strength (MPa)	110	70	115	95	105
Compressive strength (GPa)	1.3	0.95	1.0	1.0	1.0

given. The values of strengths are in the fibre direction.

3.7   Conclusions

Organic resins find wide uses as matrices for composite materials. They can be either thermosetting resins or thermoplastic polymers. Their properties depend intimately on their chemistry as does their cost and ease of incorporation as a matrix material. Polyester resins are most widely used in general purpose composites but epoxy resins represent the most important group of resin systems used in high performance composites. Other resin systems exist which have better high temperature properties than these common systems but their chemistry is generally more complex. Thermoplastic resins offer an adaptability after manufacture which is lacking in thermosetting systems and so are attractive both for the speed of composite manufacture that they allow but also as there should be fewer problems associated with recycling. Whilst thermosetting resin systems represent about two-thirds of the matrix systems used in composites, thermoplastics show two or three times their growth rate as composites find increasing uses in an ever-widening field of industrial sectors.

3.8   Revision exercises

1. Explain the difference between a thermosetting resin and a thermoplastic. What happens if they are heated? Give examples of each type of material.
2. What defines a thermostable resin?
3. The macromolecules on which resins are built are derived from monomers. Discuss how these monomers can be made into long chain macromolecules. Explain how a bifunctional system results in a thermoplastic and how crosslinking can occur with other systems.
4. Polyesters are the most commonly used thermosetting resin system but polyesters are also used in the production of fibres. Discuss and explain the signification of this observation.
5. Explain what is meant by a homopolymer and copolymerization? What are isotactic, syndotactic and atactic polymers?
6. Explain the significance of glass transition temperature.
7. Discuss the phenomena involved in the crosslinking of a thermosetting resin, explaining what is meant by gel time and how shrinkage can be a problem with these resins.
8. If the curing of a thermosetting resin is characterized by DSC it is found that there is a heat flow (φ) of 20 mW in a period of 30 s (Δt) for the specimen which weighs 60 mg (M_r). If the specific heat is

$0.1 \, \mathrm{J \, kg^{-1} \, {}^{\circ}C^{-1}}$ calculate the rise in temperature. (Answer 100 °C.)

9. All polyester resins are based on a recurring molecular group. What is this group and what is it called?
10. What is the most common crosslinking agent used with thermosetting polyesters?
11. What is the role of an initiator in the crosslinking process?
12. Discuss some of the drawbacks of polyester resin which need to be overcome if accurately dimensioned structures may be affected by fire. Discuss the relevant importance of using styrene or methyl methacrylate as the crosslinking monomer.
13. Why are epoxy resins preferred to unsaturated polyesters for high performance composites. Give the chemical formula for the epoxide group.
14. What is the most widely used epoxy resin?
15. There are several families of curing agents which are used with epoxy resins but one is particularly sensitive to degradation by water. Which one is it and why?
16. If low water uptake is required which type of resin system could be used and why must precautions be taken during its manufacture?
17. Give examples of resin systems which can be used at higher temperatures than those which are possible with epoxy systems.
18. What is the most widely produced thermoplastic and why does this present some difficulties when used as a matrix material? Nevertheless the polymer is increasingly used in composites. Why is this?
19. What is the chemical name of Nylon 6.6 and what do the numbers signify?
20. What is the melting point of saturated polyester (PET) and why does this mean that it is important to prepare the polymer under dry conditions?

References

Lee S M 1991 *International Encyclopaedia of Composites* (New York: VCH Publishers)
Daniel I M and Ishiai O 1994 *Engineering Mechanics of Composite Materials* (Oxford: Oxford University Press)
Delaware Composites Design Encyclopaedia 1990 (Technomic Publishing Co. Inc.)
Kelly A and Zweben C (eds) 2000 *Comprehensive Composite Materials* (Oxford: Pergamon)

Chapter 4

Composite manufacturing processes

4.1 Introduction

Manufacturing processes used for composite materials are often very different from those used with conventional structural materials such as metals, although in some cases similarities can be found with the processing of polymers. The materials we are considering are based on fine filaments, which are very long with respect to their diameters, so that there are many aspects of composite manufacture which are common with textile processing. Metal structures are most often made by first producing bulk metal and then machining it into the final form. However, advanced composite materials are most often two-dimensional materials in which the fibres lie parallel to the plane of the material. A carbon fibre pressure vessel can be made by winding the filaments around a preform which can become an impermeable liner. An aircraft wing may be a three-dimensional structure but it can be made by stretching a two-dimensional composite skin over a honeycomb form. The fibres in the skin may be laid up in discrete layers of parallel fibres with each layer having fibres arranged in different directions or the fibres can be woven. The directions in which the fibres are placed control the behaviour of the final product. In some cases techniques borrowed from the textile industry can place fibres at right angles to the plane of the fabric by such means as needle punching, sewing or knitting so as to improve interlaminar shear behaviour. The flexibility of fibres allows them to be woven or wound onto mandrels so that complex forms can be produced which, when impregnated with resins, and then cured, can lead directly to the final product which requires little or no extra finishing. In this way the total cost of manufacturing a component with a composite material can be reduced, compared with traditional materials. Another technique borrowed by the composite industry from traditional fibre processing is braiding. Braiding is rather like plaiting hair so that by using a number of spools of fibres virtually continuous tubes or profiles can be made at high speed. Pultrusion also produces virtually continuous tubes or profiles by combining the simultaneous pulling of the fibres through a die, which determines the

132

cross sectional shape, and the extrusion of the resin through the same die. The final product is generally unidirectional but subsequent turning the profile after the die or over winding fibres around it can overcome this limitation. Lower cost products may be made by injection or closed moulding techniques which resemble those used in the plastics industry. In these processes the fibres are of a short length, ranging from a few millimetres to one or two centimetres, and are arranged randomly either in the plane of a sheet of the composite or in three dimensions so as to give a material which can be formed in a closed mould. Alternatively fibres can be chopped and sprayed together with the resin onto an open mould. This technique is useful for large structures including boat hulls but is also used for other smaller low cost structures. By far the greatest number of composite materials are made with polymeric matrices which can be either thermosetting or thermoplastics. The manufacturing process for all of these composites must follow the following flow chart:

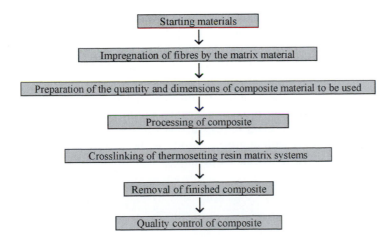

All manufacturing processes require the material to be formed into its final shape. An advantage with composites is that the final shape can often be achieved in one operation with little need for further expensive finishing processes, whereas metals often have to be machined into the final form required. However, shrinkage of the composite due to the crosslinking process can be considerable, if not countered. A polyester resin will shrink by around 8% in volume which will lead to a linear shrinkage of around 2%. Nevertheless the presence of reinforcements as well as appropriate fillers can reduce this to a fraction of a percent in volume. Often the composite is made in a mould but as the pressures and temperatures which are necessary for forming composites are low compared with say metal casting the moulds can be made of cheaper materials. For many fabrication methods the moulds which are used are themselves made from composite materials. The cost of such moulds made in metal would typically be ten times higher. This explains why composites can be competitive with metal structures for modest fabrication runs but if the rate of production is increased they may become less attractive as moulds have to be replaced more frequently or more expensive, longer-lasting moulds have to be used. For example, a production run of, say, one hundred thousand composite structures per year may be competitive but doubling this rate may mean that the composite has to be replaced by metal.

4.2   Reinforcements

Fibre reinforcements come in different forms and glass fibres present the widest range of these forms. Figure 4.1 shows a spool of continuous glass fibres. The glass fibres are typically 14 µm in diameter, which is a fifth of the diameter of a human hair, so although glass is a brittle elastic material the fineness of the fibres makes them very flexible and they can be treated in much the same way as continuous textile fibres. They can be woven or filament wound around a mandrel. The fibres are, however, sensitive to abrasion, so they are coated with a resin or size which serves to protect them, to facilitate their conversion into finished products by acting as a lubricant, to hold them together as a tow to aid handling and also to bond them to the composite matrix and in some cases as an antistatic agent. Figure 4.2 shows a woven cloth of glass fibres. In this form the fibres can be used to form composite parts with complex shapes as the weave holds the fibres in position. The use of a woven cloth means also that the position of the

Figure 4.1. Glass fibres are first wound into spools or bobbins after which they can be either used in traditional textile processes such as weaving or transformed into other forms.

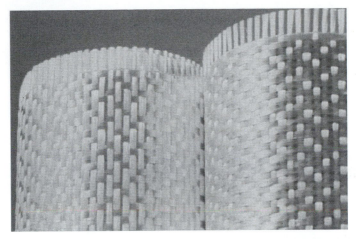

Figure 4.2. Continuous glass fibres can be woven with a variety of different weaves to produce a cloth which then can be impregnated with a resin and placed in a mould to form the composite.

fibres is controlled so that they can be placed so as to make optimum use of their properties.

Some fibres, usually carbon but also glass and aramid fibres, are available pre-impregnated in the form of a sheet and this material is known as prepreg. The continuous unidirectional fibres are coated with the uncured thermosetting resin or the thermoplastic polymer. The thickness of the prepreg sheets is around 0.3 mm and they can easily be cut to shape with a cutter and then stacked in a mould for processing. The layers of prepreg can be arranged in different directions so as to place fibres in the numbers and directions that give optimum properties to the finished composite.

Continuous or long fibres can be used to form non-woven mats of randomly arranged fibres, as shown in figure 4.3. The size is used to hold the fibres together before the mat is cut to shape and impregnated with the uncured thermosetting resin before composite manufacture. Cutting out the weaving process reduces cost but means that the fibres cannot be aligned preferentially so as to make optimum use of their properties.

Glass and sometimes other fibres, such as carbon fibres, are often chopped into short lengths varying from 5 to 75 mm for use in a variety of manufacturing processes. The widest use is in the form of sheet moulded compound (SMC) in which the fibres are cut to short lengths and allowed to fall on a moving sheet of uncured thermosetting resin. The resulting mixture of short fibres in a viscous resin can then be cut to shape, and placed in a mould for a cheap method of making a composite part. Figure 4.4 shows short chopped glass fibres.

Figure 4.3. Long glass fibres can also be randomly laid out to make a non-woven mat. The size on the fibres is used to hold them together before the mat is impregnated with the resin.

Short fibres must also be used in making composite parts by injection moulding into a closed mould. The fibres can be added separately from the thermoplastic resin in the screw extruder or they can be first prepared in the form of granules. In this latter case the continuous fibres are first extruded with the molten polymer to form a continuous ribbon of unidirectional material which is then cut into short lengths of around 20 mm. These

Figure 4.4. The glass fibres can be chopped into lengths generally between 5 and 75 mm in length and used to produce chopped strand mat or sheet moulded compound. The size on the fibres is used to hold them together.

Figure 4.5. Glass fibres can be extruded in a continuous form together with a thermoplastic such as polyamide and then chopped to lengths of around 20 mm to be used in injection moulding.

granules can then be used in the extrusion process with better control over final length. Figure 4.5 shows such granules.

4.3 Resin matrix processes

Manual production processes are the simplest for producing composite structures but they are also the slowest. They can be adapted for the production of large structures but are only of interest for short production runs. Most commonly the reinforcements which are used are glass fibres and the resin is unsaturated polyester, although higher performance fibres and epoxy resin are both sometimes employed but, as these are more costly, other techniques are usually preferred for these latter systems. The techniques often involve the use of open moulds which means that the quality of the finish of only one surface can usually be controlled as it is formed against the mould. Also the emission of styrene, which is used in the crosslinking of polyester resins, can be high. Although no direct link between styrene and health problems has been conclusively demonstrated the distinctive smell associated with its use leads to much speculation as to its effects. Legislation varies enormously with factors of five existing between different industrialized countries. A tightening of international regulations on styrene emission can be expected.

At the laboratory scale, the simplest method of producing a high performance composite flat plate is to use a process called hot pressing which makes use of a female mould of the dimensions of the desired plate. The clean mould is coated with a release agent such as PTFE or Teflon spray to allow the finished product to be removed. The fibres are taken from a bobbin, cut into lengths equal to the length of the mould which has the dimensions of the desired plate, placed in the mould and then impregnated with uncured resin, which has been premixed with the hardener, by stippling the resin into the fibres with a brush and closing the mould. The mould is designed with escape holes around the edge through which excess resin can escape. The shape of the mould controls the final dimensions of the composite plate and the quantity of fibres placed in the mould determines the fibre volume fraction. In such a process the mould is placed between the heated platens of a press, closed and taken through the heat cycle for curing the resin. If a solvent is used to control the viscosity of the resin, a dwell time to allow the solvent to evaporate may be necessary. On heating, the resin first becomes very liquid and the pressure applied forces it around the fibres. Excess resin is forced out of the mould. As the resin begins to crosslink its viscosity increases dramatically and eventually the resin gels. Crosslinking proceeds asymptotically towards complete crosslinking, as discussed in chapter 3.

As described, the resin flows towards the edge of the mould, taking with it any entrapped bubbles of air or solvent but also dragging fibres so that the finished composite plate often shows fibre misalignment, particularly in the centre. The use of a bleeder cloth placed over the composite, within the mould, allows the excess resin to be absorbed and provides a much shorter channel across the thickness of the specimen for the resin to be removed. This largely eliminates the displacement of the fibres. At the end of the curing process the mould is opened and the composite plate removed.

The process described above allows us to examine several important factors which must be taken into account when making a composite. Firstly a release agent must be used so the finished product can be removed. At the laboratory level a Teflon spray can be used, but this is an expensive product, and more often a wax, such as bees' wax, or a synthetic wax is used, often diluted with a solvent. Silicone sprays are usually avoided so as not to pollute the composite and because the finish obtained is not always smooth. Several layers of wax may be necessary but need not be replaced after each fabrication cycle. An alternative method is to use a thermoplastic film such as thermoformable polyester, PVC (polyvinyl chloride) or cellulosic films. Not all thermoplastic films are suitable as they are attacked by the styrene monomer used in the crosslinking process of polyester resins. Other films can be deposited onto the mould. These are PVA, polyvinyl alcohol dissolved in water, which has the advantage of allowing the mould to be easily cleaned.

The surface finish of the composite which is made depends on the quality of the surfaces of the mould. Many structures have a gel coat which is a surface layer of resin, usually containing other products which both act as fillers and modify the properties of the resin. Such a gel coat takes on the finish of the mould and, if this is very smooth, can provide an excellent surface finish. Most pleasure boats, and some bigger boats such as mine sweepers, are made by applying the composite to an open mould which has the shape of the hull. The attractive appearance of many of these crafts is due to the quality of the moulds and the gel coat which assures a smooth surface finish. The gel coat can be coloured to give the composite material a particularly desired appearance and is also used as a protective layer against abrasion and water ingress, although this latter characteristic is not as obvious as is often claimed since water can diffuse through such a resin layer. Environmental ageing is covered in chapter 11, which discusses long term behaviour. The gel coats vary as a function of the composite system used but are often of the same type of resin as used in the matrix. They should not be diluted with acetone as the presence of this solvent can cause problems due to outgassing, forming bubbles and porosities, so producing an unacceptable finish. In the case of polyester resins the gel coat can be diluted with styrene monomer. They can be applied to the inside of the mould with a brush or projected onto it. The gel coat has to be left to gel before the composite can be added to the mould and the mould closed.

After the manufacture of the composite it may be desirable to post-cure it, which may enhance the crosslinking process and in the case of polyester resins, for example, remove excess styrene left over from the manufacturing process.

4.3.1 Contact moulding

This technique is simple and suitable for large as well as small structures, but it is slow and only suitable for small series at a slow rate of production of up to 500 units per year. The most widely used composite system which is used with this method is glass fibre reinforced polyester. A rigid mould with a hard surface is used and is most often made with glass fibre reinforced polyester or epoxy resin, concrete, wood or plywood, plaster, aluminium or steel sheet. The fibres, in the form of mat or woven roving, are placed inside a female mould or on a male mould, depending on the desired finished article, which has previously been coated with a release agent and, if required, a gel coat with a thickness of up to 1 mm, which may be coloured. The reinforcement is then manually impregnated with the resin to which an appropriate catalysing agent has been added to promote crosslinking. Figure 4.6 shows a schematic illustration of this process. The resins used for this process are most commonly unsaturated polyesters, vinyl esters and epoxies although the latter are used less frequently than the other two. The laminate is then consolidated with a brush or roller to eliminate air bubbles, as shown in figure 4.7. To produce a thick composite the process is repeated a number of times as necessary, with a dwell time between each layer as the layers should begin to gel but not to cure completely between each application.

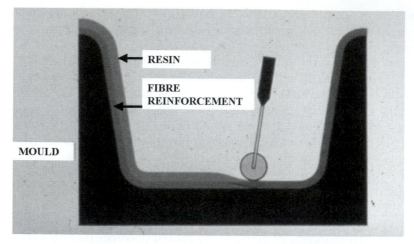

Figure 4.6. A schematic illustration of contact moulding. The reinforcement, usually in the form of random mat, is placed in the mould and impregnated with the resin to which a suitable crosslinking agent has been added. The laminate is then rolled or brushed to consolidate it and to remove entrapped air.

Finally the composite is allowed to fully cure, usually at room temperature. The finished product has a smooth appearance on the surface which is in contact with the mould, but the other surface will usually be rough and reveal the fibre reinforcements. This process can be used to make very large structures such as large boats.

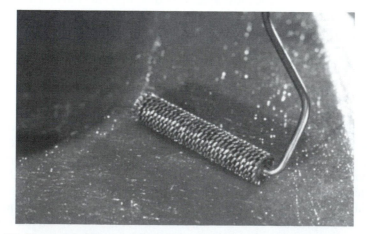

Figure 4.7. Entrapped air and volatiles are removed by rolling the composite, which also consolidates the structure.

4.3.2 Spray-up moulding

This technique has much in common with contact moulding and the two are often combined in alternating sequences. The principle of the technique is to use compressed air to simultaneously project the fibres and the resin onto the surface of the mould. The continuous fibres are brought to the spray gun, which is often hand held, from bobbins and are there chopped, to lengths between 25 and 75 mm, by a rotary chopping head before being projected together with the resin, as shown in figure 4.8. The fibres are randomly distributed on the surface of the mould. After projection the laminate has then to be rolled to eliminate entrapped air and ensure good consolidation. Successive layers can be sprayed to obtain the desired thickness and if required mat or woven layers can be included in the thickness.

Hand lay-up, contact moulding and spray moulding may be combined in the production of a composite structure, for example a boat hull, as the latter technique is unable to provide the fibre content of the former and this determines mechanical properties. Although spray lay-up can be automated, if the structure is simple, such as cylindrical tubes and pipes, and is the cheaper and more versatile process, it is restricted to a glass fibre weight content of around 35% whereas around double is possible with hand lay-up. In addition hand lay-up allows greater control over fibre alignment which again is important for mechanical properties.

4.3.3 Bag moulding and autoclaving

The previous techniques do not use any applied pressure during crosslinking of the composite. Bag moulding allows pressure to be applied evenly over the

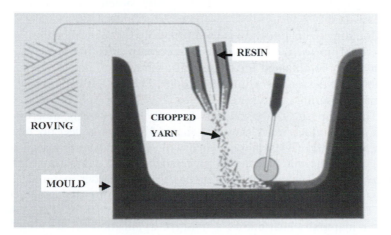

Figure 4.8. Spray moulding involves simultaneously projecting chopped fibres and resin against the mould.

Figure 4.9. A large autoclave in which is placed a composite laminate that is to be consolidated. The autoclave allows complex cure cycles to be used and the application of pressure under controlled temperature conditions.

composite structure. A flexible sheet, which can be rubber, is laid over the composite lay up, sealed at the edges and the air sucked from beneath the sheet. The depressurization needs only to be relatively slight, as too great a suction will extract styrene from polyester systems which is needed for the crosslinking process. Alternatively, pressure, in the range of 1 to 3 atmospheres is applied over the whole area of the sheet. Often both air removal and pressure is applied. The appearance of the two surfaces of the composite is better than in the two previously described techniques. Further curing can take place after this initial process by heating in an oven. This is usually only considered necessary in the case of advanced composite systems for which prepreg sheets are used.

Moulding in an autoclave is used primarily for the production of small numbers, of up to 100, high performance composite structures, as higher pressures can be obtained for consolidation but, in principle, has much in common with the bag moulding techniques. The process is time consuming with manufacturing cycles lasting one to three hours. It is therefore expensive. The composite to be made is laid up as previously described and placed in the autoclave, as shown in figure 4.9. Often a peel ply of woven Nylon, polyester or glass fibre treated with release agent is placed on top of or under the composite and is removed after fabrication for final surface finishing such as bonding or painting. A bleeder material, usually of glass fibre, is also used to absorb excess resin during the manufacturing cycle. There is then a barrier film of material such as PTFE. Then comes a breather material which allows a uniform application of the pressure by allowing entrapped air or volatiles to be removed during cure. It may be a loosely woven fabric or felt which can be easily draped over the composite to take its form. Then comes the vacuum bag. The air in the bag is removed so that it collapses onto the composite to be consolidated and allows the solvents to be removed. The curing cycle and applied pressure can then be automatically controlled. The autoclave is usually pressurized with nitrogen or a mixture of nitrogen and air in order to avoid a fire risk. The autoclave is designed so as to allow gas circulation to achieve a uniform temperature. This manufacturing process is of little interest for large volume production but is widely used in the aerospace industry.

4.3.4 Medium series moulding

Manual techniques are slow and are rarely used in industrial processing except for very large structures such as boat hulls. Medium series processes use techniques for which the impregnation of the fibres is achieved automatically and the finish of the structure is more reproducible than in the above techniques.

4.3.5 Resin transfer moulding (RTM)

This is one of the most widely used composite manufacturing techniques in industry. It can be used to produce quite large series of complex composite structures. The principle of the technique is to inject resin into a closed mould which contains the reinforcements which have previously been placed there. The pressures to force the resin into the mould are low, typically 690 kPa, which allows simple and cheap moulds to be used. The number of pieces to be made will determine the type of material used for the mould. Composite moulds with a polyester resin can be used to produce up to 2000 items but better finishes and larger production runs of up to 4000 items are obtained with epoxy composite moulds. Stamped steel moulds will allow several tens of thousands of composite structures to be made although the forms which can be made can only be simple. Composite moulds with a projected surface finish of zinc bismuth or zinc aluminium can give composite surfaces which can be either matt or shiny and allow production runs of up to 10 000 units. For runs of up to 50 000 pieces, nickel copper electro-plated onto epoxy composite moulds can be used and give excellent surface finishes. The reinforcement is placed inside the mould which is then closed and the resin subsequently injected into it under low pressure so as to avoid disturbing the placement of the fibres. The resins are of low viscosity, typically in the range of 0.1 to 1 Pa/s, fast curing resins such as polyester and epoxy, although bismaleimide and phenolic based resins find uses for high temperature applications, and the reinforcement chopped strand or continuous glass fibre mat. Mould closing pressures used are low and in the range 1–6 atm (100–600 kPa), which means that the tonnage of the presses need not be high (20 tons/m^2) which keep the cost of equipment low. Permeability of the mat is therefore important. Air is driven out as the resin impregnates the fibres, so escape holes for the air have to be provided in the mould. The resin spreads through the fibre reinforcements. If the permeability of the mat is low a bleeder material can be used to increase the speed of resin transfer. Crosslinking occurs without further heating of the composite. Styrene emissions are low. The surfaces of the mould must be non-deformable and if this is the case an excellent finish for both surfaces is achieved. Fibre volume fractions can be in the range of 0.20 to 0.55.

4.3.6 Vacuum moulding

This technique requires little investment, particularly as the mould costs are lower than in the straight RTM method as they do not have to resist the higher pressures involved in that method. A closed mould is used so that the final product has well finished surfaces, but this time both the fibres and the uncured resin are placed in the female part of the mould which is then closed and the air extracted through the male part. The air is extracted from points at the extremity of the composite item being made and a reservoir of resin is connected to the mould at its lowest point to ensure that enough resin is provided to completely impregnate the reinforcements. The two parts of the mould are forced together by atmospheric pressure and impregnation of the fibres is achieved. The points at which air is extracted are closed when resin begins to be seen. Styrene emission can be high. The moulding cycles are slow with this technique and there is a danger that the reinforcements will be displaced by the movement of the viscous resin. Fibre volume fractions achieved are typically in the range 0.25 to 0.40 and the composite can contain some porosities due to trapped air or solvent.

4.3.7 Vacuum-assisted resin injection/transfer

This technique combines both aspects of the two previously described techniques, as shown in figure 4.10. A closed mould is used with silicone seals at the periphery. If necessary the injection of the resin can be by a gravity feed into the lowest part of the mould. The air is pumped out of the mould at locations removed from the injection points. The design of the mould is more complex than in the two previous techniques and this increases cost. However, the manufactured products have very low porosity and surface finish is good. The uniform quality and lack of porosity means that the technique is used to produce composite parts for applications for which their good dielectric qualities are used. The emission of styrene is low. The fibre volume fraction which can be achieved is in the range 0.25–0.55.

Vacuum-assisted RTM has the advantage of requiring less rigid moulds than the other two processes and, as a consequence, can be both cheaper and lighter. Figure 4.11 shows such a structure during resin filling, the darker area

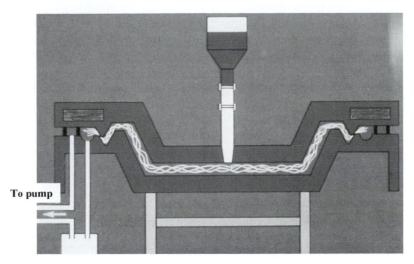

To pump

Figure 4.10. Vacuum-assisted RTM requires the mould to be filled with the reinforcement, closed and simultaneously the air pumped out and the resin allowed to flow in to impregnate the fibres.

in the centre revealing the extent of the resin flow. Very large structures can be made by this technique, as can be seen from figure 4.12.

4.3.8   Cold press moulding

This industrial process employs male and female moulds fitted to a simple hydraulic press. Large structures such as lorry cabs, roofs for caravans and

Figure 4.11. Vacuum-assisted RTM. The darkened area shows the extent of resin flow as it permeates the fibres.

Figure 4.12. Vacuum-assisted RTM can be used to fabricate very large structures.

mobile homes as well as large panels, baths, shower units and containers of various types can be produced by this method. The cost of the moulds is not high as they are not heated and only low pressures in the range of 0.35–4 atm (35–400 kPa) are used. The dry reinforcement, such as continuous glass mat, is placed in the mould and the catalysed resin is poured onto it. Polyester resins of higher reactivity are used than in contact moulding. The faster rate of crosslinking allows a relatively higher production rate. The stages of cold press moulding can be seen in figure 4.13. The manufacture of items at rates of five to ten per hour is possible. The only heating which occurs is due to the exothermic reaction of the resin curing process but temperatures of up to 50 °C can be developed.

A variation on this technique is thermoforming, in which the composite, generally of glass reinforced polyester, is sandwiched between two thermoplastic acrylic sheets of around 0.6 mm thickness. The acrylic sheets are first shaped, either at room temperature by vacuum forming or, for higher production rates, at temperature and then the reinforcements and resin, combined with a room temperature curing catalyst, are sandwiched between the acrylic sheets and the press closed during curing. The finished product has the high quality surface appearance of acrylic sheets and can be produced with a range of colours at a lower cost than if a paint cycle were necessary. The fibre volume fraction is around 0.25.

4.3.9   Large series methods

Many composite manufacturers are concerned with small series production, for which manual or other slow processing techniques may be appropriate. However, as soon as large numbers of similar items, with reproducible

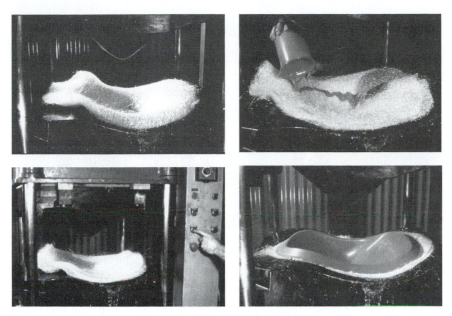

Figure 4.13. Cold press moulding brings two platens together to compress the reinforcement onto which the catalysed resin has been poured. Heating is provided by the exothermic crosslinking of the resin.

characteristics, need to be made, some of the processes so far described are not ideal. The differences may not be of quality as some industrial sectors require much lower production rates than others, as a rapid consideration of the aerospace and automobile industries illustrates. The manufacturing processes which are described below are those which are suited for high production numbers.

4.3.10   Hot press moulding

This hot compression moulding of composites (figure 4.14) is used in preference to the cold techniques described above for the manufacture of large series of an article, typically up to and above 20 000 units per year. The moulds, which are heated in this process, are made of steel and the pressures are of the order of 20–200 atm (2–20 MPa). The starting materials can be in the form of chopped or continuous fibre mat, the latter being more common, woven cloth or fibre preforms, often made by spraying the fibres onto a pre-shaped metal mesh. The fibres are held onto the preform by suction and bonded lightly together. Alternatively short chopped fibres, most often glass, can be mixed with the resin to form a dough or sheet which can then be moulded into a finished article.

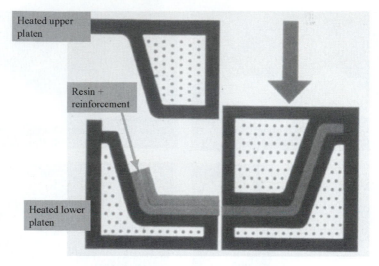

Figure 4.14. Hot press moulding brings together the two matched parts of the mould to form the composite.

4.3.11 Hot press moulding—wet process

In this process the fibres are positioned onto a heated mould, which has been preheated to the temperature at which the crosslinking of the resin will take place, the catalysed resin poured onto the fibres and the mould closed. The predetermined manufacturing cycle depends on the reactivity of the resin and the thickness of the laminate. The resins are usually filled with additives which counter shrinkage of the composite article due to the crosslinking process and higher temperatures used in manufacture. The moulds, which are simple in form, are usually heated to 100–140 °C and the pressures applied are most often between 20 and 50 atm (2–5 MPa). A production rate of up to 30 articles per hour can be achieved with this technique. Both surfaces of the finished product have excellent appearances, porosity is generally very low, although flash has to be removed and the complexity of the finished shapes which it is possible to produce is limited. A fibre volume fraction of 0.2 to 0.3 is commonly used with this technique.

4.3.12 Hot press moulding—dry process

This technique uses one of the most common forms of pre-impregnated composite material called SMC which is an abbreviation of 'sheet moulded compound'. It is made by chopping continuous glass fibre tow into lengths, typically from 25 to 50 mm. The chopped fibres fall continuously onto a moving belt onto which a polyethylene film has been laid and onto which the catalysed but uncured resin has been poured. This is shown in figure 4.15.

Figure 4.15. Sheet moulded compound (SMC) is made by allowing chopped glass fibres to fall onto a moving continuous sheet of polyethylene coated with uncured polyester resin. The SMC is finally coated with another sheet of polyethylene and then stored until its viscosity increases sufficiently to give a sheet material which can be stacked in moulds.

The result is an almost completely random arrangement of the glass fibres, in the plane of the belt, which become impregnated with the resin. As table 4.1 shows, the fibre content is usually in the range 25–40%, by weight. The resin may be coloured if required. In this continuous process the fibre resin mixture is then covered with another polyethylene film to produce a sheet of pre-impregnated fibre resin, cut to length and usually around 6 mm thick. The resins used are most commonly polyesters with styrene or acrylic used as crosslinking agents. The acrylic monomers are used to obtain low shrinkage and vinyl can be used for higher temperature strength. Peroxides such as benzoyl peroxide (BPO) and *t*-butyl perbenzoate for higher temperatures are used as catalysts. Fillers, such as calcium

Table 4.1. Typical formulations of general purpose and low-shrinkage SMC (% by weight).

	General purpose SMC (%)	Low-shrink SMC (%)
Resin	44	30
Polyethylene powder	0	15
Styrene	5	5
Catalyst (BPO)	0.4	0.7
$CaCO_3$	15.5	18.6
Mould release agent	0.1	0.7
Glass fibre	35	30
Thickeners	0	1.0

carbonate and sometimes ground (kaolin) clay, are added to reduce cost and others to modify the properties of the SMC during moulding and in its final form. Filler agents, known as thickeners, such as magnesium oxide and hydroxide, calcium oxide and hydroxides, can modify the viscosity so as to control flow in the mould, while others reduce the coefficient of thermal expansion, control the temperature rise due to the exothermic crosslinking reaction, increase hardness, rigidity and especially reduce shrinkage during crosslinking. Shrinkage is controlled by adding powdered polyethylene to the formulation. Talc is added to improve dielectric properties and resistance to humidity. As resistance to fire is a concern, aluminium hydrate, which releases 35% by weight of water during heating, is often added, as are antimony trioxide and chlorinated waxes which are also used for self-extinguishing properties. SMC compounds have also been produced for the automobile industry using vinylester and can be used for large exterior body panels and under the bonnet because of their high resistance to corrosion and ability to withstand high temperatures.

The SMC is then stored for some days during which the viscosity of the resin increases to give a sheet of material which can be handled easily. The SMC sheet can be cut into shapes required by the mould and stacked in the mould so as to achieve the desired thickness. Stearates added to the SMC migrate to the surface and act both as a mould release agent and pigmentation. An important difference between this dry technique and the wet process described above is that in the latter it is the resin which flows to fill the mould whereas in the dry process the entire compound flows. The size of the charge is therefore determined by the weight and has to completely fill the mould when it is closed. The process is generally used for large series moulding of small and medium size parts of up to one or two square metres, but sometimes it is used for larger mouldings. Moulding pressures are 30–100 atm (3–10 MPa). Production rates are of the order of 30 items per hour.

As the fibre volume fraction is not very high and the short fibres are randomly arranged in the plane of the composite, the mechanical properties are far less than in some other types of composite material but the ease of manufacture and low cost make the process attractive. Table 4.2 shows some typical properties of SMC materials.

Table 4.2. Typical properties of two types of SMC materials.

	General purpose	Low-shrinkage
Specific gravity	1.14	1.67
Flex modulus (GPa)	4.2	11.6
Flex strength (MPa)	95	150
Tensile strength	70	35

Other forms of this type of compound exist such as HMC, which is a compound with a high glass fibre content of 50–60% by weight. To achieve this volume fraction the fibres are usually chopped to lengths of 25 mm. TMC describes a form of SMC which is made in thicker sheets, hence the T in the initials and XMC which is a unidirectional reinforced SMC produced by a winding technique with a high glass content of 60–70% by weight.

A three-dimensional distribution of glass fibres in the matrix to give a sort of dough which can be compression moulded is known as DMC. Not surprisingly this stands for 'dough moulded compound'. This compound is made by the fibres, which are chopped to lengths of 6–10 mm, the resin, catalysts, pigments and fillers being mixed by a twin bladed mixer. The three-dimensional arrangement of the fibres naturally leads to fewer fibres being arranged in any given direction, when compared with SMC, so that mechanical properties of DMC are lower than those of SMC. A form of DMC with higher fibre volume fraction is BMC which is short for 'bulk moulded compound'. The fibres in BMC are chopped to shorter lengths than in DMC and are between 4 and 6 mm. Such compounds can be used in the production of small pieces even if of thick and complicated shapes.

The moulds are similar to those used in the wet process and the description of the manufacturing process is reminiscent of cold press moulding but for one important difference. In the latter process the movement of fibres is unwanted whereas it is a requisite of the hot press technique. The SMC, or DMC, flows under the effect of the applied pressure to fill the mould and as it does so the fibres also move. If the mould is of a complex shape additional pre-cut SMC may be positioned in particular parts of the mould where filling by the flow of the compound might be difficult. Moulding pressures are in the range 30–100 atm (3–10 MPa) and mould temperatures are between 140 and 160 °C. Manufacturing rates are in the range 20–40 items per hour, perhaps less for DMC pieces. The investment cost for this process are higher than in many other techniques due to the higher pressures used but the higher production rates makes it attractive.

4.3.13   Injection moulding

It is possible to inject short fibres mixed in an uncured resin into moulds in much the same way as many unreinforced plastics are moulded. Compounds with fibre lengths up to 30 mm are used and are fed manually or automatically from a roll or bulk form of the material. The steel moulds are heated in the temperature range 140–160 °C. The technique is suitable for making small thick specimens and it is possible to produce up to 60 per hour. Variations on the technique exist which differ essentially in the manner by which the compound is injected into the mould. The compound can be transferred to the principal injection chamber by a piston and then

Figure 4.16. Unsaturated polyester resin loaded with short glass fibres can be injected into a mould by the combined action of an Archimedes screw thread and a piston action.

injected into the mould by another piston. In this way the fibres are not broken down to shorter lengths, which is of great importance in maintaining optimum mechanical properties in the final composite product. Alternatively the injection can be achieved, as with the injection moulding of unreinforced plastics using a screw thread injection system, as shown in figure 4.16. This is known as ZMC. Injection pressures are high (50–100 MPa) and the parts which can be made are limited by the dimensions of the mould and the force necessary to hold it closed. The compound is introduced into the injection chamber by a piston and then a screw thread operating on the Archimedes screw principle injects it into the mould. No pressure is applied during this transportation stage and as the feed rate is kept constant, with a low rotation rate generally between 50 and 90 rpm, a minimum of damage to the fibres occurs. The injection channel is heated with the temperature increasing from 30 °C at the arrival point for the compound to 80–90 °C at the entrance to the mould, which is typically at a temperature of 150–200 °C. One difficulty with this technique is that the injection of the compound has to be stopped once the mould is filled. This can be by simply stopping the movement of the screw thread or by the use of a valve which closes off the entrance to the mould. Both these actions can introduce some variability into the quantity of compound placed in the mould and also can cause the breakage of the fibres. Overall, however, the screw thread offers advantages over a piston as production cycles can be reduced and a greater control over product quality can be achieved.

4.3.14   Injection moulding of fibre reinforced thermoplastics

The production of parts made by the injection of a molten thermoplastic into a, usually, cold mould, which after cooling of the part can be opened and the part ejected, is a rapid and well established production route suitable for high production rates. When fibres are added to the plastic the extruder has to be made so as to resist the added wear due to abrasion which occurs. Injection moulding has been adapted for fibre reinforced thermosetting resins as described above although in this case the mould has to be heated to induce crosslinking. In the case of injection moulding of fibre reinforced thermoplastics the starting materials are solid and can be in the form of separate fibres and the matrix in powder form or granules or alternatively, and increasingly, in the form of preformed pellets of fibre in the matrix. These pellets are produced initially in the form of continuous aligned fibres extruded with the molten thermoplastic and cut into short lengths of around 25 mm. The resulting fibre lengths can be short (0.1 mm for separate mixing and 4–6 mm for the pellets). The moulding pressures are some 30% higher than when no fibre reinforcement is introduced in the thermoplastic and the injection temperatures are generally higher to reduce viscosity. The fibres and thermoplastic are introduced into the injection moulding machine, either together or, if thought necessary, the introduction of the fibres can be nearer the mould so as to reduce fibre breakage. As in figure 4.16 the material to be injected is moved towards the mould by the action of a single or twin screw thread which shears the material, which in the case of some thermoplastics, such as polypropylene, is necessary to achieve adhesion with the fibres. This action, however, breaks the fibres so that their final lengths are greatly reduced to lengths measured in millimetres, but as their diameters are of the order of 0.01 mm their aspect ratios are such that reinforcement is still achieved. The fibre volume fraction is generally no greater than 0.4 and can be as low as 0.25.

The flow of the matrix in the mould leads to alignment of the fibres parallel to the flow lines and this has to be understood so as to achieve optimum properties. The composite products which are made by this process can be extremely complex in shape and high rates of production are possible although sizes are limited. Table 4.3 gives some properties of injection moulded thermoplastic composites.

4.3.15 Reinforced reaction injection moulding

Many composite manufacturing processes are too slow for really large volume fabrication such as might be needed in the automobile industry, so reinforced reaction injection moulding is of particular interest as it allows large parts to be made quickly. The composite is made in a closed heated mould into which are injected two highly reactive monomers, one of which is also used to carry the short milled glass fibres into the mould. Milled or broken fibres are generally preferred to cut fibres because of lower cost. The monomers are carefully metered and brought together in a mixing head and injected under low pressure. The interaction between the two monomers produces a crosslinked polymer which fills the mould and which is reinforced with between 8 and 25% by weight of the fibres. The reaction is generally between a liquid polyol and an isocyanate, such as diphenylmethane 4,4'-diisocyanate (MDI), to produce a polyurethane, as

Table 4.3. Properties of some injection moulded thermoplastics.

Polymer	Glass fibre content (wt%)	Glass fibre content (%)	Specific gravity	Injection temperature (°C)	Mould temperature (°C)	CTE $(10^{-5}°C)$	Flex modulus (GPa)	Tensile strength (MPa)	Failure strain (%)
PE(HD)	40	20	1.28	230	40	4.3	7.5	80	2.5
PP	40	19	1.22	240	40	2.7	7.0	103	4
PA66	40	23	1.46	295	105	2.5	9.0	210	2.5
PS	40	22	1.38	245	65	3.4	11.3	103	2.5
ABS	40	22	1.38	260	90	2.2	7.6	110	3.5
PPS	40	26	1.65	320	120	2.7	12.5	160	3
PC	40	24	1.52	310	120	1.8	10.3	145	4

PE(HD) is high density polyethylene, PP is polypropylene, PA66 is polyamide (Nylon) 66, PS is polystyrene, PPS is polyphenylsulphone, PC is polycarbonate.

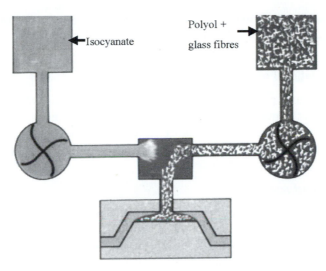

Figure 4.17. Reinforced reaction injection moulding (RRIM) is produced by mixing iso-cyanate with a polyol carrying milled glass fibres. The reaction takes place in the mould and fabrication times are short.

described in chapter 3, and the process is schematically shown in figure 4.17. It is important to avoid exposure of the monomers to water as this can produce gas during manufacture of the composite and produce poor results. Isocyanates are particularly sensitive to water which will produce a hard precipitate so hindering processing and reduce the characteristics of the composite. For this reason the isocyanates are generally stored under nitrogen. Storage of the monomers is usually controlled to maintain a temperature between 24 and 30 °C in order to keep them liquid. Fluorocarbon blowing agents can be used in small quantities to produce a microcellular structure at the centre of the composite. This means that the composite is less dense and allows for better packing of the mould. Filling the mould occurs in seconds and the entire process, including curing and removal from the mould, usually takes only two minutes. The process is cheap as the low pressures involved allow less expensive moulds to be used and little heat input is needed as the reaction is exothermic. The process can be automated and fast moulding times are possible. However, polyurethanes have high coefficients of thermal expansion and poor dimensional stability at raised temperatures. The presence of the fibres considerably reduces these problems. Tensile strengths in the range of 30 MPa and above are attainable, with flexural moduli being around 1.4 GPa. Strains to failure are high, in the region of 25%.

A variation of RRIM combined with the RTM process involves injecting the polyols and isocyanate into a mould prefilled with glass fibre mat. This allows the fibres to be placed in particular areas of the mould rather than relying on the polyol to transport them there.

4.3.16   Rotational moulding

There are several processes which involve placing the fibre reinforcements, together with the matrix material, onto a rotating mandrel in order to fabricate the composite part. Both short fibre reinforcement and continuous reinforcements can be used but they produce very different end products.

4.3.17   Centrifugal casting

A low cost process for producing cylinders and pipes involves placing both resin and fibres into the interior of a rotating circular mould. The fibres can be in the form of mats or woven rovings or chopped fibres can be projected into the rotating mould, as illustrated in figure 4.18.

Figure 4.19 shows a close up of the glass fibres being projected onto the inside of the rotating mould. The mould rotates at a sufficiently high speed for the centrifugal force to cause the glass fibres to be well impregnated with the resin and induce good compaction. This technique is the only process which gives a good finish on both sides of the composite product without the use of matching moulds and gives a very smooth finish to both surfaces. The dimensions of the products which can be made do not have an upper limit but sprayed chopped fibres cannot be used for cylinders of small diameter as a gap of 250–300 mm has to be left between the spray head and the mould to ensure even fibre distribution. The process can be used to make cylindrical or slightly tapered pipes.

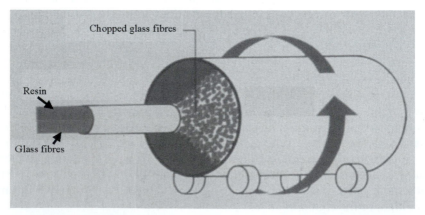

Figure 4.18. Centrifugal casting by projection of chopped glass fibres together with the resin into the interior of a rotating cylindrical mould.

Figure 4.19. Chopped glass fibres being projected onto the inside of a rotating mould.

4.3.18 Filament winding

Filament winding of continuous filaments onto a rotating mandrel, after first being impregnated with uncured thermosetting resin or coated with thermoplastic, is a means of making high performance, high reproducible, composite structures. The simplest structures which can be made by this technique are tubes, using a cylindrical rotating mandrel onto which the continuous fibres are wound from a head which repeatedly sweeps from one end of the mandrel to the other. Such cylinders can be very large, as can be seen from figure 4.20.

Wet winding, in which the fibres pass first through a bath of uncured thermosetting resin, is the most common. The fibres, impregnated with the resin, are then wound onto the mandrel and subsequently cured either at room temperature or in an oven. There is generally no external pressure applied during manufacture, except that provided by the tension in the fibres, although, in some cases, curing can take place under pressure in an autoclave. It is also possible to use prepreg rovings which can then be wound onto the mandrel without the need of a resin bath. The tackiness of the material is ensured by control of the temperature at which the winding takes place. Occasionally the fibres can be wound onto the mandrel and the resin applied afterwards.

If a thermoplastic matrix is required the fibres must first be processed to produce a ribbon impregnated with the thermoplastic. This can be achieved by passing the tow of fibres through the thermoplastic in the form of a fine powder so that it adheres to the fibres by electrostatic forces or the fibres can be co-extruded with the molten polymer. In both cases the ribbons are wound onto the mandrel as in the wet process but at the point of contact

Figure 4.20. The filament winding process used to make a large tube.

the ribbon is both heated to remelt the thermoplastic and a pressure is applied locally to achieve bonding.

The mandrels can either be dismantled or dissolved away after composite manufacture or most often remain and serve a useful role as a liner, providing an impervious barrier for whatever fluid, liquid or gas is to be held in the structure. Typically such liners are in aluminium or polyethylene.

Filament winding machines can be extremely complex and possess up to seven degrees of movement so as to enable complex shapes to be wound. Figure 4.21 shows a helically winding pattern used to make a pressure vessel. Some examples of complex winding patterns are shown in figure 4.22.

Figure 4.21. A helical pattern used to produce a filament wound pressure vessel.

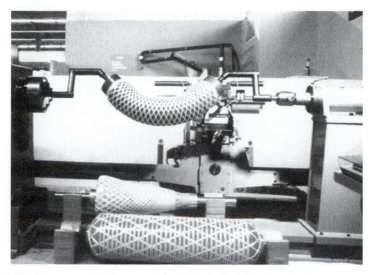

Figure 4.22. Some complex shapes produced by helical filament winding.

Polar winding, which allows spheres to be produced, involves the mandrel remaining stationary whilst the fibre feed arm rotates about the longitudinal axis which is inclined at the necessary winding angle.

In both helical and polar windings the fibre tows can be arranged so as to sit perfectly next to one another to give a complete covering in one ply, or alternatively several plies can be used to cover the mandrel entirely.

The fibres are generally wound on the mandrel following geodesic paths, which are the shortest distances between two points on a three-dimensional surface, and this ensures that no slippage occurs and no friction is required to hold the fibres in place. A geodesic path can be described by the relation

$$r \sin \alpha = \text{constant} \qquad (4.1)$$

in which r is the radius at a given point on the surface and α is the winding angle. It follows that the fibres must experience only tensile forces so that the following expressions can be used to calculate the strength of a filament wound tube.

Where S is the unit strength of the wrapped tow, S_H is the strength of the cylinder in the hoop direction, S_A is the strength of the cylinder in the axial direction, t is the wall thickness, α is the winding angle, W is the width of the tow, $L = W/\sin \alpha$, F_1 is the component of the strength of the tow in the α direction, and F_2 is the component of the strength of the tow in the hoop direction, we can write

$$F_1 = \frac{SWt}{2} \qquad (4.2)$$

$$F_2 = \frac{\sin \alpha \, SWt}{2} \tag{4.3}$$

$$S_H = \frac{2F_2}{Lt} = \frac{2 \sin \alpha \, SWt}{2W(\sin \alpha)^{-1}t} = S \sin^2 \alpha. \tag{4.4}$$

Similarly $S_A = S \cos^2 \alpha$. This means that for a cylinder with a 2:1 ratio of hoop strength to axial strength

$$\frac{S_H}{S_A} = \frac{2}{1} = \frac{S \sin^2 \alpha}{S \cos^2 \alpha} = \tan^2 \alpha. \tag{4.5}$$

This gives a value for the winding angle α of 54.75° and coincides with a minimum value, given by a more detailed analysis, of the shear forces in the composite and of the stresses tending to separate the fibres.

Filament wound composite pressure vessels require at least one end fitting so that they can be filled. These fittings are incorporated into the structure during winding. The fibres which wrap around the ends provide the necessary strength to support the end forces due to the internal pressure, but they are inevitably at an angle with respect to the hoop axis of the vessel over the central cylindrical part of the structure. Passing around the ends of the vessel can produce a too-thick layer of fibre if several windings are made. This is avoided by slightly changing the winding angle for subsequent windings. Additional fretting windings are wound around the central cylindrical section to provide the required hoop strength.

4.3.19 Pultrusion

This technique integrates fibre and resin mixing, composite forming and curing in one continuous process. The fibres, which are generally glass but can be of any continuous type, are brought together from a number of spools or bobbins and impregnated with a, usually, thermosetting resin such as polyester. The impregnated fibres then pass through a die, heated between 90 and 150 °C, which crosslinks the resin sufficiently for a sufficiently strong composite profile to be pulled from the die. The process allows profiles of extremely complex cross section to be made at speeds in the range 0.6–1.2 m/min.

Figure 4.23 illustrates the pultrusion process with the profile being pulled through the die and heating oven by two constantly rotating, cleated belts between which the cured profile is pulled and then cut to length. Figure 4.24 shows such a machine and the spools which feed the fibres into the pultruder.

The cleated caterpillar-belt pulling system has the drawback of requiring the belts to be shaped to maintain contact with the profile being drawn. Single clamps can be used which pull the required length of profile through

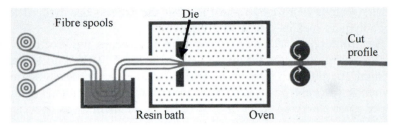

Figure 4.23. Schematic view of the pultrusion process.

the die. Improved production rates can be achieved by using a radio frequency (RF), typically at 70 MHz, as a means of dielectric heating the resin. This accelerates curing and allows the profile to be pulled through the die quicker.

Profiles can be either solid or hollow and can be made of all one type of reinforcement or a combination of continuous fibres and fibreglass mat. In this latter case the continuous fibre rovings are sandwiched between two layers of mat as the mat would not be strong enough to pass through the resin bath without the roving. This type of material is used to make flat sections or thin walled shapes.

Thermoplastics can be used as the matrix material, but the higher viscosity of the polymers is an added complication compared with thermosetting resins which have viscosities two or three orders of magnitude lower, around 1–10 Pa/s. Despite this difficulty the thermoplastics have one big advantage in that they can be remelted or shaped after fabrication of the profile. Initially the process was developed for high performance

Figure 4.24. Fibres being pulled from spools into a pultruder.

polymers such as PEEK, PEI and PPS, but the process has evolved to less costly polymers such as saturated polyester.

The profiles produced by pultrusion are intrinsically unidirectional even though a variation of the technique can induce some twist into the profile by turning the pulling mechanism. The technique can be developed into a pullwinding process in which the pultruded unidirectional profile is wrapped after emerging from the pultruder by filaments which overcomes the anisotropy of the unidirectional structure.

4.3.20   Modelling of the thermosetting process

The cure or crosslinking process is central to the manufacture of most composite materials. There are many different types of thermosetting resin together with the crosslinking agents which are used but the reactions, at the molecular level, always involve creating bonds between the molecules which produce profound effects in the molecular, chemical and physical nature of the resin. There is a maximum number of these bonds which can be made and the crosslinking process proceeds towards this saturation limit but rarely reaches it, which is why post-curing, or the further heating of the composite product after its initial manufacture, is quite common. Post-curing leads to an increase in the number of crosslinking bonds in the resin and to further stability to the properties of the composite. It is difficult to measure directly the number of bonds which have been created but the effects of crosslinking can be measured, such as changes in shear viscosity of the curing resin or other physical properties like shear or elastic modulus, changes in density or enthalpic reactions which can be measured with a differential scanning calorimeter. If one of these properties, which we can call S, is monitored so as to gain insight into the crosslinking process, the degree of crosslinking can be written as

$$u = \frac{S - S_0}{S' - S_0} = f(t) \qquad (4.6)$$

where S_0 is the measured variable or property at the beginning of the reaction, S' is the variable when the reaction is finished, and S is the current value of the variable.

The degree of the reaction therefore varies between being fully uncured, $u = 0$, to being fully cured with $u = 1$.

In the simplest situation, the degree of crosslinking (u) proceeds towards saturation in an asymptotic manner so that the parameter S, which is being followed as an indication of crosslinking, increases, initially rapidly, but tends to a steady value. This means that the rate of crosslinking, $\delta u/\delta S$, steadily decreases from a maximum at the beginning of the curing reaction to zero when the process is finished. Figure 4.25 shows what happens at different temperatures $(T_1 > T_2 > T_3 > T_4)$ when inhibiting agents are

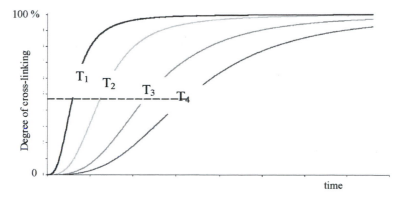

Figure 4.25. The addition of inhibitors initially slows crosslinking rates to thermosetting resins.

added. These inhibitors are added to control crosslinking at the beginning of manufacture such as during mould filling. They initially saturate the reactive intermediate products but their effects diminish with time. The result is that the crosslinking process is retarded. This behaviour begins with an S-shaped curing curve leading, asymptotically, towards saturation. The degree of reaction is asymptotic to a value which is independent of temperature and is therefore seen to be related to the number of crosslinks which are made. Equation (4.7) gives an empirical expression for this behaviour:

$$u = \frac{t^j}{(1/k_0)\,e^{T_0/T} + t^j} \tag{4.7}$$

where t is time, j is a whole number, T is the temperature in Kelvin, $T_0 = E^a/R$, E^a is the activation energy, R is the universal gas constant, and k_0 is a rate constant or factor or proportionality.

By putting $k = k_0\,e^{-(T_0/T)}$ it becomes possible to differentiate equation (4.7) and it can be shown that the reaction rate varies as a function of temperature, T in Kelvin, and time t such that

$$\frac{\delta u}{\delta t} = k_0\,e^{-(E^a/RT)} u^{2-n}(1-u)^n \tag{4.8}$$

where, k_0, E^a and the exponent n, which is the reaction exponent related to the number of molecules involved in the curing process, are three variables which are considered as being independent of the temperature and the degree of crosslinking. The reaction exponent n should be an integer but as several processes can be occurring in parallel this is not always the case. The reaction rate described by equation (4.8) is shown in figure 4.26 and it can be seen that the rate passes through a maximum which is more pronounced at higher temperatures and is shifted towards shorter reaction times. The peaks determine the points of inflection of the degree of

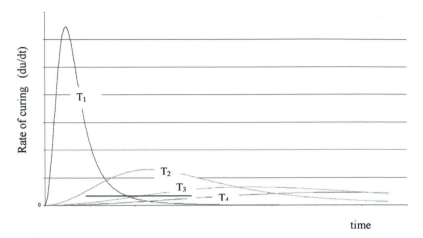

Figure 4.26. The rate of curing or crosslinking passes through a peak which is the more pronounced and occurs earlier with increasing temperature $T_1 > T_2 > T_3 > T_4$.

crosslinking which lie on the dotted line shown in figure 4.25 and which are known as the critical conversion points.

The analysis given here is typical of the crosslinking of polyester thermosetting resins.

4.3.21   Mould filling in RTM

The filling of a mould in which has been placed the fibrous reinforcements, as in the RTM technique, has to take into account the ease with which the resin can permeate into the fibrous structure. This means that the viscosity of the resin, the surface area A of the mould and the porosity of the fibrous filler have to be considered. Darcy's Law describes the flow of a resin into a fabric, which is considered to be an open porous medium. The resin is assumed to show Newtonian flow characteristics and the flow rate Q is a function of the pressure gradient (P/L) so that

$$Q = \Delta \left(\frac{A}{\mu} \right) \left(\frac{\delta P}{L} \right) \tag{4.9}$$

where Δ is Darcy's constant for permeability (1 Darcy = $0.00987\,\mathrm{m^2/s}$), μ is the Newtonian viscosity, and A is the cross-sectional area.

Darcy's constant for flow into a capillary network of parallel fibres is given by

$$\Delta = \frac{\phi^3}{CA^2(1 - \phi)^2} \tag{4.10}$$

where ϕ is the porosity, A is the surface area, and C is a constant.

In the fibre direction permeability is given by

$$\Delta_x = \frac{r_f^2(1 - V_f)^3}{(4s_x V_f^2)} \qquad (4.11)$$

where Δ_x is the permeability in the fibre direction, r_f is the fibre radius, s_x is a constant, and V_f is the fibre volume fraction.

In the direction perpendicular to the fibres the permeability constant Δ_x is given by

$$\Delta_z = \frac{r_f^2[(V_a/V_f)^{1/2} - 1]^3}{4S_x(V_a/V_f - 1)} \qquad (4.12)$$

where V_a is the critical fibre volume fraction necessary for capillary flow and S_x is a constant.

These equations are used to produce computer software packages to model mould filling in the RTM process.

4.4 Metal matrix composite manufacture

Reinforced metal composite materials (MMCs) were the first to be studied in a systematic way and many of the models and theories used for resin matrix composites were first developed on metal matrix systems. However, the use of MMCs remains limited. This is because of the cost of production and the fact that for most applications for which metals are used their intrinsic properties suffice. This is in contrast to resin matrix composites which possess much superior properties than those of the unreinforced matrix. If the reinforcement is a fibre in the MMC there is the difficulty of maintaining its integrity as many forming processes for metals would break long fibres and reduce their reinforcement value. This is also a difficulty with some resin matrix manufacturing processes such as injection moulding but cold or hot forging can be seen to be poorly adapted to a metal reinforced with long fibres. This removes an attractive characteristic of metals and adds a complexity to realization of parts from MMCs. In addition, adding fibres to a metal inevitably has to increase production costs which means that MMCs are restricted to relatively niche markets. These, however, can be important. The addition of short ceramic fibres or even ceramic particles can change soft aluminium into a hard-wearing composite material. As the reinforcements used have lower coefficients of thermal expansion (CTEs) than those of metals, MMCs can be produced to have controlled CTEs to make them compatible with other materials and, at the same time, allying the low density of aluminium with the structure. In some cases, as with carbon or magnesium reinforced with carbon fibres, the very low or negative coefficients of expansion of the reinforcements can give composite materials with a zero CTE over a wide temperature range. Alternatively the addition of ceramic fibres, as described in chapter 2, to aluminium or titanium can control their creep and increase operating temperatures of parts made from these composites. The reinforcements are ceramics, usually alumina or silicon carbide, in various forms which can be short or continuous fibres or particles.

The manufacture of MMCs requires the metal to be processed in its molten form or near to its melting point in order to achieve compaction of the composite by diffusion. As the melting points of metals used in MMCs are much higher than any processing temperature used for resin matrix composites this introduces an additional complication of compatibility between the fibres and the matrix, particularly during manufacture. It is necessary that the reinforcement is wetted by the matrix but it should not be degraded.

It is possible to produce metals which are directionally reinforced by a judicious choice of alloy. Oriented eutectics such as Ni–TaC are produced by melting the alloy, which has been produced with the optimum eutectic composition and melted and solidified in a continuous unidirectional manner. The metal solidifies into two phases and as cooling is arranged so that solidification is directional the phases are aligned in the direction of solidification. This type of oriented eutectic MMC was much studied in the 1970s but the slowness of the production process, linked to a high cost, has limited exploitation. It is possible to produce discontinuous ceramic phases by coprecipitation; for example, TiC particles of several microns in size can be precipitated in steel alloyed with titanium and carbon.

4.4.1   Casting processes

The use of a casting process to incorporate ceramic particles into light alloys, usually aluminium alloys, produces the cheapest form of MMC. Obviously the form of the particles means that they are not fibre reinforced, which is the central theme treated in this book, but these materials are generally grouped together with other types of composites. These types of MMC can be processed in much the same way as unreinforced metals, such as by forging, which makes them economically attractive. Ceramic particles, usually of SiC, around $15\,\mu m$ in size, can be mixed in molten aluminium or magnesium to give volume fractions of the particles of 10– 20%. The aluminium alloy used has to be rich in silicon to avoid the formation of Al_4C_3. Alumina (Al_2O_3) particles can be mixed with aluminium but produce an undesirable reaction in magnesium. The mixing has to be violent so as to avoid clumping of the particles. Tubular pieces can be made by a centrifugal process in which both the molten metal and the ceramic particles are poured into a rotating cylindrical mould. The centrifugal forces and the differences in densities between the metal and the ceramic particles means that the volume fraction of particles is greater at the surface of the tube produced. This effect can be put to advantage so as to produce composites with a gradation of properties through the wall thickness.

4.4.2 Squeeze casting

This process is analogous to resin transfer moulding of resin matrix composites. The fibres used are usually short and fine, such as Saffil alumina silica fibres, which are produced as a felt. The felt is compressed so as to make an open preform with the addition of a bonding agent. Typically the fibres represent only 20% of the volume of the preform. The preform is placed in the mould, which is heated, and the molten metal is introduced into the preform under pressure. The pressure can be applied by a piston to produce parts of up to 150 mm in diameter or by gas at pressures between 1 and 10 MPa. The latter technique uses lower pressures than when a piston is used and could allow larger parts to be made, but the process is slower. Complex structures such as connecting rods and piston heads can be made by this method or the MMC can be used to locally reinforce such parts.

4.4.3 Powder metallurgy

Particle or short fibre reinforced MMCs can be made by powder metallurgy processes. The sizes of the reinforcements are smaller than in the casting process, being in the range 2–5 μm, with the particles of aluminium being around five times bigger. The two types of component have to be completely and intimately mixed if a good composite material is to be produced. The MMC is made under pressure, or by extrusion, at a temperature lower than the melting point of the aluminium alloy. The interfacial reactions between the reinforcement and the alloy matrix are less than those obtained by the casting processes in which the matrix is molten. This allows aluminium alloys other than those rich in silicon to be used when SiC is the reinforcement.

4.4.4 Long fibre MMCs

The techniques described above for the manufacture of particular or short fibre MMCs usually produce composites with a random arrangement of reinforcements although the extrusion process can induce orientation. These composites also usually have a low volume fraction of the reinforcement. Maximum efficiency of reinforcement requires the fibres to be well ordered, which implies the use of long or continuous fibres. Despite the difficulties of associating long fibres with a metal matrix, some such composites have been developed and find applications.

Wires of fibre reinforced aluminium are made by continuous processes. These can then be incorporated into structures such as cables. The wish to reinforce aluminium by carbon fibres is as old as the existence of the fibres, since the combination of the high performance properties of carbon fibres coupled with the properties of the aluminium could produce a highly desirable composite. Unfortunately chemistry comes into the equation, which means that research is still continuing in order to make a completely acceptable carbon–aluminium composite. As with any composite, the fibres have to be wetted but not damaged by the matrix. In the case of carbon fibres the wetting has to be facilitated by the addition of a fluoride-based flux which reacts with the metal and dissolves the fine film of alumina associated with the aluminium bath. Wetting can also be added by agitation of the molten aluminium bath by ultrasonics or alternating electromagnetic fields. Carbon fibre reinforced wires can therefore be produced by drawing the bundle of continuous fibres through the bath of molten metal. This procedure has been used commercially by 3M in the USA to make continuous alumina-based fibre reinforced aluminium wires and in Japan by Nippon Carbon to make SiC-based fibre reinforced aluminium wires. In both cases the fibres are of small diameter, around 10–15 μm.

Large diameter SiC fibres (100–140 μm) are used to reinforce titanium with the aim of producing structures for applications in the range 400–600 °C for jet engines. The continuous fibres are wound onto a drum of a diameter of the order of 1 m. The fibres can be positioned close to one another and then bonded together either by an organic binder, which has to be removed at a later stage, or by the plasma projection of the matrix material. In both cases this allows the fibres to be removed from the drum by cutting along a line parallel to the axis, so giving a sheet of fibres. This sheet can then be cut and positioned in a mould between other sheets of the matrix material and hot pressed to form the composite. The process takes place at a temperature below that of the melting point of the matrix but sufficiently high to allow diffusion bonding. A similar process was used to make the tubes from which the frame of the American space shuttles are composed, but in this case the matrix is aluminium and the reinforcements are boron fibres.

4.5 Carbon–carbon composites

Carbon can exist in a variety of forms so that it is possible to place carbon fibres, which have an oriented microstructure, in a carbon matrix possessing a randomly arranged microstructure in order to produce a carbon–carbon composite. The carbon–carbon bond is the strongest in nature and, in the absence of oxygen which oxidizes carbon from around 400 °C, carbon–carbon composites can resist temperatures in excess of 3000 °C. This characteristic meant that these composites were first produced for rocket nozzles, nose cones and heat shields. However, the biggest market which has developed is for brake pads for aircraft and large trucks. Carbon–carbon composites can be made using carbon particles but the composites which interest us here are made from carbon fibres. The first composites of this type were made using carbon fibres made from cellulosic precursors, as in the 1950s there were no other types available. Even today these types of carbon fibre find use in these composites because of their superior heat conduction characteristics. The fibres can be arranged randomly in the form of a mat or felt, or organized by weaving, or by other means to position them in particular directions in the composite. The matrix is generally introduced into the fibrous preform by one of two methods. Pitch, which contains a very high (~90%) carbon content, can be forced into the preform under high pressure (1000 atm or 100 MPa) and high temperature of the order of 1000 °C. The high pressure infiltration of pitch produces a composite with little porosity. Several infiltration cycles are necessary, usually up to six, but the density which can be obtained is close to that which is of a fully dense carbon of around $2 \, \text{g/cm}^3$. If infiltration of pitch is used, but at room temperature, the structure remains porous with a density of around $1.65 \, \text{g/cm}^3$. Alternatively the fibrous perform can be infiltrated, under a low pressure, with a hydrocarbon gas (chemical vapour infiltration) which is then cracked at a temperature in the range 900–1200 °C to produce carbon which is deposited on the fibres. As the carbon builds up on the fibres it inevitably leads to pores becoming closed and inaccessible. This effect is greatest at the surface and it is possible to remove the surface to increase densification through further vapour deposition. However, the carbon–carbon composites made by this means inevitably are porous and values of porosity of 15% are typical. This latter technique, which requires about 100 h of vapour deposition, is cheaper than the other processes and is more widely employed.

4.6 Ceramic matrix composites

Ceramics are crystalline materials which possess some remarkable properties not found in organic or metal materials due to the atomic bonds which make up their structures. These bonds are ionic and covalent with an absence of other types of bond which are found in the other materials. These strong bonds give ceramics great hardness which is coupled with high elastic moduli, great resistance to heat and chemical resistance. The lack of plastic deformation processes, at least at room temperature, however, make ceramics brittle as once a crack starts to propagate there exists no mechanism in bulk ceramics which can stop it and failure occurs. Crack propagation can be initiated by mechanical or thermal shock. Most bulk ceramics contain grains of varying sizes and defects in their crystalline structures which cause stress raisers and which become points of weakness giving ceramics a very wide scatter in mechanical properties. This unreliability of mechanical properties has meant that bulk ceramics are unsuitable for most high performance structural applications.

The introduction of a second phase and in particular ceramic fibres can provide a means by which cracks in the ceramic can be blocked and tougher material can be produced with the attributes of bulk ceramics but with the less desirable characteristics mitigated. The failure of the interfaces between the fibres and the matrix is an important mechanism used to toughen ceramic matrix composites (CMCs) which means that the interfaces should not be too strong. Crack bridging by the ceramic fibres and their subsequent pullout is also an important mechanism in determining toughness in CMCs. The fibres which are used in ceramic matrix composites are based on SiC or alumina. They are fine with diameters of less than $15\,\mu m$ and continuous. The matrix material, in the former case, is usually SiC, deposited by techniques analogous to those used to make carbon–carbon composites, or mullite produced by powder slurry infiltration or a sol-gel process, in the latter case.

Chemical vapour infiltration (CVI) is used to introduce the matrix material into a fibrous preform at a temperature in the region of $1000\,°C$. Silicon carbide reinforced with silicon carbide based fibres (SiC–SiC) is prepared by a reaction in which methyltrichlorosilane decomposes in the presence of hydrogen:

$$CH_3SiCl_3 + H_2 \longrightarrow SiC + 3HCl + H_2. \tag{4.13}$$

As with the carbon–carbon composites, the closing of pores in the preform is a limitation to achieving maximum density, so close control of temperature and gas flow is important. However, infiltration times can be as long as several days or weeks. The final product has then to be surface ground to achieve the finish required. Typically densities of the order of 80–90% of the theoretical maximum can be achieved.

An alternative approach is to infiltrate the fibrous preform with a polymer such as polycarbosilane, as is used in the production of the fibres. Deposition of the SiC onto the fibres allows the SiC–SiC CMC to be produced, although repeated infiltration cycles are necessary. The use of polysilazanes allows a matrix of silicon nitride (Si_3N_4) to be produced. Long processing times are again a difficulty and with the high cost of the precursor materials contribute to the high cost of manufacture, even though the lower temperatures involved, which are less than $1000\,°C$, are an advantage.

Sol-gel infiltration of fibrous preforms can also be used to infiltrate fibrous preforms but again the need to repeatedly infiltrate the preform adds to the cost. The sol can be loaded with a ceramic powder which can provide seeds to encourage mullite production. It can be combined with a CVI process to achieve maximum infiltration. This process has been mainly used in the production of oxide–oxide CMCs such as Nextel 440 fibres in a mullite matrix.

Reaction bonding can be used by infiltrating the fibrous form with a slurry or liquid which, after deposition, is reacted with a gas or liquid to form the desired ceramic matrix. Preforms have been infiltrated with silicon powder slurries and, after drying, sintered in nitrogen to form reaction bonded silicon nitride. An important aspect of this technique is that the reactions usually involve an increase in volume, which provides a means of decreasing unfilled porosity.

Slurry infiltration can be used to make composites with glass matrices. The glasses which are used for this process are usually lithium aluminium silicates (LAS) and magnesium aluminium silicates (MAS). The glass in powder form is made into a slurry which is used to impregnate the ceramic fibres. As the glasses have relatively low melting points of around 1200 °C, densification can occur when pressure is applied during sintering. These vitreous ceramic composites differ from the SiC and Al_2O_3 matrix composites as the elastic moduli of the matrix material is lower than that of the reinforcing fibre.

4.7   Revision exercises

1. Explain how the nature of composite materials is different from that of most conventional materials.
2. Which well-established industry is often a source of manufacturing processes for composite materials?
3. How is it that the total cost of composite structures is reduced by the manufacturing processes used?
4. Glass fibres are used in many forms and are therefore adaptable to many manufacturing processes. Give examples of the forms of glass fibres that can be obtained and briefly indicate some corresponding processes in which they are used.
5. Composite structures made from glass mat do not give as high properties in the final structure as woven glass fibre cloth. Why is this?
6. How are pellets of glass fibres in a thermoplastic matrix made?
7. What does SMC stand for. How is it made and used to produce a composite part?
8. Describe manufacturing with an open mould. Why is a bleeder cloth desirable in this type of process? Approximately what rate of manufacture is possible with such a technique? Give an example of a product made in this way.
9. What is an autoclave? What pressures are generally involved with this process? What is the approximate rate of production with this technique?

10. Describe what the RTM process is. Discuss the choice of moulds which can be used and explain why different materials are used. What sort of pressures are used?

11. What advantages are there in vacuum-assisted RTM? What would be a typical fibre volume fraction obtained with this technique?

12. Cold press moulding is used in industry to produce reasonably large pieces. How does this technique differ from RTM and what sort of production rate is possible? What advantage does hot press moulding have over cold press moulding?

13. SMC is widely used and the composite often contains other components than just glass fibres and resin. Explain why this is so. What is the most common hardener used in these composites and what major drawbacks does it present for the manufacturer and for the use of the composite part? How can some of these problems be resolved?

14. Thermoplastic matrix composites are often made by injection moulding. Explain this technique. How does it compare with the injection of non-reinforced thermoplastics?

15. What does RRIM stand for? How it is made and what is its main attraction? Why is it particularly important to store the products involved in dry conditions? (See also chapter 3.)

16. Describe the centrifugal casting process and give an example of a structure which could be made in this way. Why is there a lower limit to the diameter which can be produced?

17. Tubes and other structures can be made by the filament winding technique. Describe this technique for manufacturing with both thermosetting and thermoplastic matrix materials. Why are the fibres automatically laid down following geodesic paths? Derive the equation for the hoop strength of a filament wound cylinder.

18. How can the inherent anisotropic structures made by the pultrusion process be modified?

19. The value equivalent to one Darcy is $0.00987\,\mathrm{m^2/s}$. Explain what this means. Darcy's constant for a flow in a capillary network of parallel fibres is given by $\Delta = \phi^3/[CA^2(1-\phi)^2]$. Explain the terms in the relationship. Permeability in the same direction is given by $\Delta_x = r_f^2(1-V_f)^3/(4s_xV_f^2)$. Explain the meaning of this equation and the significance of the parameters.

20. What are the main advantages and disadvantages of metal matrix composites compared with polymer matrix composites. Explain the squeeze casting process. Give an example of the use of a metal matrix composite for abrasion resistance and state which type of fibre is used?

21. Carbon–carbon composites suffer from oxidation from what temperature? In the absence of oxygen, to which temperatures can these composites be used? Explain the manufacturing process used to make these composites and give examples of their use.

22. What type of applications are foreseen for ceramic matrix composites. Give two ceramic matrix composites systems which have been developed. How are they made?

References

Akovali G 2001 *Handbook of Composite Fabrication* (Shawbury: RAPRA Technology Ltd)

Noakes K 1999 *Successful Composite Techniques*, 3rd edition (Oxford: Osprey Publishing)

Bader M 2002 Selection of composite materials and manufacturing routes for cost effective performance *Composites* **A33** 913–934

Chapter 5

Constitutive relations of a continuum and their application to composite behaviour

5.1   Introduction

A mathematical treatment of the behaviour of materials, including composite materials, is necessary in order to design even the simplest structures. The models have to take into account all the possible variables which determine the behaviour of the material and to be applied require that the variables are first determined experimentally as accurately as possible. Writing a constitutive law is needed to solve the equations of a continuum mechanics problem. The local equations introduce six unknown stresses σ and three unknown displacements u, v, w. That means there are nine unknown variables related by only three available equilibrium equations. Six other equations are therefore needed to determine all the nine variables. These equations which relate the behaviour between stresses and strains are called constitutive equations. As we have six stresses and strains, we get six new equations which allow the studied medium, the material, to be characterized.

The material is considered to be made up of regions which possess its characteristics, the sum of which makes up the whole material. In fact, we have to describe the response of the representative elementary volume (REV) subjected to a given loading. We shall take loading to mean all events which can lead to changes in the material response. Of course this means the effects of applying mechanical forces but also other factors such as environmental conditions of humidity and temperature. The choice of the REV, which will be further discussed in chapter 6, is very important because it will determine the nature of the constitutive law which will be applied. This means that the REV has to be statistically representative of the medium we wish to characterize. In order to make this choice the following steps to establish a constitutive law are carried out.

- The first step is an experimental analysis of samples which are sufficiently representative of the REV. As the observable physical phenomena are well known, the aim of this step is to identify non-observable phenomena which are internal and dissipative in the material. These observations need both destructive and non-destructive tools to evaluate all physical events, appreciate their relative importance and so decide which are the most important for modelling.

- The second step is a theoretical one. As the physical events to be considered have been selected, we have to formulate the most appropriate mathematical model to describe each variable state and its evolution. This formulation has to respect some theoretical framework, which we are going to develop.

- Finally the third step is an identification procedure. Like the first one, it is an experimental phase carried out on representative samples of the material. As a mathematical law to describe different physical phenomena, we have to identify the different coefficients which characterize this law and which are dependent on the material itself. In order to do this, we have to select appropriate experiments to determine specific coefficients. It is most important that the results used to measure these coefficients be as accurate and reproducible as possible.

This chapter will pay particular attention to the second step described above.

The general principles which have to be considered when formulating constitutive relations have to be coherent with the physics of all bodies in motion and be able to describe the effects of motion which differ according to the nature of the material that makes up the body. A general theory of constitutive relations has to be able to describe an infinite number of classes of materials, distinguished by properties of symmetry and invariance. We can classify these general principles as follows.

- *Principle of material frame-indifference.* This means that we regard materials properties as being indifferent to the choice of framework. Since constitutive equations are designed to express idealized material properties, we require that they shall be frame-indifferent. In popular terms this means that the responses of a material and its microstructure depend neither on their location in the world nor on the position of different observers.

- *Respect of the geometrical microstructure.* This means that the constitutive relations have to satisfy material symmetry during any mapping or change of framework.

- *To be thermodynamically admissible during any motion combined with acting forces.* Continuum mechanics provide a framework which will be considered in order to describe dynamic process as a function of time t.

Considering these three last requirements, it is first necessary to answer the following question: to state the constitutive law of a REV for a given point at a time t, is it sufficient to consider only its own history or is it necessary to take into account the history of all neighbouring particles with their own histories too? This answer will be given by the Principle of Determinism and the Principle of Local Action which we shall now study.

5.2   Principle of determinism

Modelling the response at a material point M at time t when the material is submitted to a stress field, requires answers to the following questions.

- How can the local stress field be taken into account?
- What is the influence of the surroundings on the behaviour of the studied particle?

The principle of determinism answers these two questions by assuming that the stress at the place occupied by a material point M at time t is determined by the history of the motion of the point M itself, and also of all other points surrounding M up to time t. That means the history of all points making up the body must be taken into account. The concept of material defined here reflects the common observation that many natural bodies exhibit a memory of their past experiences, sometimes continuing to respond to the effect by a change of form long after the stress field has been applied. In this case the constitutive relationships describe the past and present positions of points in the material as a function of forces acting on the studied material. These mathematical expressions have to include the variables which have been used to model observable phenomena (generally, stress σ, strain ε or temperature T) and internal variables, V_i (plasticity, damage, ageing, rupture, etc.). Changes in these mechanical or physical variables are consequences of the applied loading. In this way the constitutive expressions are able to describe the state of all the points in the material at any time up to time t.

5.3   Principle of local action

In the principle of determinism the motions of body-points that lie far away from M are allowed to affect the stress at M. This principle is too general and also too difficult to be applied. Accordingly, we assume a second constitutive axiom: only the motion of body-points at a finite distance from M and in the neighbourhood of M are allowed to affect the stress state at M. Sometimes we limit this principle to the point M itself.

5.4   Principle of material frame-indifference

The question is now to know if the constitutive laws and the associated variables are indifferent to the choice of framework during motion of the body. Considering that the material behaviour has to be objective, constitutive laws have to be universal and then indifferent to the choice of framework. That leads to the use of invariants of the stress and strain tensors.

5.5 Thermodynamically admissible processes

All the above-mentioned conditions can be verified in the frame of continuum mechanics and thermodynamics. Once the variables describing the state considered are defined, we can postulate the existence of a thermodynamic potential from which the state laws can be derived. This potential is both a function of observable variables and internal variables, such as

$$\psi = \psi(\varepsilon, T, V_1, \ldots, V_i, \ldots, V_n) \tag{5.1}$$

where ε is the strain and T is the temperature. The refinement of the description of physical phenomena controlling the material characteristics will determine the number of variables V_i and their significance. According to the principle of local action, these variables, at a point M and time t, will only depend on the point considered and its particular history up to time t. The evolution of the material is thermodynamically admissible if obeying the inequality which characterized the first and second principle [1]. In the case where there is no variation of the temperature T during the transformation, we obtain

$$\left(\sigma - \rho \frac{\partial \psi}{\partial \varepsilon}\right)\dot{\varepsilon} - \rho \sum_{i=1}^{n} \frac{\partial \psi}{\partial V_i} \dot{V_i} \geq 0 \tag{5.2}$$

where ρ is the density of the material.

If we now imagine particular transformations in which all variables are zero except only one, we obtain the law of state associated with this variable. If A_i is defined as the thermodynamic force associated with the internal variable V_i, we have

$$A_i = \rho \frac{\partial \psi}{\partial V_i}. \tag{5.3}$$

These relations which constitute the laws of state give very interesting information about the behaviour of the material, but do not describe the dissipation processes. To describe the evolution of the internal variables, a complementary formalism is needed. These laws which are able to describe these dissipation processes are called 'complementary laws'.

We are not going to discuss in detail the construction of these complementary laws as this chapter is only concerned with linear elastic media, but we can say that dissipation can be described by the introduction of a dissipation potential (or pseudo-potential) expressed within the same constraints as the thermodynamic potential: principle of material frame-indifference and the material symmetries of the media. The other condition is that the dissipation process of a body in motion given by (5.2) is never negative. With this condition the thermodynamic processes are admissible, the complementary laws describe dissipative phenomena, and the variables introduced in these dissipation potentials depend on the position of point M and its own history up to time t.

In this way the thermo-mechanical behaviour at a point M is determined with a potential and associated dissipation processes. We shall mention some of these processes when discussing damage in composite materials.

5.6 Anisotropic linear elastic materials

For reversible (or elastic) phenomena, at each instant t, the state only depends on observable variables. If we only consider isothermal transformation, the unique observable variable is the elastic strain equal in that case to the total strain ε. Then the thermodynamic potential can be written as

$$\psi = \psi(\varepsilon)$$

and the unique law of state is given by

$$\sigma = \rho \frac{\partial \psi}{\partial \varepsilon}. \tag{5.4}$$

We obtain the relation $\{\sigma\} = [C]\{\varepsilon\}$ written in a condensed form which corresponds to the generalized Hooke's law for a linear, anisotropic, elastic continuum material. $[C]$ is the stiffness matrix the components of which are the elastic moduli which may depend on the temperature. By inverting the above relation, the linear strain-stress relation becomes:

$$\{\varepsilon\} = [S]\{\sigma\} \tag{5.5}$$

where $[S]$ is the elastic compliance matrix.

C and S are fourth-order tensors, the components of which are respectively C_{ijkl} and S_{ijkl}. These tensors depend on 81 coefficients (each mark varies from 1 to 3). Further, as the stress and the strain tensors are symmetrical, their number reduces to 6 independent components instead of 9 and the linear relationship between stress and strain can be written in a general tensorial way with 36 material coefficients all referred to an orthogonal Cartesian coordinate system (x_1, x_2, x_3) fixed in the body

$$\begin{Bmatrix} \sigma_{11} \\ \sigma_{22} \\ \sigma_{33} \\ \sigma_{23} \\ \sigma_{31} \\ \sigma_{12} \end{Bmatrix} = \begin{bmatrix} C_{1111} & C_{1122} & C_{1133} & C_{1123} & C_{1131} & C_{1112} \\ C_{2211} & C_{2222} & C_{2233} & C_{2223} & C_{2231} & C_{2212} \\ C_{3311} & C_{3322} & C_{3333} & C_{3323} & C_{3331} & C_{3312} \\ C_{2311} & C_{2322} & C_{2333} & C_{2323} & C_{2331} & C_{2312} \\ C_{3111} & C_{3122} & C_{3133} & C_{3123} & C_{3131} & C_{3112} \\ C_{1211} & C_{1222} & C_{1233} & C_{1223} & C_{1231} & C_{1212} \end{bmatrix} \begin{Bmatrix} \varepsilon_{11} \\ \varepsilon_{22} \\ \varepsilon_{33} \\ 2\varepsilon_{23} \\ 2\varepsilon_{31} \\ 2\varepsilon_{12} \end{Bmatrix}.$$

The tensors $C = [C_{ijkl}]_{i,j,k,l=1,2,3}$ and $S = [S_{ijkl}]_{i,j,k,l=1,2,3}$ are symmetrical tensors so that $C_{ijkl} = C_{ijlk} = C_{jikl} = C_{jilk}$. That reduces the number of independent coefficients to 21 coefficients, which means that the most general anisotropic elastic continuum medium can be characterized with 21 coefficients.

A single subscript notation is often used for stresses and strains which is called the contracted notation or the Voigt notation with the following correspondences:

$$\sigma_1 = \sigma_{11}, \quad \sigma_2 = \sigma_{22}, \quad \sigma_3 = \sigma_{33}, \quad \sigma_4 = \sigma_{23}, \quad \sigma_5 = \sigma_{13}, \quad \sigma_6 = \sigma_{12}$$

$$\varepsilon_1 = \varepsilon_{11}, \quad \varepsilon_2 = \varepsilon_{22}, \quad \varepsilon_3 = \varepsilon_{33}, \quad \varepsilon_4 = 2\varepsilon_{23}, \quad \varepsilon_5 = 2\varepsilon_{13}, \quad \varepsilon_6 = 2\varepsilon_{12}$$

In this case the most general behaviour can be written by using Voigt notation $C_{IJ} = (I, J = 1, 6)$:

$$\begin{Bmatrix} \sigma_1 \\ \sigma_2 \\ \sigma_3 \\ \sigma_4 \\ \sigma_5 \\ \sigma_6 \end{Bmatrix} = \begin{bmatrix} C_{11} & C_{12} & C_{13} & C_{14} & C_{15} & C_{16} \\ C_{12} & C_{22} & C_{23} & C_{24} & C_{25} & C_{26} \\ C_{13} & C_{23} & C_{33} & C_{34} & C_{35} & C_{36} \\ C_{14} & C_{24} & C_{34} & C_{44} & C_{45} & C_{46} \\ C_{15} & C_{25} & C_{35} & C_{45} & C_{55} & C_{56} \\ C_{16} & C_{26} & C_{36} & C_{46} & C_{56} & C_{66} \end{bmatrix} \begin{Bmatrix} \varepsilon_1 \\ \varepsilon_2 \\ \varepsilon_3 \\ \varepsilon_4 \\ \varepsilon_5 \\ \varepsilon_6 \end{Bmatrix}. \tag{5.6}$$

In the same way, the inverse strain–stress relations are given by

$$\{\varepsilon\} = [S]\{\sigma\} \qquad \text{and} \qquad S_{IJ} = S_{JI}$$

$$\begin{Bmatrix} \varepsilon_1 \\ \varepsilon_2 \\ \varepsilon_3 \\ \varepsilon_4 \\ \varepsilon_5 \\ \varepsilon_6 \end{Bmatrix} = \begin{bmatrix} S_{11} & S_{12} & S_{13} & S_{14} & S_{15} & S_{16} \\ S_{12} & S_{22} & S_{23} & S_{24} & S_{25} & S_{26} \\ S_{13} & S_{23} & S_{33} & S_{34} & S_{35} & S_{36} \\ S_{14} & S_{24} & S_{34} & S_{44} & S_{45} & S_{46} \\ S_{15} & S_{25} & S_{35} & S_{45} & S_{55} & S_{56} \\ S_{16} & S_{26} & S_{36} & S_{46} & S_{56} & S_{66} \end{bmatrix} \begin{Bmatrix} \sigma_1 \\ \sigma_2 \\ \sigma_3 \\ \sigma_4 \\ \sigma_5 \\ \sigma_6 \end{Bmatrix} \tag{5.7}$$

with $[S] = [C]^{-1}$.

5.7 Material symmetry

As we have seen previously, the stiffness matrix and the compliance matrix are determined using 21 material coefficients. This is the general case for an anisotropic material, described as being 'triclinic' and displaying no symmetry. The response of such a characteristic REV submitted to a given loading depends on the loading direction.

The question which is now raised is whether when the REV is loaded in different directions, some of which give the same response. If it is the case, the identical responses characterize the material symmetry. According to the set of material symmetry, which characterizes the material, its material behaviour is said to be monoclinic, orthotropic, transversally isotropic or isotropic.

It is mainly the geometry of the microstructure which induces material symmetry (single crystal, fibre reinforced composite, textile, etc.). In addition other causes can modify the microstructure or change material symmetry (pre-stresses, damage, etc.) and modify the lifetime of the material.

The constitutive law has to be written to respect the material symmetry and then the number of independent elastic coefficients can be reduced. To respect these physical symmetries, we choose a thermodynamic potential defined with polynomial quantities of state variables which remain invariable with the material plane symmetry. We can show that the number of these invariants is finite and describe the integrity basis. Further, the definition of a thermodynamic potential written with invariant variables according to material symmetries allows the material behaviour to be correctly described.

5.8 Monoclinic material

A monoclinic material is a material with only one plane of symmetry. The stiffness matrix or compliance matrix do not have to be modified by a transformation with respect to a symmetry of this plane. In the case where the plane of symmetry is $(1, 2)$, the stiffness matrix is given by

$$\begin{bmatrix} C_{11} & C_{12} & C_{13} & 0 & 0 & C_{16} \\ C_{12} & C_{22} & C_{23} & 0 & 0 & C_{26} \\ C_{13} & C_{23} & C_{33} & 0 & 0 & C_{36} \\ 0 & 0 & 0 & C_{44} & C_{45} & 0 \\ 0 & 0 & 0 & C_{45} & C_{55} & 0 \\ C_{16} & C_{26} & C_{36} & 0 & 0 & C_{66} \end{bmatrix}. \tag{5.8}$$

The compliance matrix is similar. The number of independent coefficients reduces to 13.

5.9 Orthotropic material

An orthotropic material has three orthogonal planes of material symmetry; each plane is normal to the others. The existence of two of them implies the existence of the third plane. The stiffness matrix is then obtained by adding another plane of symmetry to the monoclinic material. Respecting invariance towards these three planes of material symmetry requires us to write a thermodynamic potential as a function of seven invariant polynoms of strains in the 'integrity basis'.

- Degree 1 with strains: $I_1 = \varepsilon_1$, $I_2 = \varepsilon_2$, $I_3 = \varepsilon_3$.
- Degree 2 with strains: $I_4 = \varepsilon_4^2$, $I_5 = \varepsilon_5^2$, $I_6 = \varepsilon_6^2$.
- Degree 3 with strains: $I_7 = \varepsilon_4 \varepsilon_5 \varepsilon_6$.

A strain second order thermodynamic potential can then be written as

$$\psi = \psi(I_1, I_2, I_3, I_4, I_5, I_6).$$

The law of state

$$\sigma = \rho \frac{\partial \psi}{\partial \varepsilon}$$

gives the stiffness matrix as

$$\begin{bmatrix} C_{11} & C_{12} & C_{13} & 0 & 0 & 0 \\ C_{12} & C_{22} & C_{23} & 0 & 0 & 0 \\ C_{13} & C_{23} & C_{33} & 0 & 0 & 0 \\ 0 & 0 & 0 & C_{44} & 0 & 0 \\ 0 & 0 & 0 & 0 & C_{55} & 0 \\ 0 & 0 & 0 & 0 & 0 & C_{66} \end{bmatrix}. \tag{5.9}$$

The stiffness matrix is similar and the number of independent coefficients is reduced to nine. If the material coordinate system is taken to be a rectangular coordinate system (x_1, x_2, x_3) in which the x_1 axis is parallel to the fibre direction, the compliance matrix can be given by

$$\begin{Bmatrix} \varepsilon_1 \\ \varepsilon_2 \\ \varepsilon_3 \\ \varepsilon_4 \\ \varepsilon_5 \\ \varepsilon_6 \end{Bmatrix} = \begin{bmatrix} 1/E_1 & -\nu_{21}/E_2 & -\nu_{31}/E_3 & 0 & 0 & 0 \\ -\nu_{12}/E_1 & 1/E_2 & -\nu_{32}/E_3 & 0 & 0 & 0 \\ -\nu_{13}/E_1 & -\nu_{23}/E_2 & 1/E_3 & 0 & 0 & 0 \\ 0 & 0 & 0 & 1/G_{23} & 0 & 0 \\ 0 & 0 & 0 & 0 & 1/G_{13} & 0 \\ 0 & 0 & 0 & 0 & 0 & 1/G_{12} \end{bmatrix} \begin{Bmatrix} \sigma_1 \\ \sigma_2 \\ \sigma_3 \\ \sigma_4 \\ \sigma_5 \\ \sigma_6 \end{Bmatrix} \tag{5.10}$$

where E_1, E_2, E_3 are Young's moduli in 1, 2 and 3 material directions, respectively, ν_{ij} is Poisson's ratio, defined as the ratio of transverse strain in the jth direction to the axial strain in the ith direction when stressed in the i direction, and G_{23}, G_{13}, G_{12} are the shear moduli in the 2-3, 1-3, and 1-2 planes, respectively. The symmetry of the behaviour implies that the following reciprocal relations hold:

$$\frac{\nu_{21}}{E_2} = \frac{\nu_{12}}{E_1}, \quad \frac{\nu_{31}}{E_3} = \frac{\nu_{13}}{E_1}, \quad \frac{\nu_{32}}{E_3} = \frac{\nu_{23}}{E_2} \tag{5.11}$$

or in short

$$\frac{\nu_{ij}}{E_i} = \frac{\nu_{ji}}{E_j} \quad \text{(no sum on } i, j)$$

for $i, j = 1, 2, 3$. Once again, the number of independent coefficients reduces to nine,

$$E_1, \quad E_2, \quad E_3, \quad \nu_{12}, \quad \nu_{13}, \quad \nu_{23}, \quad G_{23}, \quad G_{13}, \quad G_{12}$$

for an orthotropic material.

5.10 Transversally isotropic material

A continuum material is said to be transversally isotropic for a given axis D, if its properties remain invariant or isotropic with respect to a change in direction orthogonal to D. Often fibre-reinforced lamina are characterized as being transversally isotropic. If the fibre direction is parallel to the x_1 axis, we experimentally observe that perpendicular to this axis all the properties are isotropic. We can demonstrate that the integrity basis of such a material is characterized by four independent invariant polynoms of strains:

$$I_1 = \varepsilon_1$$

$$I_2 = \varepsilon_2 + \varepsilon_3$$

$$I_3 = \varepsilon_4^2 - \varepsilon_2\varepsilon_3$$

$$I_4 = \varepsilon_5^2 + \varepsilon_6^2 - \varepsilon_1\varepsilon_2 - \varepsilon_1\varepsilon_3.$$

Thus the constitutive law can be written with five independent material coefficients:

$$\begin{Bmatrix} \sigma_1 \\ \sigma_2 \\ \sigma_3 \\ \sigma_4 \\ \sigma_5 \\ \sigma_6 \end{Bmatrix} = \begin{bmatrix} C_{11} & C_{12} & C_{12} & 0 & 0 & 0 \\ C_{12} & C_{22} & C_{23} & 0 & 0 & 0 \\ C_{13} & C_{23} & C_{22} & 0 & 0 & 0 \\ 0 & 0 & 0 & (C_{22}-C_{23})/2 & 0 & 0 \\ 0 & 0 & 0 & 0 & C_{66} & 0 \\ 0 & 0 & 0 & 0 & 0 & C_{66} \end{bmatrix} \begin{Bmatrix} \varepsilon_1 \\ \varepsilon_2 \\ \varepsilon_3 \\ \varepsilon_4 \\ \varepsilon_5 \\ \varepsilon_6 \end{Bmatrix}$$

$$\begin{Bmatrix} \varepsilon_1 \\ \varepsilon_2 \\ \varepsilon_3 \\ \varepsilon_4 \\ \varepsilon_5 \\ \varepsilon_6 \end{Bmatrix} = \begin{bmatrix} S_{11} & S_{12} & S_{12} & 0 & 0 & 0 \\ S_{12} & S_{22} & S_{23} & 0 & 0 & 0 \\ S_{12} & S_{23} & S_{22} & 0 & 0 & 0 \\ 0 & 0 & 0 & (S_{22}-S_{23})/2 & 0 & 0 \\ 0 & 0 & 0 & 0 & S_{66} & 0 \\ 0 & 0 & 0 & 0 & 0 & S_{66} \end{bmatrix} \begin{Bmatrix} \sigma_1 \\ \sigma_2 \\ \sigma_3 \\ \sigma_4 \\ \sigma_5 \\ \sigma_6 \end{Bmatrix}$$

$$(5.12)$$

Young's moduli in direction 2 and 3 are the same, and so are the shear moduli in planes 1-2, 1-3. We obtain

$$S_{11} = \frac{1}{E_1} \qquad S_{22} = S_{33} = \frac{1}{E_2} \quad S_{44} = \frac{2(1+\nu_{23})}{E_2}$$

$$S_{12} = S_{13} = \frac{-\nu_{12}}{E_1} = \frac{-\nu_{21}}{E_2} \quad S_{23} = \frac{-\nu_{23}}{E_2} \qquad S_{55} = S_{66} = \frac{1}{G_{13}} = \frac{1}{G_{12}}.$$

$$(5.13)$$

$$\left\{ \begin{aligned}
C_{11} &= \frac{E_1^2(1-\nu_{23})}{-E_1(1-\nu_{23})+2\nu_{12}^2 E_2} = \frac{E_1(1-\nu_{23})}{1-\nu_{23}-2\nu_{12}\nu_{21}} \\
C_{22} &= \frac{E_2(\nu_{12}^2 E_2 - E_1)}{(1+\nu_{23})(E_1(\nu_{23}-1)+2\nu_{12}^2 E_2)} = \frac{E_2(1-\nu_{12}\nu_{21})}{(1+\nu_{23})(1-\nu_{23}-2\nu_{12}\nu_{21})} \\
C_{12} &= \frac{-E_1 E_2 \nu_{12}}{-E_1(1-\nu_{23})+2\nu_{12}^2 E_2} = \frac{E_2 \nu_{12}}{1-\nu_{23}-2\nu_{12}\nu_{21}} \\
&= \frac{E_1 \nu_{21}}{1-\nu_{23}-2\nu_{12}\nu_{21}} \\
C_{23} &= \frac{-E_2(E_1\nu_{23}+\nu_{12}^2 E_2)}{(1+\nu_{23})(E_1(\nu_{23}-1)+2\nu_{12}^2 E_2)} = \frac{E_2(\nu_{23}+\nu_{12}\nu_{21})}{(1+\nu_{23})(1-\nu_{23}-2\nu_{12}\nu_{21})} \\
C_{66} &= G_{12} \\
\frac{C_{22}-C_{23}}{2} &= \frac{E_2}{2(1+\nu_{23})}
\end{aligned} \right.$$

$$(5.14)$$

5.11 Isotropic materials

A material is said to be isotropic if, when stressed in all the directions, the responses remain identical. The mechanical properties are the same whatever the considered direction and so the stiffness and compliance matrices have to be invariant for any orthogonal frame transformation. In that case, we have two basic strain invariants:

$$I_1 = \text{Tr}(\varepsilon) = \varepsilon_1 + \varepsilon_2 + \varepsilon_3, \qquad I_2 = \tfrac{1}{2}\text{Tr}(\varepsilon^2).$$

This requires that the thermodynamic potential ψ be a quadratic invariant of the strain tensor as a linear combination of the square of I_1 and I_2:

$$\psi(\varepsilon) = \psi(I_1^2, I_2).$$

The number of independent coefficients reduces to two and then

$$C_{22} = C_{21}, \qquad C_{23} = C_{12}, \qquad C_{66} = \tfrac{1}{2}(C_{11}-C_{12})$$

leading to the stiffness matrix

$$
\begin{bmatrix}
C_{11} & C_{12} & C_{12} & 0 & 0 & 0 \\
C_{12} & C_{11} & C_{12} & 0 & 0 & 0 \\
C_{12} & C_{12} & C_{11} & 0 & 0 & 0 \\
0 & 0 & 0 & \frac{1}{2}(C_{11} - C_{12}) & 0 & 0 \\
0 & 0 & 0 & 0 & \frac{1}{2}(C_{11} - C_{12}) & 0 \\
0 & 0 & 0 & 0 & 0 & \frac{1}{2}(C_{11} - C_{12})
\end{bmatrix}.
$$

Generally, the stiffness coefficients are also replaced by λ and μ ($\mu = G$), Lamé's two coefficients:

$$
C_{12} = \lambda, \qquad \frac{C_{11} - C_{12}}{2} = \mu
$$

and so $C_{11} = \lambda + 2\mu$. The stress–strain relation becomes

$$
\begin{Bmatrix}
\sigma_1 \\
\sigma_2 \\
\sigma_3 \\
\sigma_4 \\
\sigma_5 \\
\sigma_6
\end{Bmatrix}
=
\begin{bmatrix}
\lambda + 2\mu & \lambda & \lambda & 0 & 0 & 0 \\
\lambda & \lambda + 2\mu & \lambda & 0 & 0 & 0 \\
\lambda & \lambda & \lambda + 2\mu & 0 & 0 & 0 \\
0 & 0 & 0 & \mu & 0 & 0 \\
0 & 0 & 0 & 0 & \mu & 0 \\
0 & 0 & 0 & 0 & 0 & \mu
\end{bmatrix}
\begin{Bmatrix}
\varepsilon_1 \\
\varepsilon_2 \\
\varepsilon_3 \\
\varepsilon_4 \\
\varepsilon_5 \\
\varepsilon_6
\end{Bmatrix}.
$$

By inverting the above relations, or by differentiating a dual potential, we find the compliance matrix given as

$$
\begin{Bmatrix}
\varepsilon_1 \\
\varepsilon_2 \\
\varepsilon_3 \\
\varepsilon_4 \\
\varepsilon_5 \\
\varepsilon_6
\end{Bmatrix}
=
\begin{bmatrix}
1/E & -\nu/E & -\nu/E & 0 & 0 & 0 \\
-\nu/E & 1/E & -\nu/E & 0 & 0 & 0 \\
-\nu/E & -\nu/E & 1/E & 0 & 0 & 0 \\
0 & 0 & 0 & 1/G & 0 & 0 \\
0 & 0 & 0 & 0 & 1/G & 0 \\
0 & 0 & 0 & 0 & 0 & 1/G
\end{bmatrix}
\begin{Bmatrix}
\sigma_1 \\
\sigma_2 \\
\sigma_3 \\
\sigma_4 \\
\sigma_5 \\
\sigma_6
\end{Bmatrix}.
$$

The Young's modulus E and the Poisson's ratio ν are related to Lamé's coefficients by well-known relations:

$$
E = \frac{\mu(3\lambda + 2\mu)}{\lambda + \mu}, \qquad \nu = \frac{\lambda}{2(\lambda + \mu)}.
$$

Coefficients E and ν are generally mostly used because they can directly be obtained from experiments (tensile tests, for example).

5.12 Plane stress assumption

A plane stress state is defined to be one in which all transverse stresses are negligible. This is the case of most laminates which are thin. For a ply in the $x_1 x_2$ plane, the out of plane stress components are σ_{33}, σ_{13} and σ_{23} (figure 9.1, page 299). In the case where these components are neglected, the constitutive equations must be modified to account for the fact that

$$\sigma_3 = \sigma_{33} = 0, \qquad \sigma_4 = \sigma_{23} = 0, \qquad \sigma_5 = \sigma_{13} = 0$$

and the constitutive law is obtained from the three-dimensional compliance matrix:

$$\begin{Bmatrix} \varepsilon_1 \\ \varepsilon_2 \\ \varepsilon_3 \\ \varepsilon_4 \\ \varepsilon_5 \\ \varepsilon_6 \end{Bmatrix} = \begin{bmatrix} S_{11} & S_{12} & S_{13} & S_{14} & S_{15} & S_{16} \\ S_{12} & S_{22} & S_{23} & S_{24} & S_{25} & S_{26} \\ S_{13} & S_{23} & S_{33} & S_{34} & S_{35} & S_{36} \\ S_{14} & S_{24} & S_{34} & S_{44} & S_{45} & S_{46} \\ S_{15} & S_{25} & S_{35} & S_{45} & S_{55} & S_{56} \\ S_{16} & S_{26} & S_{36} & S_{46} & S_{56} & S_{66} \end{bmatrix} \begin{Bmatrix} \sigma_1 \\ \sigma_2 \\ 0 \\ 0 \\ 0 \\ \sigma_6 \end{Bmatrix} \Rightarrow \begin{Bmatrix} \varepsilon_1 \\ \varepsilon_2 \\ \varepsilon_6 \end{Bmatrix}$$

$$= \begin{bmatrix} S_{11} & S_{12} & S_{16} \\ S_{12} & S_{22} & S_{26} \\ S_{16} & S_{26} & S_{66} \end{bmatrix} \begin{Bmatrix} \sigma_1 \\ \sigma_2 \\ \sigma_6 \end{Bmatrix}. \tag{5.15}$$

By inverting the above equation, we get

$$\begin{Bmatrix} \sigma_1 \\ \sigma_2 \\ \sigma_6 \end{Bmatrix} = \begin{bmatrix} Q_{11} & Q_{12} & Q_{16} \\ Q_{12} & Q_{22} & Q_{26} \\ Q_{16} & Q_{26} & Q_{66} \end{bmatrix} \begin{Bmatrix} \varepsilon_1 \\ \varepsilon_2 \\ \varepsilon_6 \end{Bmatrix} \tag{5.16}$$

where $Q_{\alpha\beta}$ are the plane stress-reduced stiffnesses in its material coordinate system.

$$Q_{\alpha\beta} = S_{\alpha\beta} - \frac{S_{\alpha 3} S_{\beta 3}}{S_{33}} \qquad (\alpha, \beta = 1, 2, 6) \tag{5.17}$$

We first have to point out that we directly get the plane stress compliance matrix (5.12) as a part of the three-dimensional compliance matrix which is not the case for the stiffness in-plane matrix which is not a restriction of three-dimensional stiffness matrix.

Further, ε_3 does not appear in plane–stress behaviour. The reason is that generally this component is not useful. That does not mean it is equal to zero. If we develop the stress–strain behaviour, we get

$$\varepsilon_3 = S_{13}\sigma_1 + S_{23}\sigma_2 + S_{16}\sigma_6. \tag{5.18}$$

If now we consider an orthotropic material which verifies the plane stress assumption, we obtain

$$
\left\{
\begin{array}{c}
\sigma_1 \\
\sigma_2 \\
\sigma_6
\end{array}
\right\}
=
\left[
\begin{array}{ccc}
Q_{11} & Q_{12} & 0 \\
Q_{12} & Q_{22} & 0 \\
0 & 0 & Q_{66}
\end{array}
\right]
\left\{
\begin{array}{c}
\varepsilon_1 \\
\varepsilon_2 \\
\varepsilon_6
\end{array}
\right\}
\tag{5.19}
$$

and then $Q_{\alpha\beta}$ as function of 'Engineer coefficients':

$$
Q_{11} = \frac{E_1}{1 - \nu_{12}\nu_{21}}, \qquad Q_{12} = \frac{\nu_{12}E_2}{1 - \nu_{12}\nu_{21}}, \qquad Q_{22} = \frac{E_2}{1 - \nu_{12}\nu_{21}}, \qquad Q_{66} = G_{12}.
$$

$$
\tag{5.20}
$$

5.13 Transformation of stresses and strains

A laminate is the result of fibre-reinforced laminae which are stacked with their x_1x_2 planes parallel but each having different fibre directions. Let us call (x_1^L, x_2^L, x_3^L), the local material coordinate system with the x_1^L-axis parallel to the fibre direction and (x_1^G, x_2^G, x_3^G) the global coordinate system of the laminate (figure 9.1, page 299). If the x_3^G coordinate is oriented normal to the plane of the laminate, the x_3^L coordinate of each lamina will coincide with x_3^G. Thus we only have a transformation in the plane of the laminate by rotating by an angle θ about the x_3^G, x_3^L axis. The coordinates of a material point in the coordinate systems are related by $(x_3^G = x_3^L)$ by the following equations:

$$
\left\{
\begin{array}{c}
x_1^L \\
x_2^L \\
x_3^L
\end{array}
\right\}
=
\left[
\begin{array}{ccc}
\cos\theta & \sin\theta & 0 \\
-\sin\theta & \cos\theta & 0 \\
0 & 0 & 1
\end{array}
\right]
\left\{
\begin{array}{c}
x_1^G \\
x_2^G \\
x_3^G
\end{array}
\right\}
= [A]
\left\{
\begin{array}{c}
x_1^G \\
x_2^G \\
x_3^G
\end{array}
\right\}.
\tag{5.21}
$$

The inverse of equation (5.21) is

$$
\left\{
\begin{array}{c}
x_1^G \\
x_2^G \\
x_3^G
\end{array}
\right\}
=
\left[
\begin{array}{ccc}
\cos\theta & -\sin\theta & 0 \\
\sin\theta & \cos\theta & 0 \\
0 & 0 & 1
\end{array}
\right]
\left\{
\begin{array}{c}
x_1^L \\
x_2^L \\
x_3^L
\end{array}
\right\}
= [A]^{\mathrm{T}}
\left\{
\begin{array}{c}
x_1^L \\
x_2^L \\
x_3^L
\end{array}
\right\}.
\tag{5.22}
$$

Note that the inverse of $[A]$ is equal to its transpose, $[A]^{-1} = [A]^{\mathrm{T}}$. Next we consider the relationship between the components of stress and stress in the two coordinate systems. Let σ_i^L $(i = 1, 6)$ be the components of stress in the material coordinate (x_i^L) and σ_i^G $(i = 1, 6)$ be the components in the laminate (or global) coordinate system (x_i^G). The following relations among the two sets of components can be established.

$$
\begin{Bmatrix} \sigma_1^G \\ \sigma_2^G \\ \sigma_3^G \\ \sigma_4^G \\ \sigma_5^G \\ \sigma_6^G \end{Bmatrix} = \begin{bmatrix} \cos^2\theta & \sin^2\theta & 0 & 0 & 0 & -\sin 2\theta \\ \sin^2\theta & \cos^2\theta & 0 & 0 & 0 & \sin 2\theta \\ 0 & 0 & 1 & 0 & 0 & 0 \\ 0 & 0 & 0 & \cos\theta & \sin\theta & 0 \\ 0 & 0 & 0 & -\sin\theta & \cos\theta & 0 \\ \frac{1}{2}\sin 2\theta & -\frac{1}{2}\sin 2\theta & 0 & 0 & 0 & \cos^2\theta - \sin^2\theta \end{bmatrix} \begin{Bmatrix} \sigma_1^L \\ \sigma_2^L \\ \sigma_3^L \\ \sigma_4^L \\ \sigma_5^L \\ \sigma_6^L \end{Bmatrix}.
$$

$$(5.23)$$

The inverse relationship between $\{\sigma_i^L\}$ and $\{\sigma_i^G\}$ is given by

$$
\begin{Bmatrix} \sigma_1^L \\ \sigma_2^L \\ \sigma_3^L \\ \sigma_4^L \\ \sigma_5^L \\ \sigma_6^L \end{Bmatrix} = \begin{bmatrix} \cos^2\theta & \sin^2\theta & 0 & 0 & 0 & \sin 2\theta \\ \sin^2\theta & \cos^2\theta & 0 & 0 & 0 & -\sin 2\theta \\ 0 & 0 & 1 & 0 & 0 & 0 \\ 0 & 0 & 0 & \cos\theta & -\sin\theta & 0 \\ 0 & 0 & 0 & \sin\theta & \cos\theta & 0 \\ -\frac{1}{2}\sin 2\theta & \frac{1}{2}\sin 2\theta & 0 & 0 & 0 & \cos^2\theta - \sin^2\theta \end{bmatrix} \begin{Bmatrix} \sigma_1^G \\ \sigma_2^G \\ \sigma_3^G \\ \sigma_4^G \\ \sigma_5^G \\ \sigma_6^G \end{Bmatrix}.
$$

$$(5.24)$$

We obtained (5.24) from (5.23) by replacing θ with $-\theta$.

Similarly, the transformation relations for strain components in the material coordinate system (ε_i^L) and the laminate coordinate system (ε_i^G) can be established as being

$$
\begin{Bmatrix} \varepsilon_1^G \\ \varepsilon_2^G \\ \varepsilon_3^G \\ \varepsilon_4^G \\ \varepsilon_5^G \\ \varepsilon_6^G \end{Bmatrix} = \begin{bmatrix} \cos^2\theta & \sin^2\theta & 0 & 0 & 0 & -\sin\theta\cos\theta \\ \sin^2\theta & \cos^2\theta & 0 & 0 & 0 & \sin\theta\cos\theta \\ 0 & 0 & 1 & 0 & 0 & 0 \\ 0 & 0 & 0 & \cos\theta & \sin\theta & 0 \\ 0 & 0 & 0 & -\sin\theta & \cos\theta & 0 \\ \sin 2\theta & -\sin 2\theta & 0 & 0 & 0 & \cos^2\theta - \sin^2\theta \end{bmatrix} \begin{Bmatrix} \varepsilon_1^L \\ \varepsilon_2^L \\ \varepsilon_3^L \\ \varepsilon_4^L \\ \varepsilon_5^L \\ \varepsilon_6^L \end{Bmatrix}.
$$

$$(5.25)$$

These relationships are a bit different from (5.23) because, if we use the Voigt notation, there is a ratio of 2 with shear strains given as

$$\varepsilon_4 = 2\varepsilon_{23}, \qquad \varepsilon_5 = 2\varepsilon_{13}, \qquad \varepsilon_6 = 2\varepsilon_{12}.$$

The inverse relationship is given by

$$
\begin{Bmatrix} \varepsilon_1^L \\ \varepsilon_2^L \\ \varepsilon_3^L \\ \varepsilon_4^L \\ \varepsilon_5^L \\ \varepsilon_6^L \end{Bmatrix} = \begin{bmatrix} \cos^2\theta & \sin^2\theta & 0 & 0 & 0 & -\sin\theta\cos\theta \\ \sin^2\theta & \cos^2\theta & 0 & 0 & 0 & \sin\theta\cos\theta \\ 0 & 0 & 1 & 0 & 0 & 0 \\ 0 & 0 & 0 & \cos\theta & \sin\theta & 0 \\ 0 & 0 & 0 & -\sin\theta & \cos\theta & 0 \\ \sin 2\theta & -\sin 2\theta & 0 & 0 & 0 & \cos^2\theta - \sin^2\theta \end{bmatrix} \begin{Bmatrix} \varepsilon_1^G \\ \varepsilon_2^G \\ \varepsilon_3^G \\ \varepsilon_4^G \\ \varepsilon_5^G \\ \varepsilon_6^G \end{Bmatrix}.
$$

$$(5.26)$$

5.14   Transformed constitutive law of a lamina

As we know the constitutive law of a lamina in the material coordinate system, we now want to evaluate its behaviour in the global coordinate system of the laminate. This will be useful when evaluating the constitutive law of the laminate itself. As

$$\{\sigma^L\} = [C]\{\varepsilon^L\} \qquad (5.27)$$

and we replace

$$\{\sigma^L\} = [T_\sigma]^t\{\sigma^G\}, \qquad \{\varepsilon^L\} = [T_\varepsilon]^t\{\varepsilon^G\} \qquad (5.28)$$

in equation (5.27), we get the stress–strain relationship in the global coordinate system:

$$\{\sigma^G\} = [T_\sigma][C][T_\varepsilon]^t\{\varepsilon^G\} \qquad (5.29)$$

and then the constitutive law of each lamina in the global coordinate system:

$$\{\sigma^G\} = [C(\theta)]\{\varepsilon^G\} \qquad \text{with } [C(\theta)] = [T_\sigma][C][T_\varepsilon]^t. \qquad (5.30)$$

The stiffness matrix which is obtained has the form

$$
\begin{bmatrix} C_{11}(\theta) & C_{12}(\theta) & C_{13}(\theta) & 0 & 0 & C_{16}(\theta) \\ C_{12}(\theta) & C_{22}(\theta) & C_{23}(\theta) & 0 & 0 & C_{26}(\theta) \\ C_{13}(\theta) & C_{23}(\theta) & C_{33}(\theta) & 0 & 0 & C_{36}(\theta) \\ 0 & 0 & 0 & C_{44}(\theta) & C_{45}(\theta) & 0 \\ 0 & 0 & 0 & C_{45}(\theta) & C_{55}(\theta) & 0 \\ C_{16}(\theta) & C_{26}(\theta) & C_{36}(\theta) & 0 & 0 & C_{66}(\theta) \end{bmatrix}.
$$

$$(5.31)$$

We find, after transformation (rotation of θ) that the constitutive law in the global coordinate system has a similar form to the constitutive law of a monoclinic material:

$$C_{11}(\theta) = C_{11} \cos^4 \theta + 2(C_{12} + 2C_{66}) \sin^2 \theta \cos^2 \theta + C_{22} \sin^4 \theta$$

$$C_{12}(\theta) = (C_{11} + C_{22} - 4C_{66}) \sin^2 \theta \cos^2 \theta + C_{12}(\sin^4 \theta + \cos^4 \theta)$$

$$C_{13}(\theta) = C_{12} \cos^2 \theta + C_{23} \sin^2 \theta$$

$$C_{14}(\theta) = 0$$

$$C_{15}(\theta) = 0$$

$$C_{16}(\theta) = (C_{11} - C_{12} - 2C_{66}) \sin \theta \cos^3 \theta + (C_{12} - C_{22} + 2C_{66}) \sin^3 \theta \cos \theta$$

$$C_{22}(\theta) = C_{11} \sin^4 \theta + 2(C_{12} + 2C_{66}) \sin^2 \theta \cos^2 \theta + C_{22} \cos^4 \theta$$

$$C_{23}(\theta) = C_{12} \sin^2 \theta + C_{23} \cos^2 \theta$$

$$C_{24}(\theta) = 0$$

$$C_{25}(\theta) = 0$$

$$C_{26}(\theta) = (C_{11} - C_{12} - 2C_{66}) \sin^3 \theta \cos \theta + (C_{12} - C_{22} + 2C_{66}) \sin \theta \cos^3 \theta$$

$$C_{33}(\theta) = C_{22} \tag{5.32}$$

$$C_{34}(\theta) = 0$$

$$C_{35}(\theta) = 0$$

$$C_{36}(\theta) = (C_{12} - C_{23}) \sin \theta \cos \theta$$

$$C_{44}(\theta) = \frac{C_{22} - C_{23}}{2} \cos^2 \theta + C_{66} \sin^2 \theta$$

$$C_{45}(\theta) = \left(C_{66} - \frac{C_{22} - C_{23}}{2} \right) \sin \theta \cos \theta$$

$$C_{46}(\theta) = 0$$

$$C_{55}(\theta) = \frac{C_{22} - C_{23}}{2} \sin^2 \theta + C_{66} \sin^2 \theta$$

$$C_{56}(\theta) = 0$$

$$C_{66}(\theta) = (C_{11} + C_{22} - 2C_{12} - 2C_{66}) \sin^2 \theta \cos^2 \theta + C_{66}(\sin^4 \theta + \cos^4 \theta)$$

5.15 Conclusion

This presentation describes a general frame to build constitutive laws according to the microstructures of materials. We are now going to apply these concepts to composite materials. First at the scale of the microstructure in chapter 6 which deals with micromechanical models of composite behaviour and in chapter 8 dealing with the ply behaviour and laminate theory.

5.16   Revision exercises

1. Give the relationships between the stresses $\{\sigma^G\}$ expressed in the global reference system and the stresses $\{\sigma^L\}$ expressed in the local reference system.
2. Same question for the strains $\{\varepsilon^G\}$ and $\{\varepsilon^L\}$.
3. Can you explain the differences between these two relationships?
4. The ply behaviour is generally said to be transversally isotropic for the fibre axis. In this case, what is the number of independent constants which are necessary to describe this behaviour?

Reference

[1] Lemaitre J and Chaboche J L 1990 *Mechanics of Solid Materials* (Cambridge: Cambridge University Press)

Chapter 6

Micromechanical models of composite behaviour

6.1   Introduction

It is now increasingly seen to be inadequate to model the behaviour of composite materials only using a macroscopic approach. This is for two main reasons. The first is that an approach which only considers phenomena on a macroscopic scale cannot detect if a given macroscopic event can be generated by a single or several physical events. The second reason is that computers are now more and more powerful and allow the behaviour of materials to be studied on different scales and so to analyse the influence of local defects on the global behaviour or lifetime of materials and structures. Homogenization becomes the appropriate tool for such a multi-scale approach.

Our objective here is not to present all the existing homogenization methods, but mainly to analyse some of them and to point out their particularities and their differences.

Suppose that the studied material has a heterogeneous microstructure resulting from the assembly of different constituents which are themselves considered to be homogeneous. If we intend to understand physical events inside the microstructure (local stresses, onset and propagation of cracks and their influence on the macroscopic behaviour) the analysis has to be carried out at an appropriate scale—that means at the scale of the microstructure.

So, more than the homogenization process which allows the behaviour of the heterogeneous material to be homogenized or considered as a continuum, we need a reverse procedure, called localization, with which to evaluate the local stresses and strains at the micro-scale. These two complementary but reverse procedures, homogenization and localization, are the basis of a multi-scale approach to understanding composite behaviour.

The purpose of the homogenization method is to evaluate, in a non-experimental way, the behaviour of the representative elementary volume

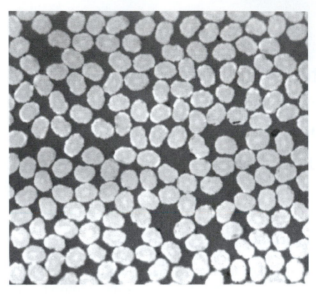

Figure 6.1. Fibre distribution in a carbon–epoxy composite.

(REV) of the material. This volume is defined as being characteristic of the heterogeneous material and is the key to using continuum mechanics for modelling the behaviour of composites. This elementary volume has to be large enough to be statistically representative of the studied material and to provide reproducible information concerning the behaviour of the material. Further, the choice of the REV depends on the refinement or detail which is required to characterize the phenomena occurring inside the material.

If the arrangement of constituents inside the REV is periodic, we can assume that the medium is periodic. In this case, it is equivalent to studying the entire REV, the behaviour of which is similar to the whole composite, or just one cell of the periodic pattern. Strictly speaking it is clear that the periodic case is very rare because of irregular distribution of the components in the composite due to the manufacturing process, or to local fluctuations which lead to a scatter in properties throughout the material.

Figures 6.1 and 6.2 illustrate the meaning of a statistically representative volume element. We can easily see that the arrangement of fibres on figure 6.1 is non-periodic but a large enough volume allows representative average values of the behaviour of the material to be obtained.

The choice of the REV is not therefore necessarily unique but only one is enough if it is relevant to the studied materials and if it can give reproducible results. In that case, the choice of the REV does not need any periodicity even if in many cases, for convenient modelling, periodicity is assumed. Figure 6.2,

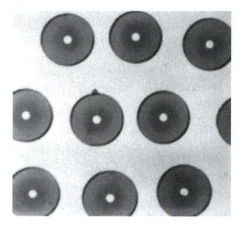

Figure 6.2. Fibre distribution in a SiC–titanium composite.

in contrast to figure 6.1, shows a material the microstructure of which is very close to a periodic pattern.

6.2   Principles of homogenization

First, let us remember that a problem in continuum mechanics has a unique solution for all stresses and strains if the following quantities can be found without ambiguity and verified:

- local equations and constitutive relations,
- boundary conditions in terms of force distributions and displacement conditions, or both.

Then it can be shown that

- the solution for the stress field is unique,
- the solution for the displacement field exists but is not necessarily unique.

In order to formulate correctly our multi-scale approach, we have to be consistent with the above considerations. To be more precise, we have to solve the problem of calculating the structure of the REV, the constitutive law of which is linearly elastic and supposed to be known in each point (or constant per constituent). The boundary conditions are given by the structure. The purpose is to evaluate the average behaviour of the REV and the local state of materials according to the overall, macroscopic loading. As the REV represents an element, of infinitesimally small volume of the structure, we assume that the macroscopic stress is the average value of stresses in the domain:

$$\langle \sigma \rangle = \Sigma \qquad (6.1)$$

In a similar way, if E represents the macroscopic strain:

$$\langle \varepsilon \rangle = E \tag{6.2}$$

Then, the problem to solve is

$$\begin{cases} \overrightarrow{\operatorname{div}}\, \sigma(x) = \vec{0} \text{ in each point of the REV} \\ \sigma = C(x)\varepsilon(\vec{u}(x)) \\ \text{given boundary conditions } \Sigma \text{ or } E \\ \text{as } \Sigma = \langle \sigma \rangle \text{ or } E = \langle \varepsilon \rangle. \end{cases}$$

This problem is not well enough defined because of the multiple boundary conditions which can obey $\langle \sigma \rangle = \Sigma$ and $\langle \varepsilon \rangle = E$, and various solutions are possible. In order to improve on this, we need to define better the boundary conditions by making a subjective choice. Classical homogenization methods use uniform displacements or forces as boundary conditions. The compatibility between them and the average values of stresses and strains can be demonstrated as follows (using tensorial notation):

(a) If we apply $\sigma \vec{n} = \Sigma \vec{n}$ on the boundary with Σ uniform (div $\sigma = 0$ in v), we get

$$\int_v \sigma_{ij}\,\mathrm{d}v = \int_v (\sigma_{ik} x_j)_{,k}\,\mathrm{d}v = \int_s \sigma_{ik} x_j n_k\,\mathrm{d}s = \int_s \Sigma_{ik} x_j n_k\,\mathrm{d}s = \Sigma_{ik} \int_s x_j n_k\,\mathrm{d}s$$

$$= \Sigma_{ik} \int_v x_{j,k}\,\mathrm{d}v = \Sigma_{ik} \delta_{jk} v = \Sigma_{ij} v.$$

Then we get

$$\langle \sigma \rangle = \frac{1}{v} \int_v \sigma\,\mathrm{d}v = \Sigma.$$

(b) If we apply $u = Ex$ on the boundary, with E uniform, we get

$$v\langle \varepsilon_{ij} \rangle = \int_v \varepsilon_{ij}\,\mathrm{d}v = \frac{1}{2}\int_v (u_{i,j} + u_{j,i})\,\mathrm{d}v = \frac{1}{2}\int_s (u_i n_j + u_j n_i)\,\mathrm{d}s$$

$$= \frac{1}{2}\int_s (E_{ik} x_k n_j + E_{jk} x_k n_i)\,\mathrm{d}s = \frac{E_{ik}}{2}\int_s x_k n_j\,\mathrm{d}s + \frac{E_{jk}}{2}\int_s x_k n_i\,\mathrm{d}s$$

$$= \frac{1}{2} E_{ik} \delta_{jk} v + \frac{1}{2} E_{jk} \delta_{ik} v = E_{ij} v.$$

Then we get $\langle \varepsilon \rangle = E$.

As E or Σ are imposed, the solution of the problem can provide the homogenized stiffness or compliance tensors as

$$\langle \sigma \rangle = C^{\mathrm{hom}} \langle \varepsilon \rangle \tag{6.3}$$

$$\langle \varepsilon \rangle = S^{\mathrm{hom}} \langle \sigma \rangle. \tag{6.4}$$

The purpose of a homogenization problem is to determine the homogenized behaviour C^{hom} and S^{hom} between the macroscopic quantities Σ and E.

There are three methods which can be used in order to obtain the elastic coefficients of an anisotropic homogeneous material equivalent to the composite.

1. The 'effective moduli' based methods. In these methods the purpose is to determine the mechanical strain and stress concentration factors A and B which express the microscopic variables as functions of the macroscopic variables as

$$\varepsilon = AE \quad \text{and} \quad \sigma = B\Sigma. \tag{6.5}$$

An extensive body of literature exists on the evaluation of these mechanical concentration factors but we can say that the estimations obtained depend on the type of mechanical model used to describe the cell. Among these methods, we shall analyse the following.

- The 'rule of mixtures' laws which use the average of all local parameters describing the REV. That means that concentration factors are constant in each phase. Generally the results are poor but these analyses are regarded as first estimates and consider that all constituents are in parallel or in series. We shall introduce the best known law which, from a chronological point of view, are the earliest steps of homogenization.
- The 'self-consistent method' which does not consider an inclusion or the REV as isolated free-bodies. The essence of this method is to place these volumes in an infinite medium which is already homogenized and the properties of which have to be found. In the case of composites, the REV can be a fibre surrounded by the matrix. This REV is itself surrounded by the homogenized composite. Several variations of this approach are possible according to the surrounding homogenized medium. At the beginning, the properties of the composite are assumed so that the stress and strain fields in the REV can be computed. When the REV is homogenized to represent the composite, the resulting material properties in the REV must match those assumed previously for the composite. Iterations of the process would then converge to a solution. This approach, which considers an infinite medium with a single REV, cannot take into consideration interactions between the constituents. As a result the solutions obtained are good for low volume fractions of heterogeneities. For large variations of structure, it is better to use other techniques.

2. 'Bounding methods' are those which use the energy theorems of elasticity with potential energy or complementary potential energy. The choice of an acceptable field, displacement fields and stress fields leads to upper

and lower estimations of the elasticity tensors. These methods can be used with only a partial knowledge of the microstructure of the composite material. Of course, the scatter band bounded by the predicted limits is smaller if the microstructure is better known.

3. The 'averaging' method for a periodic medium belongs to the family of 'effective moduli' methods but the results are independent of the geometry of the periodic cell representing the REV. It must be noted that this is the only method which allows the rigidity matrix C^{hom} and the compliance matrix S^{hom} to be calculated, with a displacement or force formulation, and verify:

$$[S^{\text{hom}}]^{-1} = [C^{\text{hom}}].$$

6.3   Theory of 'effective moduli' methods

These methods are based on the resolution of a localization problem. The use of average parameters for the REV allows the homogenized behaviour of the equivalent medium to be determined. The examples given below are the 'rule of mixture' laws and those based on inclusion problems as 'self-consistent' methods.

Let us consider a heterogeneous material composed of n different homogeneous constituents (figure 6.3).

The subscript m stand for matrix. Other phases are indicated as i, with $1 \leq i \leq n - 1$. Volume fraction of the constituents are defined as f_m and f_i as $f_m = v_m/v$ and $f_i = v_i/v$, where v_m and v_i are respectively the volumes of the phase (m) and phase (i) and $v = v_m + \Sigma v_i$ is the total volume.

In each phase i, we defined the average volume of local fields

$$\langle \sigma^i \rangle = \int_{v_i} \sigma \, dv \quad \text{and} \quad \langle \varepsilon^i \rangle = \int_{v_i} \varepsilon \, dv. \tag{6.6}$$

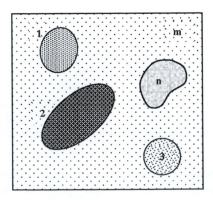

Figure 6.3. Heterogeneous medium.

At the macroscopic level, suppose that the REV is subjected to an overall strain E, using averaging relations (6.6) we obtain

$$E = f_{\mathrm{m}}\langle \varepsilon^{\mathrm{m}} \rangle + \sum_{i=1}^{n-1} f_i \langle \varepsilon_i \rangle$$

$$\Sigma = f_{\mathrm{m}}\langle \sigma^{\mathrm{m}} \rangle + \sum_{i=1}^{n-1} f_i \langle \sigma^i \rangle = f_{\mathrm{m}} C^{\mathrm{m}} \langle \varepsilon^{\mathrm{m}} \rangle + \sum_{i=1}^{n-1} f_i C^i \langle \varepsilon^i \rangle$$

and then

$$\Sigma = C^{\mathrm{m}} E + \sum_{i=1}^{n-1} f_i (C^i - C^{\mathrm{m}}) \langle \varepsilon^i \rangle. \tag{6.7}$$

In the same way, the inverse relations give

$$E = S^{\mathrm{m}} \Sigma + \sum_{i=1}^{n-1} f_i (S^i - S^{\mathrm{m}}) \langle \sigma^i \rangle. \tag{6.8}$$

Within the elastic domain, the local fields in the phases can be expressed in terms of a mechanical influence function as

$$\langle \varepsilon^i \rangle = A^i E \quad \text{or} \quad \langle \sigma^i \rangle = B^i \Sigma \tag{6.9}$$

where A^i and B^i are the concentration factors of strains and stresses defined at each point of the REV by $\varepsilon = AE$ and $\sigma = B\Sigma$ averaged over the volume v_i.

We then find C^{hom} and S^{hom} using (6.7), (6.8) and (6.9) as

$$C^{\mathrm{hom}} = C^{\mathrm{m}} + \sum_{i=1}^{n-1} f_i (C^i - C^{\mathrm{m}}) A^i \tag{6.10}$$

$$S^{\mathrm{hom}} = S^{\mathrm{m}} + \sum_{i=1}^{n-1} f_i (S^i - S^{\mathrm{m}}) B^i. \tag{6.11}$$

The overall average volume leads to:

$$\langle \varepsilon \rangle = \langle AE \rangle = \langle A \rangle E, \qquad \langle \sigma \rangle = \langle B\Sigma \rangle = \langle B \rangle \Sigma.$$

The concentration factors A and B must satisfy

$$\langle A \rangle = I \quad \text{and} \quad \langle B \rangle = I \quad (I \text{ is the identity matrix}).$$

In the elastic field, the constitutive relations of the phases are written as $\sigma = C\varepsilon$ and $\varepsilon = S\sigma$. It follows that

$$\Sigma = \langle \sigma \rangle = \langle C\varepsilon \rangle = \langle CAE \rangle = \langle CA \rangle E$$

and so the stiffness tensor of the effective medium is

$$C^{\mathrm{hom}} = \langle CA \rangle. \tag{6.12}$$

Figure 6.4. Representative cells for different types of two-phase composites.

In a similar way, using stress concentration relation, we obtain

$$E = \langle \varepsilon \rangle = \langle S\sigma \rangle = \langle SB\Sigma \rangle = \langle SB \rangle \Sigma$$

then we can obtain the compliance tensor of the effective medium

$$S^{\text{hom}} = \langle SB \rangle. \tag{6.13}$$

The major question is how to evaluate the concentration factors. 'effective moduli methods' as 'rule of mixtures methods' and 'self-consistent methods' answer this question. An extensive body of literature exists on this evaluation. We shall limit our investigation to two-phase composites: a matrix m and a fibrous reinforcement f (figure 6.4). The quantities f_{f} and f_{m} are the volume fractions of fibres and matrix.

The purpose of micromechanical modelling is to evaluate overall properties of such a composite in terms of microstructural geometry, phase volume fractions and constituent properties.

This problem is not easy and the results depend on the reliability of the mechanics model used. All make similar assumptions:

- fibres and matrix are isotropic and homogeneous. If this assumption is acceptable for the matrix, it is not for fibres which generally are transversally isotropic,
- interfaces between fibres and matrix are perfectly bonded.

6.4   Effective moduli: the rule of mixtures

6.4.1   Voigt and Reuss model

Relations (6.10) and (6.11) give the elastic material coefficients as far as concentration factors A^{i} and B^{i} are known. Considering two-phase composites, we can write

$$C^{\text{comp}} = C^{\text{m}} + f_{\text{f}}(C^{\text{f}} - C^{\text{m}})A^{\text{f}}, \qquad S^{\text{comp}} = S^{\text{m}} + f_{\text{f}}(S^{\text{f}} - S^{\text{m}})B^{\text{f}}. \quad (6.14)$$

The problem is now to determine A^{f} and B^{f}. The first estimations which answer this question are the strain field approach of Voigt [11] and the stress dual approach of Reuss [7]. The former considers that the strain is the same over all the composite and so is equal in each phase. This means that $\langle \varepsilon^{\text{m}} \rangle = \langle \varepsilon^{\text{f}} \rangle = E$. The concentration factors then reduce to the identification of tensors in each phase ($A^{\text{f}} = I$) and we obtain the effective tensor of the composite by arithmetical averaging of the rigidity tensors for each phase:

$$C^{\text{comp}} = \langle C \rangle = f_{\text{m}} C^{\text{m}} + f_{\text{f}} C^{\text{f}}. \quad (6.15)$$

Reuss's approach supposes constant stresses over all the composite and so in each phase. $\langle \sigma^{\text{f}} \rangle = \langle \sigma^{\text{m}} \rangle = \Sigma$ leading to:

$$S^{\text{comp}} = \langle S \rangle = f_{\text{m}} S^{\text{m}} + f_{\text{f}} S^{\text{f}}. \quad (6.16)$$

Such an assumption of constant strains or constant stresses in the REV are only valid when the constituents are in parallel or in series. Using similar assumptions, Ekvall [1] introduced a thin parallelepiped elementary cell (figure 6.5).

This describes the situation which is close to that seen with unidirectional plies in composite structures. In this case, the homogenized behaviour is supposed to show plane stress between the averaged strain and stress. If we call L the longitudinal fibre direction, T the transverse direction of fibres and

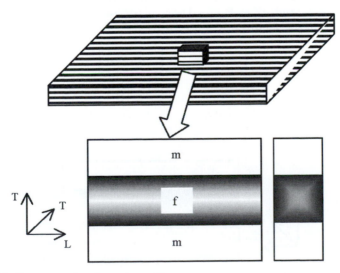

Figure 6.5. Representative elementary cell.

C the axial-transversal shear, we obtain

$$S^{\text{hom}} = \begin{bmatrix} 1/E_L & -\nu/E_L & 0 \\ -\nu/E_L & 1/E_T & 0 \\ 0 & 0 & 1/G \end{bmatrix}. \tag{6.17}$$

Thus our goal is to determine the four components of the composite compliance matrix S^{hom}: the Young's longitudinal modulus E_L, the transversal Young's modulus E_T, the Poisson's ratio ν when the ply is loaded in the fibre direction and finally the in-plane shear modulus. The assumption of isotropy for the fibre and the matrix gives the constitutive relations S^{f} and S^{m}, respectively for the fibre and the matrix as a function of Young's moduli and Poisson's ratios of the fibres and matrix respectively noted $E_{\text{f}}, \nu_{\text{f}}, E_{\text{m}}, \nu_{\text{m}}$.

$$S^{\text{f}} = \begin{bmatrix} 1/E_{\text{f}} & -\nu_{\text{f}}/E_{\text{f}} & 0 \\ -\nu_{\text{f}}/E_{\text{f}} & 1/E_{\text{f}} & 0 \\ 0 & 0 & 1/G_{\text{f}} \end{bmatrix}$$

$$S^{\text{m}} = \begin{bmatrix} 1/E_{\text{m}} & -\nu_{\text{m}}/E_{\text{m}} & 0 \\ -\nu_{\text{m}}/E_{\text{m}} & 1/E_{\text{m}} & 0 \\ 0 & 0 & 1/G_{\text{m}} \end{bmatrix}. \tag{6.18}$$

Determination of the longitudinal Young's modulus

Let us suppose a uniform macroscopic strain $\bar{E}$ applied in the fibre direction (figure 6.6). The fibre and the matrix are in parallel, leading to constant strains in both constituents. The stress–strain relations in the constituents give $\sigma_L^{\text{f}} = E_{\text{f}}\varepsilon_L^{\text{f}}$ and $\sigma_L^{\text{m}} = E_{\text{m}}\varepsilon_L^{\text{m}}$. Then, if we use the average stress in the fibre direction we obtain

$$\langle \sigma_L \rangle = \frac{1}{v} \int_v \sigma_L \, dv = \frac{1}{v} \left(\int_{v_{\text{f}}} \sigma_L^{\text{f}} \, dv + \int_{v_{\text{m}}} \sigma_L^{\text{m}} \, dv \right)$$

$$\langle \sigma_L \rangle = \frac{v_{\text{f}}}{v} \langle \sigma_L^{\text{f}} \rangle + \frac{v_{\text{m}}}{v} \langle \sigma_L^{\text{m}} \rangle = f_{\text{f}} \langle \sigma_L^{\text{f}} \rangle + f_{\text{m}} \langle \sigma_L^{\text{m}} \rangle = (f_{\text{f}} E_{\text{f}} + f_{\text{m}} E_{\text{m}}) \bar{E}.$$

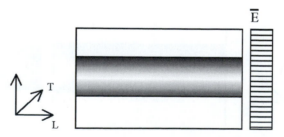

Figure 6.6. Volume element under longitudinal loading.

Therefore, the longitudinal Young's modulus becomes

$$E_L = f_f E_f + f_m E_m. \tag{6.19}$$

Evaluation of Poisson's ratio

Next, the transverse composite strain is related to the longitudinal composite strain by

$$\langle \varepsilon_T \rangle = f_f \langle \varepsilon_T^f \rangle + f_m \langle \varepsilon_T^m \rangle = -(\nu_f f_f + \nu_m f_m) \bar{E}.$$

The longitudinal Poisson's ratio is thus

$$\nu = f_f \nu_f + f_m \nu_m. \tag{6.20}$$

Transversal Young's modulus

To determine the transverse Young's modulus E_T, we apply the following boundary conditions (figure 6.7):

$$\Sigma_L = 0, \qquad \Sigma_T \neq 0, \qquad \Sigma_C = 0.$$

The fibres and matrix are now in series and the compatibility between the applied load and averaged quantities leads to

$$\langle \sigma_L^f \rangle = \langle \sigma_L^m \rangle = 0, \qquad \langle \sigma_T^f \rangle = \langle \sigma_T^m \rangle = \Sigma_T, \qquad \langle \sigma_C^f \rangle = \langle \sigma_C^m \rangle = 0.$$

The transverse strain is given by

$$\langle \varepsilon_T \rangle = f_f \langle \varepsilon_T^f \rangle + f_m \langle \varepsilon_T^m \rangle = \left(\frac{f_f}{E_f} + \frac{f_m}{E_m} \right) \Sigma_T.$$

Thus the transverse Young's modulus is obtained from

$$\frac{1}{E_T} = \frac{f_f}{E_f} + \frac{f_m}{E_m}. \tag{6.21}$$

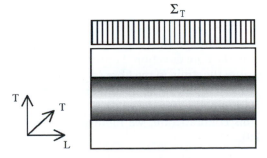

Figure 6.7. Volume element under the transverse loading.

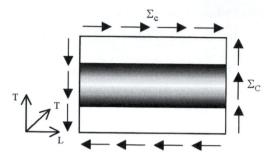

Figure 6.8. Determination of the shear modulus.

Evaluation of the shear modulus

Finally, the boundary conditions for the determination of the shear modulus are (figure 6.8):

$$\Sigma_L = 0 \qquad \Sigma_T = 0 \qquad \Sigma_C \neq 0.$$

The procedure to be followed next is similar to those for E_T. Therefore, the shear modulus is given by

$$\frac{1}{G} = \frac{f_f}{G_f} + \frac{f_m}{G_m}. \tag{6.22}$$

It is interesting to note that knowledge of the exact stress and strain fields within the composite was not required because only the average values in each constituent were necessary. Therefore, the results are exact for the chosen representative cell with the given boundary conditions.

Nevertheless the above homogenization provides an accurate prediction for the fibre direction, and gives the longitudinal Young's modulus E_L and Poisson's ratio ν. The above approach is not so accurate when applied to the transverse direction as a difference of sometimes 40% appears when the results are compared with observed experimental results or exact solution calculations.

Lateral constraint for strain compatibility

The boundary conditions to be imposed on the representative cell must simulate the *in situ* state of stress as closely as possible. When Σ_T was applied, we assumed no stress in the fibre direction. However, such a boundary condition is not realistic because the strain in the fibre $\langle \varepsilon_L^f \rangle$ is not the same as $\langle \varepsilon_L^m \rangle$ in the matrix unless $\nu_f/E_f = \nu_m/E_m$. The resulting difference in displacements cannot be sustained in real composites.

To remedy this contradiction, we modify the boundary conditions as follows (figure 6.9):

$$\langle \varepsilon_L^f \rangle = \langle \varepsilon_L^m \rangle = \bar{E}, \qquad \langle \sigma_T^f \rangle = \langle \sigma_T^m \rangle = \Sigma_T$$

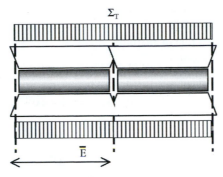

Figure 6.9. Longitudinal strain compatibility.

where $\bar{E}$ is to be determined from the condition $\Sigma_L = 0$. These boundary conditions introduced in constitutive relations of each constituent give

$$\bar{E} = -\frac{\nu}{E_L}\Sigma_T$$

with

$$\langle\sigma_L^f\rangle = \frac{\nu_f E_m - \nu_m E_f}{f_f E_f + f_m E_m} f_m \Sigma_T, \qquad \langle\sigma_L^m\rangle = \frac{\nu_m E_f - \nu_f E_m}{f_f E_f + f_m E_m} f_f \Sigma_T.$$

The transverse Young's modulus E_T is then obtained from

$$\frac{1}{E_T} = \frac{f_f}{E_f} + \frac{f_m}{E_m} - f_f f_m \frac{\nu_f^2 E_m/E_f + \nu_m^2 E_f/E_m - 2\nu_f\nu_m}{f_f E_f + f_m E_m}. \qquad (6.23)$$

The first two terms on the right-hand side are the same as the 'traditional' rule of mixtures based on a uniaxial state of average stress. The third term represents the effect of lateral constraint imposed by the strain compatibility and leads to a higher transverse modulus.

Other models which are similar to the basic 'rule of mixtures model' lead to substantial improvements. Such an approach is described by Puck's model, which is semi-empirical, using different cells to provide better estimates of moduli than are possible with the simple rule-of-mixtures equations. A correlation between several composite systems gives the following equations:

$$E_L = f_f E_f + f_m E_m$$

$$\nu = f_f \nu_f + f_m \nu_m$$

$$E_T = E_m^0 \frac{1 + 0.85 f_f^2}{(1 - f_f)^{1.25} + f_f(E_m^0/E_f)} \qquad (6.24)$$

$$G = G^m \frac{1 + 0.6\sqrt{f_f}}{(1 - f_f)^{1.25} + f_f(G_m/G_f)} \qquad \text{with } E_m^0 = \frac{E_m}{1 - \nu_m^2}$$

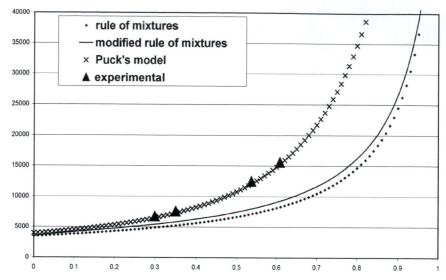

Figure 6.10. Comparison of different 'rule of mixtures'.

The results of the above models are plotted on figure 6.10 and are compared with experimental results. Equations of homogenized transverse Young's moduli are shown graphically versus different fibre volume fractions for a glass/epoxy composite, the constituents of which have the following properties: $E_f = 73\,100\,\text{MPa}$, $E_m = 3500\,\text{MPa}$, $\nu_f = 0.22$, $\nu_m = 0.35$.

6.4.2 Hopkins and Chamis' model

The above models do not take into account the inhomogeneous distribution of fibres which can lead to considerable scatter in experimental results. More accurate theoretical results can be obtained with different patterns of fibres. First consider the square fibre array in figure 6.11, the surface of which is equivalent to the circular fibre of the considered composite through the relation

$$s_f^2 = \frac{\pi d^2}{4}.$$

Then the fibre volume fraction is given by $f_f = s_f^2/s^2$.

The REV is divided into three parts. The first part, B, is homogenized using equation (6.21). We obtain the effective moduli E_T^B as

$$\frac{1}{E_T^B} = \frac{1}{E_f} f_f^B + \frac{1}{E_m} (1 - f_f^B) \qquad (6.25)$$

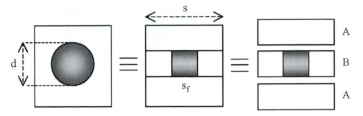

Figure 6.11. Hopkins and Chamis' representative element.

with $f_f^B = s_f/s$, the fibre volume fraction of the part B. Using $s_f/s = \sqrt{f_f}$, we obtain E_T^B as

$$E_T^B = \frac{E_m}{1 - \sqrt{f_f}(1 - E_m/E_f)}. \tag{6.26}$$

Then we homogenize part A and, by using the rule of mixtures,

$$E_T = E_T^B f_B + E_m(1 - f_B) \tag{6.27}$$

with $f_B = s_f/s = \sqrt{f_f}$. Introducing (6.26) into (6.27) we obtain

$$E_T = E_m \left[(1 - \sqrt{f_f}) + \frac{\sqrt{f_f}}{1 - \sqrt{f_f}(1 - E_m/E_f)} \right]. \tag{6.28}$$

A similar demonstration, with appropriate boundary conditions, gives the shear modulus:

$$G = G_m \left[(1 - \sqrt{f_f}) + \frac{\sqrt{f_f}}{1 - \sqrt{f_f}(1 - G_m/G_f)} \right]. \tag{6.29}$$

6.4.3   Cells with circular fibres

In the case of a circular fibre packing array, the state of stress is not homogeneous. If we consider a unidirectional ply in which the centres of the fibres are distributed according a square pattern, we can choose the cell shown in figure 6.12. $2p$ is the distance between the centre lines, R is the radius of the fibres and the fibre volume fraction is given by $f_f = \pi R^2/p^2$.

Ekvall [1] and Spencer [10] have used this cell for the calculation of the transverse Young's modulus and the axial-transverse shear modulus as there is good agreement with values obtained with the basic 'rule of mixtures' for the axial Young's modulus and Poisson's ratio. First consider a tensile or compressive stress transverse to the fibre.

Assume the cell is divided into parallel elements of width $d\delta$ and that the mean stress σ_δ is constant along each element. The stress distribution in the cell is given in figure 6.12.

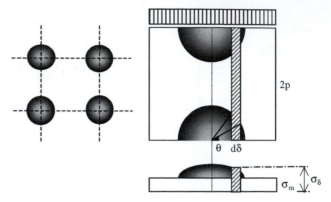

Figure 6.12. Cell with circular fibres.

As the elements are in parallel, the resulting modulus is

$$E_T = \sum_{nb}^{bandes} E_\delta \frac{d\delta}{p} = \int_0^R E_\delta \frac{d\delta}{p} + \int_R^p E_m \frac{d\delta}{p}.$$

As the last element is only made of matrix, we obtain

$$E_T = \int_0^{\pi/2} E_\delta R \frac{\sin\theta\, d\theta}{p} + \frac{p-R}{p} E_m$$

with, for each element,

$$\frac{1}{E_\delta} = \frac{R\sin\theta}{p} \frac{1}{E_f} + \frac{p - R\sin\theta}{p} \frac{1}{E_m}.$$

Then the following relationship can be written:

$$E_\delta = E_f \frac{1}{\dfrac{E_f}{E_m} + \dfrac{R}{p}\sin\theta\left(1 - \dfrac{E_f}{E_m}\right)}$$

and for the transverse Young's modulus

$$E_T = \frac{p-R}{p} E_m + \frac{R}{p} E_f \int_0^{\pi/2} \frac{\sin\theta\, d\theta}{\dfrac{E_f}{E_m} + \dfrac{R}{p}\sin\theta\left(1 - \dfrac{E_f}{E_m}\right)}. \tag{6.30}$$

If we introduce $\gamma = p/R$ and $h = 1 - (E_m/E_f)$, we can correlate the results of Ekvall with Spencer's approach:

$$\frac{E_T}{E_m} = \frac{\gamma - 1}{\gamma} + \int_0^{\pi/2} \frac{\sin\theta\, d\theta}{\gamma - h\sin\theta}. \tag{6.31}$$

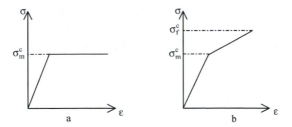

Figure 6.13. Stress–strain relationship of (a) resin (b) composite.

6.4.4 Description of non-linear behaviour

Schaffer [9] developed the stress–strain relationships of fibre reinforced plastics in linear and non-linear fields by assuming that the resin obeys a perfectly plastic yield condition (figure 6.13a). The stress–strain relationship is linear in the elastic range, where it obeys Hooke's law and may deform at constant stress once its yield stress has been reached. Since the moduli of fibres are greater than that of the resin, the stresses in the fibres are greater than those in the resin as long as both elements remain elastic and the behaviour is in agreement with the 'rule of mixtures': $E_L = f_f E_f + f_m E_m$.

As soon as the stress in the matrix reaches its yield stress the equivalent modulus of the composite (figure 6.13b) equals the fibre modulus multiplied by the fibre volume fraction.

$$E_L = f_f E_f \tag{6.32}$$

Transverse moduli evaluation is made with a hexagonal array of fibres. The cell contains a uniform distribution of fibres of diameter d, the centres are separated by a distance p (figure 6.14). The ratio of the area occupied by the fibre over the surface occupied by the triangle gives the fibre volume fraction of the composite:

$$f_f = \frac{\pi d^2}{2\sqrt{3} p^2}.$$

If the distance s between lines tangential to the fibres is positive, the fibre volume fraction is less than 68%. In that case the cell can be simplified to two parallel bars joined to rigid end plates. One is of pure matrix and the other consists of the fibre and matrix in series. Such a model is shown in figure 6.15.

The transverse modulus of elasticity E_T is determined in two steps. First the two bars in series which determine the modulus can be written as

$$E_c = \frac{E_m}{1 - \dfrac{\pi d}{4p}\left(1 - \dfrac{E_m}{E_f}\right)}.$$

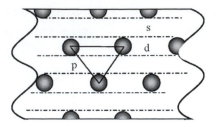

Figure 6.14. Cross section of the composite.

The second step is to evaluate the equivalent behaviour of two parallel bars. Thus we obtain

$$E_T = \frac{E_m}{d+s} \frac{d+s\left\{1-\left(\frac{\pi d}{4p}\right)\left[1-\left(\frac{E_m}{E_f}\right)\right]\right\}}{1-\frac{\pi d}{4p}\left[1-\frac{E_m}{E_f}\right]}. \tag{6.33}$$

This expression is valid as long as the yield of the matrix is not reached.

As the modulus of the equivalent bar (fibre–matrix in series) is larger than that of the matrix ($E_f > E_m$), the stress in the equivalent bar is also greater. This bar will be the first to yield in the matrix region. After yielding, the same demonstration can be carried out by introducing the matrix yield stress as the state of stress of the equivalent bar. Thus we obtain the equivalent modulus as

$$E_T = E_m \frac{s}{d+s}. \tag{6.34}$$

When f_f is greater than 68%, s vanishes and the cell reduces to only one bar with the fibre and matrix in series. Then E_T is given by

$$E_T = \frac{E_m}{1-f_f[1-(E_m/E_f)]}. \tag{6.35}$$

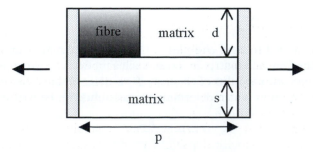

Figure 6.15. Fibre normal to an applied force.

This model gives the composite behaviour when the matrix obeys an elastic perfectly plastic behaviour.

6.4.5 The limit of the 'rule of mixtures' models

We limit our analysis to the description of the 'rule of mixtures' type models. The reader can imagine that many other models based on equivalent hypotheses can be found in the literature but they rapidly become very complicated and then their governing mathematical equations can only be solved by using numerical methods such as 'finite difference' energy methods or 'finite element' analyses. The calculations can become unnecessarily complex with little improvement in the evaluation of some characteristics of the composite.

In contrast to these models, which always associate constituents in parallel or in series to obtain an estimation of the elastic characteristics, other types of model consider the composite to be made of inclusions or inhomogeneities embedded in an infinite elastic medium. The estimations of effective stiffnesses or effective compliances always require the determination of the mechanical influence tensors relating to strains and stresses. The principles of these models and those which are most often used will be described in the next section.

6.5 'Self-consistent' models

6.5.1 General considerations

Several authors have attempted to estimate the properties of a composite by using 'self-consistent' methods. We are only going to study two of them: the 'auto-coherent scheme' and the 'Mori and Tanaka' method. Both of them are based on the Eshelby equivalent inclusion method in homage of its author who proposed to transform an elastic inhomogeneity into a 'stress free' inclusion made of the surrounding homogeneous medium.

6.5.2 The Eshelby inclusion method

The originality of this idea outlined by Eshelby [2] in the 1950s has only recently been appreciated. The original idea concerned a hypothetical situation of an infinite matrix, a part of which is submitted to a 'stress free' strain. The different steps of the transformation are described with schematic illustrations in figure 6.16.

First suppose that the matrix extends to infinity. In this unstrained medium, a cut is made with the shape of an inclusion which has the same elastic constants as the matrix. We imagine that the inclusion undergoes a

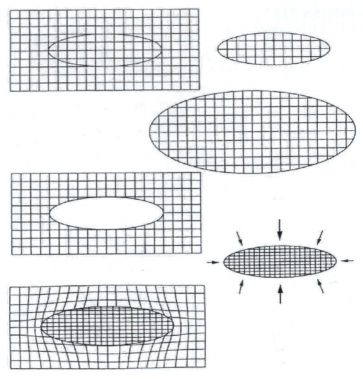

Figure 6.16. Stress free strain in an homogeneous medium.

transformation specified by a uniform stress-free strain while constrained by the matrix, and try to evaluate the resulting elastic field. Such a strain could be the case of a local heating or local plastic strain of the matrix. First remove the inclusion from the matrix. Then the transformation of the medium is allowed to occur. As the inclusion is removed from the matrix, it is now free to change in size and shape (a 'stress-free' strain).

If now the inclusion is constrained to return to its original dimensions so as to be replaced in the hole from where it came, it is then subjected to surface forces:

$$\mathrm{d}\vec{f} = -C\varepsilon^{\mathrm{L}}\,\mathrm{d}s\,\vec{n}$$

where C is the stiffness matrix of the considered medium.

In this way the continuity of the medium is maintained. If we relax the surface forces $\mathrm{d}\vec{f}$, the body adopts a new shape by distorting the matrix.

If ε^{c} is the final deformation in the inclusion from the beginning of the transformation, the stress σ in each point of the inclusion is

$$\sigma = C(\varepsilon^{\mathrm{c}} - \varepsilon^{\mathrm{L}}). \tag{6.36}$$

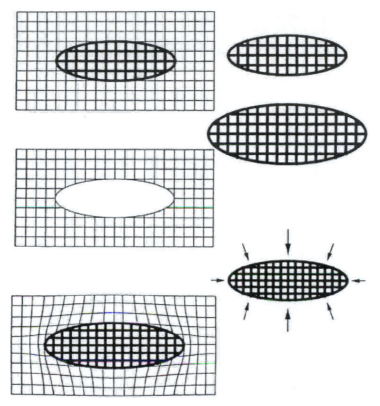

Figure 6.17. Rigid inclusion embedded in matrix.

This above simple 'thought experiment' can be interesting if we now make the analogy between an inclusion having the same elastic constants as the matrix and an inhomogeneous inclusion (fibre) embedded in the matrix. In this way, the Eshelby method can provide a stress state solution in the case of heterogeneous materials and so composite materials. In this case, the foregoing steps can be made with the dark lines of the inclusion to denote its higher stiffness (figure 6.17).

If we consider the special case where the inclusion or inhomogeneity takes the form of an ellipsoid with three unequal axes as for short fibres and if α_f and α_m respectively characterize the thermal expansion coefficients of the fibres and matrix, a stress-free strain can be written in the case of differential thermal expansion on heating by ΔT:

$$\varepsilon^L = \Delta T(\alpha_f - \alpha_m). \tag{6.37}$$

Since, in most cases, $\alpha_f < \alpha_m$, a change in temperature produces a change in the shape of the inclusion. If, as previously, we remove an inclusion from the matrix, each constituent can now change freely. After replacement of the inclusion in the cavity, it adopts a new size and shape, distorting the surrounding matrix as it does so itself. This new size and shape represents a constrained strain ε^c relative to the original state. Then the stress in the inclusion can be written as

$$\sigma^f = C^f(\varepsilon^c - \varepsilon^L) \tag{6.38}$$

where C^f is the stiffness tensor of the inclusion (fibre) and the term $\varepsilon^c - \varepsilon^L$ represents the net strain relative to the final state and the 'stress-free' state.

The link between a local free strain in a homogeneous medium and the problem of an inclusion such as a fibre in a heterogeneous medium is made by adaptation of the free strain with the induced strain in the real composite. In this way, we get identical physical phenomena. The contribution made by Eshelby was to establish that the constrained strain ε^c for this equivalent homogeneous inclusion is only related to the transformation strain by the expression

$$\varepsilon^c = S^E \varepsilon^L \tag{6.39}$$

and in the case of an ellipsoid ε^c is constant over all the inclusion. S^E is the 'Eshelby tensor' which is a function of the ellipsoid axis ratio (geometry of fibre) and the Poisson's ratio of the matrix. Readers interested in different expressions of Eshelby tensors should consult Eshelby [2].

The analysis of the equivalent inclusion allows behaviour homogenization problem to be solved by substitution of a problem with heterogeneity into a problem of a homogeneous medium submitted to a localized free strain.

6.5.3   Eshelby method with applied macroscopic strain

The problem is now to apply a 'background strain' E, on to a medium of infinite matrix with a given inclusion. Suppose the stiffnesses of the matrix and inclusion to be C^m and C^f. Far from inclusion, the displacement u can be written $u = Ex$. As previously, Eshelby supposed an equivalence between this problem and an inclusion submitted to a stress-free strain ε^L with E superimposed on this strain. The equivalence is made if the same strain and stress state is generated in both problems.

In the elastic domain the problem of an inclusion with a strain E at infinity is equivalent to solving the problem outlined in section 6.5.2, with a superimposed problem of an infinite medium subjected to a background strain E. This is the purpose of problems (a) and (b) in figure 6.18.

The strain field of problem (a) is the result of the local prestrained field ε^c and the infinite medium. Then we get

$$\sigma = C^f(\varepsilon^c + E) = C^f(S^E \varepsilon^L + E), \qquad \varepsilon = \varepsilon^c + E, \qquad \langle \varepsilon \rangle = E \tag{6.40}$$

for problem (a).

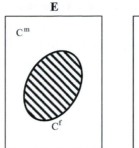 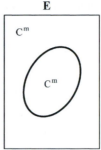

Figure 6.18. Infinite medium submitted to a background strain E: (a) fibre embedded is an infinite medium of matrix, (b) stress-free strain is an infinite medium of matrix.

The related problem (b) of a 'stress-free strain' in an infinite homogeneous medium gives

$$\sigma = C^{\mathrm{m}}(\varepsilon^{\mathrm{c}} - \varepsilon^{\mathrm{L}} + E) = C^{\mathrm{m}}[(S^{\mathrm{E}} - I)\varepsilon^{\mathrm{L}} + E], \quad \varepsilon = \varepsilon^{\mathrm{c}} - \varepsilon^{\mathrm{L}} + E, \quad \langle \varepsilon \rangle = E \tag{6.41}$$

for problem (b).

Both problems are equivalent if the strain and stress fields are the same at each point. That leads to the equality of stresses and then

$$\varepsilon^{\mathrm{L}} = [(C^{\mathrm{m}} - C^{\mathrm{f}})S^{\mathrm{E}} - C^{\mathrm{m}}]^{-1}(C^{\mathrm{f}} - C^{\mathrm{m}})E. \tag{6.42}$$

If we now introduce this relation into the strain field for problem (a)

$$\varepsilon = \varepsilon^{\mathrm{c}} + E = S^{\mathrm{E}}\varepsilon^{\mathrm{L}} + E$$

or for problem (b)

$$\varepsilon = \varepsilon^{\mathrm{c}} - \varepsilon^{\mathrm{L}} + E$$

we get the concentration tensor A^{f} as

$$\varepsilon^{\mathrm{f}} = [I + S^{\mathrm{E}}(C^{\mathrm{m}})^{-1}(C^{\mathrm{f}} - C^{\mathrm{m}})]^{-1}E. \tag{6.43}$$

To evaluate the stiffness matrix, Eshelby assumes that a fibre in the composite medium behaves in a same way as a single fibre surrounded by an infinite medium submitted to a macroscopic strain E. Then the strain concentration factor is given by the last relationship. According to relation (6.10), which describes a two phase composite, we get

$$C^{\mathrm{comp}} = C^{\mathrm{m}} + f_{\mathrm{f}}(C^{\mathrm{f}} - C^{\mathrm{m}})[I + S^{\mathrm{E}}(C^{\mathrm{m}})^{-1}(C^{\mathrm{f}} - C^{\mathrm{m}})]^{-1}. \tag{6.44}$$

This supposes the same concentration factor for all fibres and consequently the same geometry and the same orientation for all fibres.

The main drawback of this approach is to neglect the interactions between fibres, since only a single fibre is considered. It is then clear that the error due to this method depends on the volume fraction of fibres.

6.5.4   Eshelby method with an applied macroscopic stress

The problem is similar to the foregoing problem in figure 6.18 with now a background stress Σ. In this way we intend to determine the concentration factor B^f. If we apply E^* as $\Sigma = CE^*$, we get an equivalent equation of (6.43) by writing

$$\varepsilon^f = [I + S^E(C^m)^{-1}(C^f - C^m)]^{-1}E^* \tag{6.45}$$

then, introducing it into (6.40), we obtain the stress in the fibre as

$$\sigma^f = C^f[I + S^E(C^m)^{-1}(C^f - C^m)]^{-1}E^*$$

or

$$\sigma^f = C^f[I + S^E(C^m)^{-1}(C^f - C^m)]^{-1}S\Sigma.$$

The stress concentration factor is then

$$B^f = C^f[I + S^E(C^m)^{-1}(C^f - C^m)]^{-1}S. \tag{6.46}$$

With an identical hypothesis as used in the last paragraph, equation (6.11) gives the compliance of the equivalent homogeneous composite. The same remarks can be made concerning the drawbacks of this method when the fibre volume fraction is large.

6.5.5   'Auto-coherent' scheme

In the 'auto-coherent' scheme or 'self-consistent' scheme only one single fibre is used, surrounded by an infinite homogeneous medium equivalent to the real composite. This body is always submitted to a background stress Σ or strain E. We can then better consider interactions between fibres. The average fibre strain is then calculated by replacing C^m by C^{comp} in relation (6.43):

$$\varepsilon^f = [I + S^E(C^{comp})^{-1}(C^f - C^{comp})]^{-1}E \tag{6.47}$$

and then

$$A^f = [I + S^E(C^{comp})^{-1}(C^f - C^{comp})]^{-1} \tag{6.48}$$

using (6.10), we obtain

$$C^{comp} = C^m + f_f(C^f - C^m)[I + S^E(C^{comp})^{-1}(C^f - C^{comp})]^{-1}. \tag{6.49}$$

S^E always characterizes the Eshelby tensor, which has to be calculated by a different method. The equivalent inclusion of the Eshelby problem is now surrounded with an infinite composite medium with identical fibres oriented in the same direction wrapped in a matrix which is supposed to be isotropic. As the fibres are supposed to be ellipsoidal, the elastic behaviour of composite material is transversally isotropic around fibre axes.

Relation (6.49) gives an implicit expression of composite behaviour because C^{comp}, which is unknown, is on both sides of the equality. The resolution is then iterative by introducing as initial value of C^{comp}, the result of the Eshelby method. The process converges with less than a 1×10^{-6} approximation in about 20 iterations.

The hypothesis to replace the matrix medium by a homogeneous composite is not completely satisfactory because the fibre is surrounded with a stiffer medium than the real matrix and does not correctly consider the fibre–matrix interface which is weaker than the homogeneous equivalent medium.

6.5.6   The Mori–Tanaka model

This method which has been introduced by Mori and Tanaka [4] allows us to better describe the real strain state in the matrix. As before, we consider a composite consisting of identical fibres oriented in the same direction and submitted to a background strain E. As there are a great number of fibres, the average matrix strain $\langle \varepsilon^{\text{m}} \rangle$ can be divided into the macroscopic strain E and a fluctuation strain $\tilde{\varepsilon}^{\text{m}}$ due to the neighbouring fibres. The average matrix strain can be then written as

$$\langle \varepsilon^{\text{m}} \rangle = \tilde{\varepsilon}^{\text{m}} + E.$$

In a given fibre, the average strain $\langle \varepsilon^{\text{f}} \rangle$ can be written as

$$\langle \varepsilon^{\text{f}} \rangle = E + \tilde{\varepsilon}^{\text{m}} + \tilde{\varepsilon}^{\text{f}} = \langle \varepsilon^{\text{m}} \rangle + \tilde{\varepsilon}^{\text{f}}$$

where $\tilde{\varepsilon}^{\text{f}}$ is fluctuating around $\langle \varepsilon^{\text{m}} \rangle$.

To determine $\langle \varepsilon^{\text{f}} \rangle$ we use the Eshelby method where the elementary cell is a single fibre surrounded by an infinite matrix medium submitted to a background macroscopic strain $\langle \varepsilon^{\text{m}} \rangle$.

If we then apply equation (6.43), we obtain

$$\langle \varepsilon^{\text{f}} \rangle = [I + S^{\text{E}}(C^{\text{m}})^{-1}(C^{\text{f}} - C^{\text{m}})]^{-1} \langle \varepsilon^{\text{m}} \rangle$$

We finally obtain C^{comp}, which is now explicit and does not need any numerical treatment.

$$C^{\text{comp}} = C^{\text{m}} + f_{\text{f}}(C^{\text{f}} - C^{\text{m}})[I + S^{\text{E}}(C^{\text{m}})^{-1}(C^{\text{f}} - C^{\text{m}})]^{-1}$$
$$\times \{f_{\text{m}}I + f_{\text{f}}[I + S^{\text{E}}(C^{\text{m}})^{-1}(C^{\text{f}} - C^{\text{m}})]^{-1}\}^{-1}. \qquad (6.50)$$

6.5.7   Limit of self-consistent type models

A well known application of the self-consistent method in the field of composite materials is Cox's model, which is described in chapter 7. Even if such a model is an approximation of the real microstructure, it is very useful for understanding the physical mechanisms. When the microstructure of composites is more complicated, the REV is also more complicated. When the material is damaged, damage and flaws can be introduced. The general principle remains valid for all these variations, although it invariably causes complications in computation. The main drawback is that it does not consider real cases with small longitudinal and transversal distances between fibres or even consider partial overlapping reinforcements. In reality, we cannot neglect interactions between fibres and consider only constant elastic fields.

6.6 'Bounding' methods

6.6.1 General considerations

These methods are based on limit theorems, the upper bound theorem using admissible strain fields and the lower bound theorem using admissible stress fields.

The above methods were able to determine concentration strain factors A or concentration stress factors B. The methods we are now going to describe allow, without enough knowledge about these concentration factors, to propose bounding of the behaviour of the considered material. Again we consider a composite material made of fibres f wrapped in matrix m. We always assume these constituents to be homogeneous, linearly elastic and isotropic obeying Hooke's law.

6.6.2 Theorem 1: principle of potential energy

The kinematically admissible displacement field u, which is the solution of our problem and which is related through some constitutive law to a stress field satisfying the equilibrium in a body acted on by statically compatible external loads, must minimize the potential energy with respect to all other kinematically admissible displacement fields u':

$$\pi(u) \leq \pi(u') \tag{6.62}$$

with

$$\pi(u') = \frac{1}{2} \int_v \varepsilon' C \varepsilon' \, \mathrm{d}v - \int_{s_f} T^\mathrm{d} u' \, \mathrm{d}s - \int_v f^\mathrm{d} u' \, \mathrm{d}v. \tag{6.63}$$

The last two integrals characterize external virtual work. T^d and f^d are statically compatible force systems respectively surface density on part s_f of the surface and volume density which satisfy equilibrium everywhere in the body.

Then when we apply homogeneous macroscopic strains as boundary conditions (6.63) reduces to

$$\pi(u') = \frac{v}{2} \langle \varepsilon' C \varepsilon' \rangle. \tag{6.64}$$

Further, if a macroscopic stress is now applied, we get

$$\pi(u') = \frac{v}{2} \langle \varepsilon' C \varepsilon' \rangle - v \Sigma \langle \varepsilon' \rangle \tag{6.65}$$

since in tensorial notation

$$\int_{S_f} F_i^{d} u_i' \, ds = \int_{S_f} \Sigma_{ij} n_j u_i' \, ds = \Sigma_{ij} \int_v u_{i,j}' \, dv = \Sigma_{ij} \int_v \varepsilon_{ij}' \, dv = \Sigma_{ij} (v \langle \varepsilon_{ij}' \rangle).$$

6.6.3   Theorem 2: principle of complementary potential energy

The principle of total complementary energy has the following meaning: 'A statically admissible field σ, which is the solution of a given problem and which is related through some constitutive law to a kinematically compatible strain field, must maximize the complementary potential energy with respect to all other statically compatible stress fields σ^*.

$$\pi^*(\sigma^*) \le \pi^*(\sigma) \tag{6.66}$$

where $\pi^*(\sigma^*)$ is the complementary potential energy defined such that

$$\pi^*(\sigma^*) = -\frac{1}{2} \int_v \sigma^* S \sigma^* \, dv + \int_{S_u} T^* u^{d} \, ds \tag{6.67}$$

where the second integral is related to the external work for a given displacement u^{d} applied on part s_u of the surface, in any virtual statically compatible surface force field T^*.

If the boundary conditions reduce to a homogeneous stress, $\pi^*(\sigma^*)$ is defined as

$$\pi^*(\sigma^*) = -\frac{v}{2} \langle \sigma^* S \sigma^* \rangle. \tag{6.68}$$

Alternatively if the boundary conditions define a homogeneous strain E, it follows that

$$\pi^*(\sigma^*) = -\frac{v}{2} \langle \sigma^* S \sigma^* \rangle + v E \langle \sigma^* \rangle \tag{6.69}$$

since $u_i^{d} = E_{ik} x_k$, we obtain (using tensorial notations)

$$\int_{S_u} u_i^{d} \sigma_{ij}^* n_j \, ds = E_{ik} \int_{S_u} x_k \sigma_{ij}^* n_j \, ds = E_{ik} \int_v x_k \sigma_{ij,j}^* \, dv$$

$$= E_{ik} \int_v (x_k \sigma_{ij}^*)_{,j} \, dv = E_{ik} \int_v \sigma_{ij}^* \delta_{kj} \, dv$$

$$= E_{ik} \int_v \sigma_{ik}^* \, dv = E_{ik} v \langle \sigma_{ik}^* \rangle.$$

Thus, if we consider these two extremum principles, the potential energy of a kinematically admissible field u' is always superior or equal to the complementary potential energy of a statically admissible field σ^*. For the solution $\sigma = C\varepsilon(u)$ it follows that

$$\pi^*(\sigma^*) \leq \pi^*(\sigma) = \pi(u) \leq \pi(u').$$ (6.70)

6.6.4 Voigt and Reuss approximation

If we consider uniform admissible fields in all the body (strains and stresses), we obtain upper and lower estimations of the stiffness and compliance tensors. First consider the following boundary homogeneous stress conditions $\sigma^* = \Sigma$ and then $\varepsilon' = S^{\text{hom}}\Sigma$. This statically admissible field obeys the boundary conditions and is an exact solution when the constituents are in series. Considering the chosen fields:

$$-\tfrac{1}{2}\langle \sigma^* S\sigma^* \rangle \leq -\tfrac{1}{2}\Sigma S^{\text{hom}}\Sigma = \tfrac{1}{2}EC^{\text{hom}}E - \Sigma E \leq \tfrac{1}{2}\langle \varepsilon' C\varepsilon' \rangle - \Sigma\langle \varepsilon' \rangle.$$

The previous inequality is then written

$$-\tfrac{1}{2}\Sigma\langle S\rangle\Sigma \leq -\tfrac{1}{2}\Sigma S^{\text{hom}}\Sigma = \tfrac{1}{2}EC^{\text{hom}}E - \Sigma E \leq \tfrac{1}{2}\langle \Sigma S^{\text{hom}} CS^{\text{hom}}\Sigma \rangle - \Sigma S^{\text{hom}}\Sigma.$$

The two first relations imply

$$\Sigma(\langle S\rangle - S^{\text{hom}})\Sigma \geq 0 \quad (\forall \Sigma) \qquad \text{(Reuss)}.$$ (6.71)

The Reuss compliance tensor is an upper bound estimation of the effective compliance tensor.

The two latter relations imply

$$E(\langle C\rangle - C^{\text{hom}})E \geq 0 \quad (\forall E) \qquad \text{(Voigt)}.$$ (6.72)

The Voigt stiffness tensor is an upper bound estimation of the effective stiffness tensor.

We obtain identical bounds with boundary homogeneous strains as $\varepsilon' = E$ and $\sigma^* = C^{\text{hom}}E$ at each point. This estimated field is an exact solution when the constituents are in parallel.

6.6.5 Example

Consider a long fibre unidirectional composite submitted to a macroscopic stress σ in the fibre direction. Each constituent fibre or matrix is assumed to be homogeneous, linearly elastic isotropic and to obey Hooke's law.

The equivalent anisotropic homogeneous material submitted to a uniaxial stress may be considered uniformly distributed over a volume and which includes a great number of fibres. In this case, we can consider such a stress distribution as a macroscopically uniform distribution of stress. In the immediate neighbourhood of a fibre the stress distribution is, of course, inhomogeneous, but an average value of stress over a large area must be equal to the value of the applied macroscopic stress σ. Similarly, the strain distribution must be non-uniform on a small scale but mainly uniform on a large scale. This uniform macroscopic strain will be denoted ε. It is the ratio of the macroscopic quantities σ and ε which is measured during a tensile test and which defines the equivalent homogeneous elastic modulus of the composite material $E = \sigma/\varepsilon$. As σ is the only non-vanishing component of macroscopic stress, the total potential energy and the total complementary energy can be written as

$$\pi(\varepsilon) = \tfrac{1}{2} E\varepsilon^2 v - \Sigma\varepsilon v, \qquad \pi^*(\sigma) = -\frac{1}{2}\frac{\sigma^2}{E}v.$$

In order to predict an approximate solution for E, we can find a lower and upper bound by means of a particular statically admissible stress field σ^* and kinematically admissible displacement u'.

Lower bound

The total complementary energy of the statically admissible stress field σ^* can be written as

$$\pi^*(\sigma^*) = -\frac{1}{2}\int_v \sigma^* S \sigma^* \, \mathrm{d}v = -\left(\frac{1}{2}\int_{v_\mathrm{f}} \frac{\sigma^{*2}}{E_\mathrm{f}} \, \mathrm{d}v + \frac{1}{2}\int_{v_\mathrm{m}} \frac{\sigma^{*2}}{E_\mathrm{m}} \, \mathrm{d}v\right)$$

where the integration is performed over the volume v. Introducing the volume fraction respectively for the fibres and matrix, f_f and f_m, results in

$$\pi^*(\sigma^*) = -\frac{\sigma^{*2}}{2}\left[\frac{f_\mathrm{f}}{E_\mathrm{f}} + \frac{f_\mathrm{m}}{E_\mathrm{m}}\right]v.$$

The principle of total complementary energy leads to the inequality $\pi^*(\sigma^*) \leq \pi^*(\sigma)$. This gives

$$\frac{1}{E} \leq \frac{f_\mathrm{f}}{E_\mathrm{f}} + \frac{f_\mathrm{m}}{E_\mathrm{m}}.$$

Then the lower bound is

$$\frac{1}{(f_\mathrm{f}/E_\mathrm{f}) + (f_\mathrm{m}/E_\mathrm{m})} \leq E.$$

Upper bound

In order to find an upper bound for E, a suitable strain field for the fibres and matrix compatible with displacement boundary condition is

$$\{\varepsilon'\} = \{\varepsilon'; -m\varepsilon'; -m\varepsilon'; 0; 0; 0\}$$

where m is an unspecified constant.

According to this strain field, the strain energy can be formulated as

$$w(\varepsilon') = \frac{\varepsilon'^2}{2} \int_v \frac{1 - v - 4mv + 2m^2}{1 - v - 2v^2} E \, dv$$

$$= \frac{\varepsilon'^2}{2} v \left[\frac{1 - v_f - 4v_f m + 2m^2}{1 - v_f - 2v_f^2} \right] f_f E_f$$

$$+ \frac{\varepsilon'^2}{2} v \left[\frac{1 - v_m - 4v_m m + 2m^2}{1 - v_m - 2v_m^2} \right] (1 - f_f) E_m.$$

Using the principle of total potential energy, the inequality (6.62) results in the following upper bound for E.

$$E \leq \frac{1 - v_f + 2m(m - 2v_f)}{1 - v_f - 2v_f^2} E_f f_f + \frac{1 - v_m + 2m(m - 2v_m)}{1 - v_m - 2v_m^2} E_m (1 - f_f). \quad (6.74)$$

Although the inequality (6.74) is valid for any choice of m, the best results will be obtained when $w(\varepsilon')$ is a minimum, that means $\partial w(\varepsilon')/\partial m = 0$. Using the fact that $v \leq \frac{1}{2}$, $\partial^2 w(\varepsilon')/\partial m^2 > 0$, and therefore $w(\varepsilon')$ has a single minimum according as $w(\varepsilon')$ is quadratic. The value of m is given by

$$m = \frac{v_f(1 + v_m)(1 - 2v_m)f_f E_f + v_m(1 + v_f)(1 - 2v_f)(1 - f_f)E_m}{(1 + v_m)(1 - 2v_m)f_f E_f + (1 + v_f)(1 - 2v_f)(1 - f_f)E^m}.$$

In the special case where $v_f = v_m = v$, it follows that $m = v$. For this special case the inequality (6.74) reduces to

$$E \leq f_f E_f + (1 - f_f) E_m. \quad (6.75)$$

It is of interest to note that the deviation of the upper bound, depending on values of v_f and v_m, is imperceptible. This indicates that the upper bound is always very close to the straight line obtained with $v = v_f = v_m$ and so to the inequality (6.75).

Figure 6.19 shows experimental data and an approximate solution given by upper and lower bounds. In order to analytically predict the behaviour

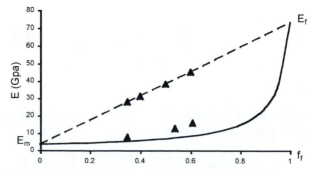

Figure 6.19. Upper and lower bounds for a glass fibre epoxy composite.

of a glass fibre composite, it is necessary to know the elastic constants of its constituents. The following values were taken: $E_f = 73\,100$ MPa; $E_m = 3500$ MPa; $\nu_f = 0.22$; $\nu_m = 0.35$.

If we now assume our cell to be loaded with a shear stress considered uniformly distributed over the volume, we get similar bounds as previously:

$$\frac{1}{(f_f/G_f) + [(1 - f_f)/G_m]} \leq G \leq f_f G_f + (1 - f_f) G_m.$$

6.7   The 'averaging method' for a periodic medium

6.7.1   Generality

The above methods give an estimation of elastic stiffnesses and compliances by using 'effective moduli methods' or upper and lower bound relationships. As we have already mentioned, the results obtained are very dependent on the localization method or the static and kinematic fields which are chosen. We can also show that the homogenized solution is also dependent on the geometry of the cell. It is of interest to point this out, because the choice of the cell is not unique. Several patterns can rebuild the real material (figure 6.20). The solution obtained will be dependent on this choice. The method we are now going to study, which is exact in a sense that we are going to further explain, gives results which do not depend on the choice of the cell and its geometry, both concerning the behaviour of the material and the evaluation of local parameters. It should be noted that the first

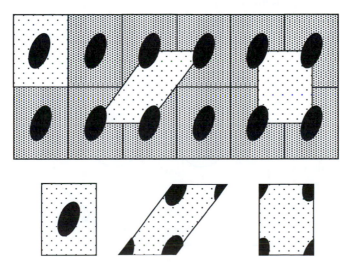

Figure 6.20. Different possible cells according to a given periodic pattern.

term of the asymptotic expansion of u according to the size of the cell gives the same results as the method we are going to present [8]. This method, which for a long time was limited to the research field, is now commonly used for the calculation of composite materials behaviour as a standard numerical method.

6.7.2   Periodic strain and stress fields

The 'averaging method' is used for heterogeneous periodic media. This period, which is characteristic of the material, is referred to as Y in the following. This medium is said to be Y-periodic. As the structure of the REV is periodic, we can limit the study of the homogenized behaviour to the basic cell.

   According to homogenization rules introduced in section 6.2, to obtain the stiffness of the equivalent homogenized material over the cell Y, it is necessary to know the localization tensor when applying a boundary macroscopic strain E or stress Σ. Since u represents the displacement field, our purpose is to establish the resulting local strain field $\varepsilon(u)$ or the local stress field $\sigma(u)$.

   If we consider a REV which is large enough to be composed of many cells and which is bounded with the displacement condition $u = Ex$, we find, by averaging over the whole volume, the applied macroscopic strain E. But locally, due to periodic heterogeneities, the local strain and stress fields 'oscillate and vary' in a periodic manner about their mean value E and Σ.

   However, the suitable local strain field $\varepsilon(u)$ is Y-periodic and can be decomposed as a uniform strain distribution E as $\langle \varepsilon(u) \rangle_Y = E$, which would be the strain of the medium if it were homogeneous, and a Y-periodic fluctuation $\varepsilon(v)$ due to local non-homogeneities of the material, the mean value of which is equal to zero, $\langle \varepsilon(v) \rangle = 0$.

   $\varepsilon(u)$ can then be written (figure 6.21)

$$\varepsilon(u) = E + \varepsilon(v). \tag{6.77}$$

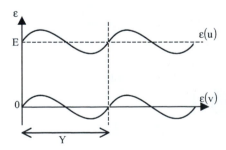

Figure 6.21. Strain decomposition over Y.

Finally, we have to characterize the displacement fields u and v which have been introduced. It is shown that the set of Y-periodic displacement u, the strain of which are also Y-periodic, can be decomposed in the following manner:

$$u(x) = Ex + v \tag{6.78}$$

where v is a Y-periodic displacement.

Further, the local stress field, which is periodic, must also be in equilibrium with the medium. Then it has to satisfy equilibrium equations ($\text{div}\,\sigma = 0$) all over the volume of the cell and boundary equilibrium conditions at the interface of the cell with neighbouring cells. This last condition associated with the periodicity of the stress field means that, on opposite sides of the cell, stress vectors σn are anti-periodic (since σ is periodic and n characterizes the normal which have opposite values on opposite sides).

Because of these conditions, the problem of elasticity which has to be solved over Y can be written as a problem of calculating a structure, the body forces of which are zero. The boundary conditions are either macroscopic imposed strains or stresses. Considering imposed macroscopic strains, the problem to solve then becomes the following.

Find (σ, u) defined over Y, with $\varepsilon(u)$ Y-periodic and satisfying

$$P_Y \begin{cases} \text{div}\,\sigma = 0 \\[4pt] \sigma = C\varepsilon(u) \text{ in } Y \\[4pt] u = Ex + v \text{ with } v \text{ Y-periodic} \\[4pt] \langle \varepsilon(u) \rangle = E. \end{cases} \tag{6.79}$$

Such a problem has a unique solution (σ, u) (within a rigid-body motion). This solution allows $\langle \sigma \rangle$ to be calculated and then the homogenized stiffnesses can be obtained according to the relation

$$\langle \sigma \rangle = C^{\text{hom}} E. \tag{6.80}$$

6.7.3   Linear sub-cell problems

Problem (6.79) is linear as a function of the applied strain E. We can then write any macroscopic strain E as a linear combination of elementary macroscopic strains $E = E^{kh} I^{kh}$, and then solve six problems P_Y^{kh} by submitting the cell with $E = I^{kh}$. I^{kh} is the elementary macroscopic strain in direction kh. u^{kh} is the displacement solution of problem P_Y^{kh}. Such a displacement field can be decomposed as

$$u^{kh} = I^{kh}x + v^{kh}. \tag{6.81}$$

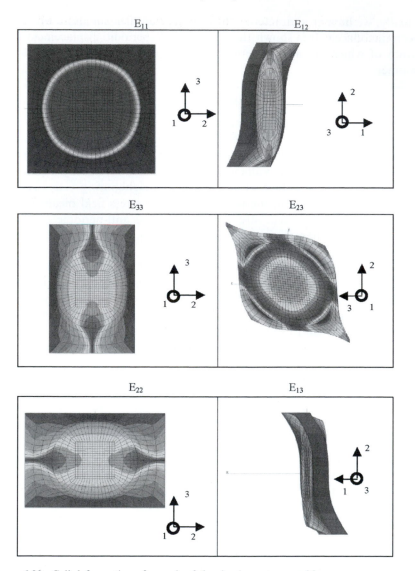

Figure 6.22. Cell deformations for each of the six elementary problems.

u^{11} represents a microscopic (or local) displacement over Y, when the cell is submitted to a pure macroscopic tension in the 1 direction. Considering u^{12}, the macroscopic strain is sheared in the 1-2 plane. Figure 6.22 illustrates cell-deformations for each of six elementary problems (as an example, a parallelepiped cell has been chosen).

Submitted to a macroscopic strain I^{kh}, the stress solution of the cellular problem can be written σ^{kh}. The problem P_Y reduces to six problems P_Y^{kh},

leading to the solution of the six following problems ($hk = 1$ to 6):
Find (σ^{kh}, u^{kh}) as

$$
(P_Y^{kh}) \left\{
\begin{array}{l}
\text{div } \sigma^{kh} = 0 \text{ in Y} \\[4pt]
\sigma^{kh} = C\varepsilon(u^{kh}) \text{ in Y} \\[4pt]
\sigma^{kh} \text{ and } \varepsilon(u^{kh}) \text{ are Y-periodic} \\[4pt]
\langle \varepsilon(u^{kh}) \rangle = I^{kh} = \{\tfrac{1}{2}(\delta_{ik}\delta_{jh} + \delta_{ih}\delta_{jk})\}_{i,j,k,h=1,2,3}.
\end{array}
\right.
\tag{6.82}
$$

When the six foregoing problems are solved, we can then calculate the homogenized coefficients.

$$
\langle \sigma_{ij}^{kh} \rangle = C_{ijpq}^{\mathrm{hom}} \langle \varepsilon_{pq}(u^{kh}) \rangle = C_{ijpq}^{\mathrm{hom}} I_{pq}^{kh}
$$

leading then to the stiffness tensor C^{hom}:

$$
C_{ijkh}^{\mathrm{hom}} = \langle \sigma_{ij}^{kh} \rangle.
\tag{6.83}
$$

Each problem P_Y^{kh} can be numerically solved by using finite element computation. The decomposition of the displacement field (6.81) suggests a solution for the problem with v^{kh}, the periodic part of u^{kh}. Then the calculation is made by applying two types of boundary conditions over the discretized cell:

1. Conditions of periodicity for the field v^{kh} by imposing identical values for each nodal components of v^{kh} which are opposite according to opposite edges of the cell. Numerically this condition of periodicity can be imposed by penalties on considered nodes.
2. Average strains over the cell by the means of global degree of freedom I^{kh}. This is made by introducing generalized strains $\{I^{kh}\}$ in each node of the mesh of the structure. Technically speaking we get generalized degrees of freedom which are common to all elements of the mesh.

The resolution is then made according two types of degrees of freedom: periodic degrees of freedom $\{v^{kh}\}$, which describe local fluctuation of the displacement field, and generalized degrees of freedom, which correspond to the macroscopic strains $\{I^{kh}\}$ which are applied over the cell.

After calculation, an integral of stresses made in each element and a summation over all elements of the mesh of the cell give the average stress value $\langle \sigma^{kh} \rangle$ and then the corresponding term kh of stiffness matrix C^{hom}. In this way, we obtain the homogenized stiffness behaviour when solving the six elementary problems.

6.8   Conclusion

In conclusion of this presentation of the 'averaging method' for periodic media, we have to note the existence of a dual method based on localization stress tensors. In that case, a macroscopic stress is imposed as boundary conditions on the cell. Then we determine the homogenized compliance tensor S^{hom}. The results obtained by using macroscopic strain or macroscopic stress are not dependent on the choice of the cell (geometry, basic pattern, etc.). Both methods give

$$C^{\text{hom}} = [S^{\text{hom}}]^{-1} \tag{6.84}$$

which is not the case for other methods presented. This is the reason why this method is the only one which can be said to be exact.

6.9   Revision exercises

1. Can you briefly explain the notion of representative elementary volume (REV)?
2. Can you mention different methods to describe the REV of the homogenized behaviour of a heterogeneous material as a composite material?
3. By using the rule of mixtures, evaluate the equivalent homogeneous behaviour of a carbon–epoxy composite, the mechanical properties of which are for the fibre f and the matrix m:

$$E_f = 230\,000\,\text{MPa}$$

$$E_m = 4000\,\text{MPa}$$

$$\nu_f = 0.22$$

$$\nu_m = 0.34$$

References

[1] Ekvall J C 1961 Structural behaviour of monofilament composites, in: *AIAA/ASME 7th Structures, Structural Dynamics and Materials Conference*, Palm Springs, California, 1966, p 250
[2] Eshelby J D 1957 The determination of the elastic field of an ellipsoidal inclusion and related problems *Proc. Roy. Soc.* **A241** 376–396
[3] Hopkins D A and Chamis C C 1988 A unique set of micromechanics equations for high temperature metal matrix composites, in: *Testing Technology of Metal Matrix Composites* ed P R DiGiovanni and N R Adsit, ASTM STP 964 (American Society for Testing and Materials) pp 159–176
[4] Mori T and Tanaka K 1973 Average stress in matrix and average elastic energy of materials with misfitting inclusions *Acta Metallurgica* **21** 571–574
[5] Paul B 1960 Prediction of elastic constants of multiphase materials *Trans. Metallurgy Soc. AIME* **218** 36–41
[6] Puck L B 1965 Zum Deformationsverhalten und Bruchmechanismus von unidirektionalem und orthogonalem G.F.K. *Kunststoffe* **55** 12

[7] Reuss A 1929 Berechnung der Fliessgrenze von Mischkristallen auf Grund der Plastizitätsbedingung für Einkristalle *Zeit. Angewandte Math. und Mechanik* **9** 49–58

[8] Sanchez-Palencia E 1980 Nonhomogeneous media and vibration theory *Lecture Notes in Physics* **127** (Berlin: Springer)

[9] Schaffer B W 1964 Stress–strain relations of reinforced plastics parallel and normal to the internal filaments *AIAA J.* **2** 348

[10] Spencer A 1986 The transverse moduli of fibre-composite material *Composite Sci. Technology* **27** 93–109

[11] Voigt W 1889 Über die Beziehung zwischen den beiden Elastizitätskonstanten isotroper Korper *Wiedemanns Annalen* **38** 573–587

Chapter 7

Micromechanisms of reinforcement and failure

7.1 Fibre–matrix load transfer

The most usual type of composite which is encountered consists of stiff fibres embedded in a more deformable matrix. This is the case of fibre reinforced resins and also fibre reinforced metals. The mechanism of reinforcement is through the transfer of the loads applied to the composite to the reinforcing fibres by the deformation of the matrix material around the fibres. As it is assumed that a good bond is maintained between the fibres and the matrix it can readily be seen that the matrix around the fibres will be constrained by their greater rigidity and will suffer shear. It is the shear of the matrix which is the dominant process in determining the composite properties. To illustrate this it is only necessary to imagine a deformable sheet on which two sets of equally spaced lines are drawn perpendicular to one another. This gives a regular pattern of identical squares. If now we imagine that the sheet is uniformly stretched in the direction of one of the sets of lines the squares become identical rectangles, retaining the 90° angles in each corner of the original squares. However if a long and thin rigid material, representing a fibre, is stuck to the original, undeformed sheet and then the sheet is stretched the results are illustrated in figure 7.1. Near the

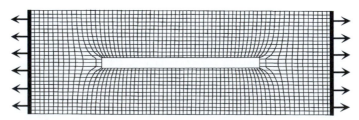

Figure 7.1. A fibre embedded in a deformable limits its strain in the neighbourhood of the fibre but the sheet can deform freely farther away. This leads to the introduction of shear into the matrix material and load transfer through the matrix to the reinforcing fibres of a composite material.

228

model fibre the sheet is obliged to follow the deformation of the stiffer fibre whilst further away the sheet can deform more. The result is that there are no longer intersections of the lines at right angles and the effect is greatest in the neighbourhood of the fibre. The presence of the fibre has introduced shear into the matrix.

This can be seen from the following brief analysis which is due to Cox and which shows how the induced shear stresses in the matrix result in a load transfer from the matrix to the reinforcing fibres.

7.1.1   Cox's analysis

Any shape of inclusion could be put into a matrix but a long thin form provides the most efficient reinforcement. H L Cox was interested in modelling the behaviour of paper and other fibrous materials. The model makes assumptions which must be understood as they determine the limits within which the model can be applied. Both the fibres and the matrix are considered to be perfectly linearly elastic. The fibres are considered to be of finite length and stiffer, that is to say, to possess Young's moduli greater, than that of the matrix. Loads are transferred from the matrix to the fibres at the fibre–matrix interface without yielding or sliding, which implies perfect adhesion between the two components. The model considers the local deformation of the fibres and the matrix and the average deformation of the composite material. The average deformation, ε_c, of the composite is considered to be attained at a point half way between the fibres in a regular array of straight and parallel fibres. This distance is denoted R and defines the radius of interaction around each fibre which is considered by the model. The model considers a single fibre of length $2L$ having a circular cross section of radius, r. In the absence of the fibre any applied load on the specimen would produce a uniform deformation of the matrix. Under load the matrix near the fibre is limited in deformation by the lower strain produced in the fibre whereas at a distance R from the fibre the matrix deforms to the macroscopic strain, ε_c, of the composite.

Let us consider an elementary section, of length dx of the specimen as shown in figure 7.2. A closer examination of this elementary section is shown in figure 7.3. The shear forces parallel to the fibre surface on a cylinder of radius z from the fibre axis must be equal to shear forces at the surface of the fibre. If the shear stress at the fibre surface is written as τ_f and the shear stress at a distance z is τ_z, we can write

$$2\pi z \tau_z \, dx = 2\pi r \tau_f \, dx \qquad (7.1)$$

which gives

$$\tau_z = r\tau_f / z. \qquad (7.2)$$

We can also write the relationship between the shear modulus of the matrix, G_m, and the displacement of the matrix, u_z, at a distance z from the fibre axis

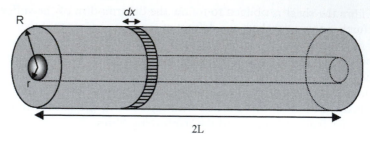

Figure 7.2. A fibre of length $2L$ and diameter $2r$ embedded in a cylinder of matrix of diameter $2R$.

which gives a shear strain of du_z/dz

$$\tau_z = G_m \frac{du_z}{dz}.$$
(7.3)

From equations (7.2) and (7.3) we can see that

$$\frac{du_z}{dz} = \frac{\tau_f r}{G_m z}$$
(7.4)

so that by integrating over the whole elementary volume we obtain

$$\int_{u_f}^{u_R} du_z = \frac{\tau_f r}{G_m} \int_r^R \frac{dz}{z}$$
(7.5)

where u_f is the displacement of the fibre and u_R is the displacement of the matrix at distance R in the elementary volume, so that

$$u_R - u_f = \frac{\tau_f r}{G_m} \ln\left(\frac{R}{r}\right)$$
(7.6)

which gives, for the shear stress,

$$\tau_f = \frac{G_m(u_R - u_f)}{r \ln(R/r)}.$$
(7.7)

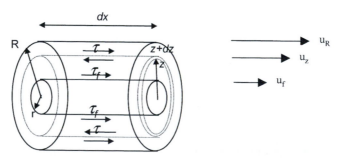

Figure 7.3. An elementary section of length dx of the specimen shown in figure 7.2.

In each element of the fibre the shear forces at the fibre surface are balanced by the tensile forces in the fibre, so that

$$2\pi r \tau_f \, dx + \pi r^2 \, d\sigma_f = 0. \tag{7.8}$$

Replacing τ_f in equation (7.7), from equation (7.8) we obtain:

$$\frac{d\sigma_f}{dx} = -\frac{2G_m(u_R - u_f)}{r^2 \ln(R/r)}. \tag{7.9}$$

Defining the strains in the fibre and in the matrix at the distance R from the fibre axis as

$$\frac{du_f}{dx} = \varepsilon_f = \frac{\sigma_f}{E_f} \quad \text{or} \quad \frac{du_R}{dx} = \varepsilon_R = \varepsilon_c. \tag{7.10}$$

Differentiating (7.9) and using the expressions from (7.10) we obtain

$$\frac{d^2\sigma_f}{dx^2} = \frac{2G_m[(\sigma_f/E_f) - \varepsilon_c]}{r^2 \ln(R/r)}. \tag{7.11}$$

Putting

$$n^2 = \frac{2G_m}{E_f \ln(R/r)} \tag{7.12}$$

which is dimensionless, we can write

$$\frac{d^2\sigma_f}{dx^2} = \frac{n^2}{r^2}(\sigma_f - E_f\varepsilon_c) \tag{7.13}$$

which is a differential equation which can be solved by the substitution

$$\sigma_f = E_f\varepsilon_c + P\sinh(nx/r) + Q\cosh(nx/r) \tag{7.14}$$

where P and Q are constants which can be determined using the boundary conditions, considering that no stress is transferred across the fibre ends, so that at $x = L$ and $x = -L$, $\sigma_f = 0$. From this we obtain $P = 0$ and $Q = -E_f\varepsilon_c/\cosh(nL/r)$. Defining the aspect ratio of the fibre as the ratio of its length over its diameter as $s = 2L/d = L/r$, we can then express the variation of the tensile stress at each point, x, along the fibre surface, and assuming that the stress is uniform across the fibre section, in the fibre, by

$$\sigma_f = E_f\varepsilon_c\{1 - \cosh(nx/r)/\cosh(ns)\}. \tag{7.15}$$

From equation (7.8)

$$\frac{d\sigma_f}{dx} = -\frac{2\tau_f}{r}. \tag{7.16}$$

Differentiating equation (7.15) and substituting from equation (7.16) we obtain the expression for the shear stress at each point, x, along the fibre

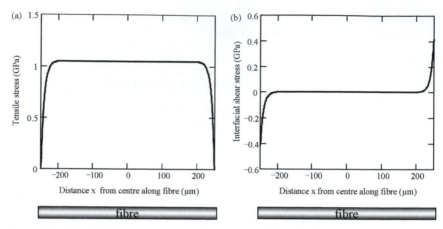

Figure 7.4. (a) Tensile stress supported by a glass fibre (length $2L = 1$ mm, $2r = 14$ μm, $E_f = 70$ GPa) embedded in a polyester resin ($G_m = 1.38$ GPa), assuming a hexagonal fibre packing array and $V_f = 0.6$. (b) Shear stress generated at the fibre–matrix interface for the same system.

surface:

$$\tau_f = \tfrac{1}{2} n E_f \varepsilon_c \sinh(nx/r)/\cosh(ns). \tag{7.17}$$

The value of n depends on the shear modulus of the matrix, the Young's modulus of the fibre and R/r, as can be seen from equation (7.12).

The ratio R/r is determined by the fibre packing and volume fraction in the composite and we can write the general expression $(2R/r)^2 = P_f/V_f$ where $P_f = \pi$ for a square packing arrangement and $P_f = (\sqrt{3}/2)\pi$ for a hexagonal arrangement.

The variations of tensile stress transferred to the fibre and shear stress in the matrix at the interface, which are given by equations (7.15) and (7.17), are shown in figure 7.4. It should be noted that the shear stresses at each end of the fibre are in opposite directions because the matrix in these two regions is sheared in opposite directions, as is illustrated in figure 7.4(b).

7.1.2 Physical interpretation of Cox's model

The above analysis shows clearly that the difference in rigidity of the fibres and the matrix leads to the development of shear stresses in the matrix, assuming that there is a good interfacial bond. This shear has the effect of transferring load from the matrix to the fibres. The model considers that all components are elastic and that the short fibres are positioned parallel to the load direction. It can be seen that the tensile load in the fibres increases from zero at the ends to attain an almost constant value approaching the stress that a continuous fibre would support for the same composite strain, $\sigma_{f\infty} = E_f \varepsilon_c$, if the aspect ratio is high enough.

Ineffective fibre length

At a given length δ along the fibre from the break, or $x = L - \delta$ from the centre, the ratio between the stress supported by the broken fibre $\sigma_{f(L-\delta)}$ and the stress that a continuous fibre would support σ_f is called the efficiency reinforcement factor Ψ:

$$\Psi = \frac{\sigma_{f(L-\delta)}}{\sigma_{f\infty}}. \tag{7.18}$$

From equation (7.15) and knowing that $\sigma_{f\infty} = E_f \varepsilon_c$ we can write

$$\Psi = 1 - \frac{\cosh[(n/r)(l-\delta)]}{\cosh(ns)} \tag{7.19}$$

$$\Psi = 1 - \cosh\left(\frac{n}{r}\delta\right) + \tanh(ns)\sinh\left(\frac{n}{r}\delta\right). \tag{7.20}$$

For most of the composites used (carbon, glass, fibres in polymeric matrices with volume fraction of 60% or more) $\tanh(ns)$ can be assimilated to 1 for fibres longer than 50 µm so that

$$\cosh\left(\frac{n}{r}\delta\right) = \frac{1 + (1-\Psi)^2}{2(1-\Psi)}. \tag{7.21}$$

As a consequence

$$\delta = \frac{r}{n}\cosh^{-1}\left[\frac{1 + (1-\Psi)^2}{2(1-\Psi)}\right].$$

Rosen took as the ineffective length δ that which gave an efficiency factor of 0.9. This value will be employed when the failure of unidirectional composites is treated. For a glass reinforced polyamide composite with a volume fraction of 60%, the ineffective fibre length is 20 µm which means that, at 20 µm from the fibre ends, the stress supported by the fibre is at least equal to 90% of the stress supported by a continuous fibre.

From equation (7.15) the mean stress supported by a short fibre embedded in a matrix is given by

$$\bar{\sigma} = \frac{1}{2L}\int_{-L}^{L} E_f \varepsilon_c \left\{1 - \frac{\cosh(nx/r)}{\cosh(ns)}\right\} dx \tag{7.23}$$

which can be re-written as

$$\bar{\sigma} = \sigma_{f\infty}\left\{1 - \frac{\tanh(ns)}{ns}\right\}. \tag{7.24}$$

It can be seen from figure 7.5 that as the length of the fibre increases the average stress supported by it increases and tends to that supported by a continuous fibre. Reducing the fibre diameter increases the aspect ratio

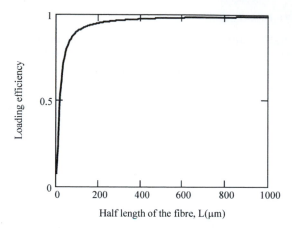

Figure 7.5. Ratio of the mean tensile stress supported by short glass fibres to the stress supported by a continuous glass fibre in a polyamide matrix, as a function of the half-length L of the short fibre.

and this results in a more rapid increase in average stress as the fibre length is increased and a greater loading efficiency. In figure 7.5 it can be seen that for a glass fibre of length $2L = 2\,\mathrm{mm}$ the efficiency is almost the same as for continuous fibres.

The load transfer length depends on the value of the shear modulus of the matrix (G_m). It can be seen that for a given type of fibre a decrease in matrix shear modulus leads to a longer load transfer length, such as might occur if a polyamide matrix were saturated with water as illustrated in figure 7.6. The load transfer length increases as the Young's modulus of the fibre increases. However, the average stress supported increases for stiffer fibres, as shown in figure 7.7(a). The rate of load transfer can be seen from figure 7.7(b) to be increased with the higher Young's modulus carbon fibre.

The plastic case is simpler than the elastic situation considered above as it can be assumed that the amplitude of the shear stress in the matrix is constant and equal to that of the yield stress, τ_y, so that equation (7.16) can be directly integrated giving

$$\sigma_f = -\frac{2 \cdot \tau_y}{r}(L + x) \qquad \text{for } -L \le x \le 0 \tag{7.25}$$

and

$$\sigma_f = \frac{2 \cdot \tau_y}{r}(L - x) \qquad \text{for } 0 \le x \le L. \tag{7.26}$$

Equations (7.25) and (7.26) show that there is a linear build up of stress from the fibre ends when it is embedded in a plastic matrix. In the case of a long fibre, however, this build up of stress is limited by the overall strain

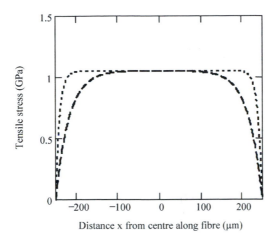

Figure 7.6. Comparison of tensile stresses supported by glass fibres, embedded in a dry polyamide (- - -) and in a polyamide saturated with water (———). The effect of absorbed water is to greatly reduce the shear modulus of the matrix (from $G_m = 1.1\,\text{GPa}$ to $G_m = 0.1\,\text{GPa}$) and so increase the load transfer length.

applied to the composite ε_c. In this case the central section of the fibre experiences a constant stress, $E_f \varepsilon_c$ equal to that which a continuous fibre embedded in the composite subjected to the same applied strain would experience, as shown in figure 7.8(a). If the stress applied to the composite is increased the load transfer length is increased and the central constant stress region reduces. Fracture will occur in the central section if the applied stress reaches the fibre failure stress. If the fibre is long, or continuous, and the fibre volume fraction V_f is not high, multiple fractures can occur. There clearly exists a length of fibre, measured from the fibre end, for which the maximum stress which can be attained at the mid-point is equal to the fibre failure stress. This is known as the critical length l_c, as shown in figure 7.8(b). At the critical length equation (7.26) can be rewritten so that

$$\frac{l_c}{r} = \frac{\sigma_{fu}}{2 \cdot \tau_y} \tag{7.27}$$

where σ_{fu} is the failure stress of the fibre.

It can be seen therefore that the ratio l_c/r is the important factor in determining the efficiency of a fibre reinforcement in the classical case of stiff fibres embedded in a deformable matrix. This is the case of glass, carbon or other high modulus fibres used to reinforce resin or metal matrices.

When a single, long fibre is embedded in such a matrix, shear of the matrix around an initial fibre break will ensure that the fibre is reloaded so that at a distance l_c the load it supports is unchanged. The fibre can therefore break again and again as shown in figure 7.8(c), until the distance between the fibre breaks becomes less than $2l_c$ [see figure 7.8(d)].

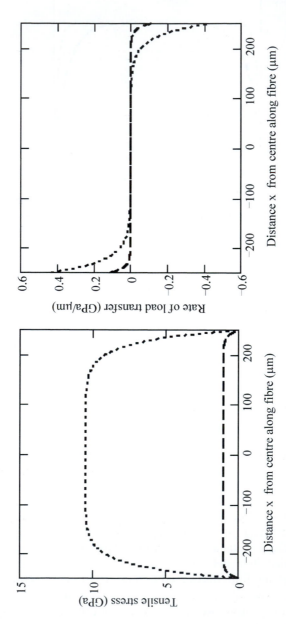

Figure 7.7. Comparison of tensile stresses supported by a glass fibre ($E_f = 70\,\text{GPa}$, $2r = 11\,\mu\text{m}$) (--) in a polyamide matrix. The effect of increasing the fibre Young's modulus is to increase the load transfer length, the rate of load transfer ($d\sigma_f/dx$) and the mean stress supported by the fibre.

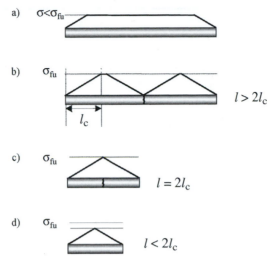

Figure 7.8. Fibre multicracking and critical length.

7.2 Unidirectional reinforcement

The preceding analyses have shown that the most efficient reinforcement is provided by continuous fibres aligned in the direction of the applied load. The fibre arrangement in the matrix influences greatly the final properties of the composite. The most spectacular properties are obtained when measured parallel to the fibres with a unidirectional composite in which all the fibres are aligned in one direction.

It is usually the properties of the fibres which are of interest so that a dense fibre packing (high fibre volume fraction) is often desirable. An ideal hexagonal arrangement for circular fibres could give a maximum fibre volume fraction greater than 90% whereas a square packing array could give more than 75%. In practice volume fractions of 60% are usual as surface tension forces and the natural viscosity of the matrix material limits what can be pushed between the fibres. In any event it would not be desirable to have the fibres touching as the points of contact would be lines of weakness in the composite. Fibre arrangements are rarely ideal in composites so that locally high fibre packing may occur and other areas are matrix rich.

When a unidirectional composite is loaded parallel to the fibres it deforms and if we assume good bonding between the fibre and matrix the strain in both will be identical and equal to the overall composite strain, so that

$$\varepsilon_c = \varepsilon_f = \varepsilon_m \qquad (7.28)$$

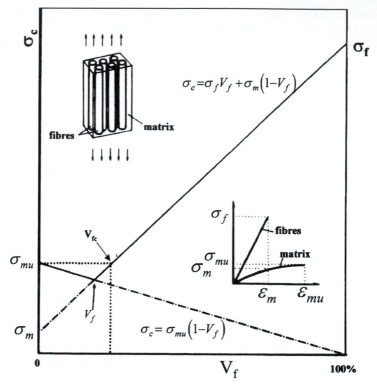

Figure 7.9. The law of mixtures for a unidirectional composite loaded in the fibre direction for which a good interfacial bond gives an increasing composite strength with increasing fibre content but a fall in strength if there is no bond.

where the suffixes c, f and m stand respectively for composite, fibre and matrix.

The applied load is shared between the fibres and the matrix, as illustrated in the inset view of a unidirectional composite in figure 7.9, so that

$$\sigma_c A_c = \sigma_f A_f + \sigma_m A_m \qquad (7.29)$$

where A is the total cross-sectional areas of the components. As the length of the components is that of the overall length of the composite we can write

$$\frac{A_f}{A_c} = V_f, \qquad \frac{A_m}{A_c} = V_m, \qquad \frac{V_f}{V_m} = 1 \qquad (7.30)$$

where V stands for volume fraction.

Equations (7.29) and (7.30) show that the stress supported by the composite is given by

$$\sigma_c = \sigma_f V_f + \sigma_m (1 - V_f) \qquad (7.31)$$

and making use of equation (7.28) we see that the Young's modulus of the unidirectional composite is given by

$$E_c = E_f V_f + E_m (1 - V_f). \tag{7.32}$$

This equation is known as the 'law of mixtures' and although simple and simplistic it is the most fundamental equation for use with composites, particularly unidirectional composites. Analogous reasoning produces the following equations for the composite shear modulus G_c and Poisson's ratio ν_c.

$$G_c = G_f V_f + G_m (1 - V_f) \tag{7.33}$$

$$\nu_c = \nu_f V_f + (1 - V_f) \nu_m. \tag{7.34}$$

When the composite is loaded perpendicularly to the fibre direction the stresses supported by the two phases are the same as that supported by the composite, $\sigma_c = \sigma_f = \sigma_m$, but they experience different strains. If we consider that the composite consists of layers of matrix and fibres perpendicular to the applied load, the total length of the composite l is the sum of the total widths of the fibre and matrix layers, denoted l_f and l_m respectively. The total elongation Δl_c of the composite is

$$\Delta l_c = \Delta l_f + \Delta l_m \tag{7.35}$$

where Δl_f and Δl_m are the total elongations of the matrix and fibre layers. The composite strain is given by

$$\varepsilon_c = \frac{\Delta l_c}{l_c} = \frac{\Delta l_f}{l_c} + \frac{\Delta l_m}{l_c}. \tag{7.36}$$

The volume fractions of the fibres and matrix can be defined as $V_f = l_f/l_c$ and $V_m = l_m/l_c$ so that

$$\varepsilon_c = V_f \frac{\Delta l_f}{l_f} + V_m \frac{\Delta l_m}{l_m}, \qquad \varepsilon_c = V_f \varepsilon_f + V_m \varepsilon_m. \tag{7.37}$$

The Young's modulus of the composite is then obtained from this equation by dividing by σ_c which is equal to σ_f and σ_m, giving

$$\frac{1}{E_c} = \frac{V_f}{E_f} + \frac{V_m}{E_m} \tag{7.38}$$

or

$$E_c = \frac{E_f E_m}{E_f V_m + E_m V_f}. \tag{7.39}$$

A more general approach describing the properties of unidirectional laminates which allows the properties of crossplied laminates to be calculated is given in chapter 8.

The law of mixture breaks down at low fibre volume fractions, for if the curve it describes is extrapolated back to zero volume fraction the value given is found not be the tensile strength of the matrix, as shown by figure 7.9. The value σ_m is, rather, the stress supported by the matrix at the breaking strain of the fibres. The stress–strain curves for the fibres and matrix which are shown as an inset in figure 7.9 reveal that when the unreinforced resin is tested it fails at a higher stress σ_{mu}. When reinforced by a considerable number of strong fibres it is their failure which provokes the complete failure of the composite. This is because the load which the fibres support before their failure is transferred to the matrix when they break and this additional load is greater than that which the matrix can withstand. This is not so at very low volume fractions in which the total load supported by the fibres can be transferred to the matrix without provoking its failure. It is possible to consider another law of mixtures which considers that the fibres do not contribute to reinforcing the matrix. This would be the case if there were no adhesion between the fibres and the matrix. In this case the fibres can be considered as voids in the matrix so that the strength falls as the fibre volume fraction is increased so that

$$\sigma_c = \sigma_{mm}(1 - V_f). \tag{7.40}$$

If a series of unidirectional composite specimens are made with a range of fibre volume fractions it is found that at very low values of V_f the composite strength falls so as to pass through a minimum after which the strength increases. The intersection of equations (7.31) and (7.40) gives the fibre volume fraction at which the composite strength is a minimum. It is necessary to reach a critical volume fraction V_{fc} before the composite strength exceeds the unreinforced matrix strength. The value of V_{fc} for most fibre–matrix mixtures is less than 5%. This dependence of the strength and the failure processes involved will be treated in greater detail in the section on the ACK model.

7.3   Off-axis loading

There are three possible mechanisms which control the failure of uni-directional composites when they are loaded at an angle θ with respect to the fibre direction. A simple consideration of the maximum applied stresses which the composite can support shows that, when the direction of loading is parallel or nearly parallel to the reinforcement, it is fibre failure which dominates. If this is the mechanism which controls composite fracture we can write that failure will occur if the applied stress in the fibre direction x attains the failure criterion X in that direction. Similarly if the limiting case at right angles to the fibres is Y and S for in-plane shear, the stresses in the composite must obey the following relationships if failure is

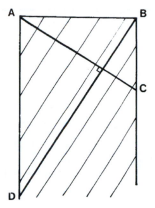

Figure 7.10. If the composite is loaded vertically, parallel to AD, but the fibres are parallel to DB, the failure processes will differ as the angle between the two directions is altered.

to occur:

$$\sigma_X \leq X, \qquad \sigma_Y \leq Y, \qquad \sigma_S \leq S. \qquad (7.41)$$

The failure stress of the composite for an orientation of the fibres is defined as σ_1. In order to calculate which mechanism controls composite failure it is necessary to determine the loads parallel to and at right angles to the fibre direction, as shown in figure 7.10. If the load on the composite applied over a section A, normal to the direction AD, is $\sigma_1 A$, these load components are $\sigma_1 A \cos \theta$ and $\sigma_1 A \sin \theta$. These loads act respectively upon sections of the composite given by $A/\cos \theta$ and $A/\sin \theta$. The component of the stress normal to the section AC and parallel to the fibre is therefore

$$\frac{\sigma_1 A \cos \theta}{A/\cos \theta} = \sigma_1 \cos^2 \theta.$$

If the composite is to fail by fibre breakage then $\sigma_1 \cos^2 \theta = X$. This means that the failure stress of the composite is given by

$$\sigma_1 = X/\cos^2 \theta. \qquad (7.42)$$

This result implies that as the angle is increased the composite strength also increases. This is clearly wrong and is due to the assumptions of the calculation not being respected in the real behaviour of the composite. As soon as the angle increases it is the shear of the composite rather than fibre failure which will control composite failure, so this situation has to be envisaged.

Shear of the composite can occur through the stresses acting parallel to the fibre direction but acting over the surface, BD, parallel to the fibres. Similar reasoning to that given above but resolving the forces on the section DB parallel to the fibre direction for shear failure gives

$$\sigma_1 = S/\sin \theta \cos \theta. \qquad (7.43)$$

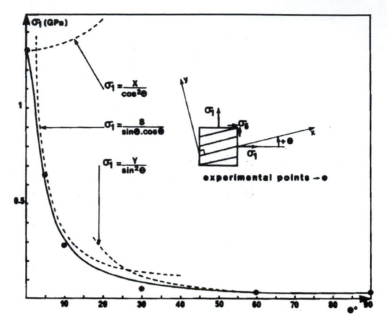

Figure 7.11. A unidirectional composite is extremely anisotropic.

In this case the composite strength is governed by the ultimate strength in shear of the composite. To simplify the calculation the shear strength will be written as σ_S and is equivalent to τ_f in equation (7.1).

Further increasing the off-axis angle produces tensile failure of the matrix (or failure of the interface or fibre splitting, as can occur with boron and Kevlar fibres) so that the composite strength is given by the stress normal to the fibres acting over the surface DB giving

$$\sigma_1 = Y/\sin^2 \theta. \tag{7.44}$$

Figure 7.11 shows how the above analysis corresponds quite well to experimental results obtained with unidirectional carbon fibre reinforced composites and emphasizes the extreme anisotropy of such a fibre arrangement. At small off-axis angles the composite strength falls greatly as it is no longer the fibre strength which determines behaviour. Such unidirectional composites do have their uses where the applied stress acts only in one direction. For example, a unidirectional composite arrangement wound around a fast rotating machine acts as a lightweight but strong and stiff fretting ring to counteract the tendency of the machine to fly apart at high speeds.

Usually however the anisotropy of unidirectional composites is unacceptable as multidirectional forces are involved.

7.4   Crossplied composites

The anisotropy of unidirectional composites is unacceptable for many applications so the solution adopted for wood when it is made into plywood is employed and crossplied laminates are used. Unidirectional layers are piled on top of one another with the reinforcement of the layers being in more than one direction. Laminate theory is considered in greater detail in chapter 8 but it can be seen from figure 7.12 that a 0°, 90° symmetrical arrangement begins to overcome the difficulties of anisotropy but at the expense of reducing the composite failure stress in the fibre directions, as in these directions half of the reinforcements play no role in strengthening the composite. Additional fibres arranged in the ±45° directions produce a material which is quasi-isotropic.

   The calculation of crossplied composite behaviour and of first ply failure are more complex than in the unidirectional case, as it is necessary to know the components of compliance and modulus and the Poisson's ratio of the unidirectional layers. It is necessary to express the behaviour of each ply by a matricial equation involving Poisson's ratio ν, the stresses σ, strains ε, Young's moduli E and shear modulus G of the ply when loaded in the longitudinal L, fibre, direction and transverse, T, to the fibre direction. This gives:

$$
\begin{bmatrix} \varepsilon_x \\ \varepsilon_y \\ \varepsilon_{xy} \end{bmatrix} = \begin{bmatrix} 1/E_L & -\nu_{TL}/E_T & 0 \\ -\nu_{LT}/E_L & 1/E_T & 0 \\ 0 & 0 & G_{LT} \end{bmatrix} \begin{bmatrix} \sigma_x \\ \sigma_y \\ \sigma_{xy} \end{bmatrix}. \qquad (7.45)
$$

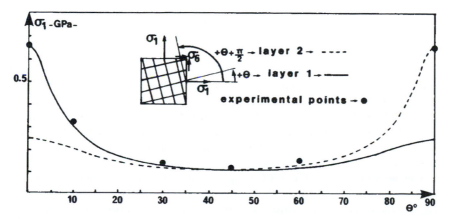

Figure 7.12. A crossplied composite with half the reinforcements in one direction and the other half at right angles goes some way to overcoming the anisotropy of unidirectional composites.

The summation of the behaviour of each layer in a crossplied composite now becomes possible and will be treated in greater detail in chapter 8.

7.5   Reinforcement efficiency of a crossplied composite

The efficiency of reinforcement can readily be calculated for any arrangement of fibres, as was shown by Krenchel. If we consider a section yy through a composite which contains fibres arranged in differing directions, the ratio of the total fibre cross-sectional areas A_{fy} to the total composite section A_c gives the fibre volume fraction.

$$\frac{A_{fy}}{A_c} = V_f. \tag{7.46}$$

The total fibre cross-sectional area is the sum of the fibre cross-sectional areas of layers of fibres striking the plane of the section, and as these are different for the various layers. For a layer denoted i this area is noted A_{fyi} and we must write

$$\frac{A_{fy}}{A_c} = \sum_i \frac{A_{fyi}}{A_c} = \sum_i V_{fi} = V_f. \tag{7.47}$$

We now consider the ratio between the cross-sectional area of any one fibre layer to the total fibre cross-sectional area so that

$$\frac{A_{fyi}}{A_{fy}} = \frac{V_{fi}}{V_f}. \tag{7.48}$$

Reinforcement efficiency (η_i) of a fibre layer i will depend on the fibre alignment with respect to the applied load. Fibres parallel to the applied load will have maximum efficiency so that $\eta_i = 1$. Fibres lying at right angles to the load will play no reinforcing role so that $\eta_i = 0$. In this way we can consider that the fibre layers crossing a given load-bearing section have an equivalent reinforcing area A'_{fy} which is only equal to A_{fy} when all the fibres in all layers are parallel to the applied load so that

$$A'_{fy} = \left(\sum_i \eta_i \right) A_{fy} = \eta A_{fy}. \tag{7.49}$$

Let us now consider a block of composite material made of several layers i, each layer containing fibres aligned in two directions at $\pm \phi_i$ with respect to the x axis as shown in figure 7.13. Assuming small strain elastic behaviour and ignoring the effect of Poisson's ratio, an external load acting in the direction of the x axis will produce a strain ε_c in the x direction. A fibre (OA) having an initial unit length and extending to the section y–y is now

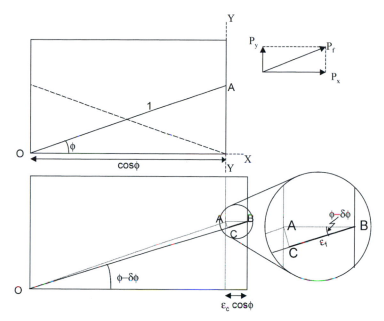

Figure 7.13. Deformation of one layer in a crossplied composite ($\pm\phi$), ignoring the effects of Poisson's ratio.

considered. The external load induces an increase in length of the fibre denoted ε_f. The internal fibre force required for this deformation is parallel to the fibre axis and is denoted P_f. P_f may be resolved into two components, P_x and P_y. The transverse force P_y can be ignored as it is balanced by the corresponding force component in the fibres aligned at $-\phi_i$ and because Poisson's ratio is ignored.

As a unit length of fibre AB is considered, the composite dimension in the x direction is $\cos\phi_i$, so that the strain ε_c induces an increase in length in the x direction to $\varepsilon_c \cos\phi_i$. When the composite is strained the fibre orientation changes slightly from ϕ_i to $\phi_i - \delta\phi_i$.

The strain ε_f induced in the fibre is given by

$$\varepsilon_f = (\varepsilon_c \cos\phi_i)\cos(\phi_i - \delta\phi_i).$$

As $\delta\phi_i$ is very small compared with ϕ_i it can be neglected in this calculation so that

$$\varepsilon_f = \varepsilon_c \cos^2\phi_i. \tag{7.50}$$

If the cross-sectional area of one fibre is S_f, the internal force in the fibre is given by

$$P_f = E_f\varepsilon_f S_f \tag{7.51}$$

and

$$P_x = P_f \cos \phi_i. \tag{7.52}$$

If the area of the fibre, in the layer i crossed by the section yy is denoted by S_{fy} we see that

$$S_{fy} = S_f / \cos \phi_i \tag{7.53}$$

so that from equations (7.50) to (7.53) we can write

$$P_x = E_f \varepsilon_c S_{fy} \cos^4 \phi_i. \tag{7.54}$$

The total internal forces, P_x, in the fibres crossing the section in a layer i consisting of fibres in the directions ϕ and $-\phi$ is obtained by summation over the section. The component of these forces in the y direction is zero as the y components in the two directions ϕ and $-\phi$ cancel, so the total force P_x is obtained by the summation, over the section, of P_x,

$$P_x = E_f \varepsilon_c \sum S_{fy} \cos^4 \phi_i. \tag{7.55}$$

Knowing that $\sum S_{fy} = A_{fi}$ this expression has the same form as Hook's law:

$$P_x = E_f \varepsilon_c A_{fyi} \cos^4 \phi_i \tag{7.56}$$

so that the equivalent fibre cross-sectional area is given by

$$A'_{fy} = A_{fyi} \cos^4 \phi_i. \tag{7.57}$$

From equation (7.57) it can be seen that the reinforcing efficiency, η_i, of each fibre layer at ϕ_i is

$$\eta_i = \frac{A_{fyi} \cos^4 \phi_i}{A_{fy}}. \tag{7.58}$$

The ratio of the surface areas of fibres in a particular direction to the total fibre cross-sectional area gives the proportion of fibres in that direction and is denoted as

$$\frac{A_{fyi}}{A_{fy}} = a_i$$

then

$$\eta = \sum_i a_i \cos^4 \phi_i. \tag{7.59}$$

Several examples will be considered of the calculation of efficiency factors for different fibre lay ups, as shown in figure 7.14 with the load applied vertically.

1. Unidirectional composites loaded in the fibre direction [see figure 7.14(a)]: in this case $\phi = 0$ and $a_1 = 1$ so that maximum efficiency is achieved and equation (7.59) gives $\eta = 1$.

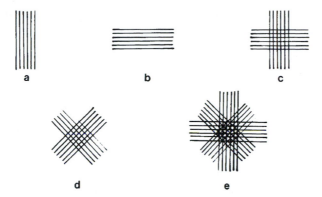

Figure 7.14. Various possible lay-ups for crossplied composites.

2. Unidirectional composites loaded at right angles to the fibres [see figure 7.14(b)]: in this case $\phi = \pi/2°$ and $\eta = 0$.
3. Symmetrical cross ply composite (0°, 90°) containing equal numbers of fibres in each direction and loaded in one of the fibre directions [see figure 7.14(c)]. In this case $\phi_1 = 0$, $\phi_2 = \pi/2$, $a_1 = \frac{1}{2}$, $a_2 = \frac{1}{2}$ so that

$$\eta = \tfrac{1}{2}\cos^4 0 + \tfrac{1}{2}\cos^4 \pi/2, \qquad \eta = \tfrac{1}{2}$$

4. Cross ply composite as in example 2 but loaded at $\phi = \pi/4$ [see figure 7.14(d)]. In this case $\phi_1 = +\pi/4$, $\phi_2 = -\pi/4$, $a_1 = \frac{1}{2}$, $a_2 = \frac{1}{2}$ so that

$$\eta = \tfrac{1}{2}\cos^4 \pi/4 + \tfrac{1}{2}\cos^4(-\pi/4) = \tfrac{1}{4}.$$

5. Symmetrical ply composite (0°, 90°, ±45°), with each layer containing the same number of fibres and loaded as in figure 7.14(e). In this case $\phi_1 = 0$, $\phi_2 = \pi/2$, $\phi_3 = \pi/4$, $\phi_4 = -\pi/4$ and $a_1 = a_2 = a_3 = a_4 = \frac{1}{4}$ so that

$$\eta = \tfrac{1}{4}(1 + 0 + \tfrac{1}{4} + \tfrac{1}{4}) = \tfrac{3}{8}.$$

If the cross ply composite considered in example 5 is loaded at a half angle between two fibre directions, the value of η is again found to be $\frac{3}{8}$. In the general case where the composite is made up of parallel layers of straight fibres distributed in all directions with an angular difference of $\delta\phi$, the value of $a_\phi = \delta\phi/\pi$. The efficiency factor is therefore given by

$$\eta = \frac{1}{\pi}\int_{-\pi/2}^{\pi/2} \cos^4 \phi\, d\phi = \frac{3}{8}.$$

The reinforcing efficiency of the (0°, 90°, ±45°) composite laminate is therefore shown to be identical to that obtained with a uniform fibre distribution in the plane of the composite and so justifies this arrangement being called quasi-isotropic.

The above analysis makes some assumptions in order to simplify the calculation, such as ignoring Poisson's ratio. The results are, however, generally in good agreement with more exact discussions of the problem—although for particular complex cases transverse deformations must be considered for the analysis in order to correspond exactly to the behaviour of the crossplied composite.

7.6 Compressive behaviour

There have been far fewer studies of the compressive failure of composites than there have been of tensile failure. This is because it is extremely difficult to produce perfect reproducible specimens and compressive strength is very sensitive to variations in composite make up. Composite strength is affected by fibre alignment, poor fibre arrangement, local resin-rich areas, porosity, poor fibre–matrix adhesion exacerbated by differing Poisson's ratio of the matrix and fibre and the viscoelastic behaviour of the matrix. In addition both fibres and matrix can be anisotropic, as is the case with Kevlar fibres which fail plastically in compression at a stress only a fifth of their tensile breaking stress. In practice the fineness of the fibres can also present a problem as the contraction of the resin which occurs during curing and cooling can cause the fibres to buckle. Such buckled fibres are not able to withstand applied compressive loads. This effect does not arise with large diameter fibres, such as boron fibres which have diameters of 140 μm compared with 7 μm for carbon fibres.

Compressive failure in unidirectional composites is due to induced fibre buckling. Two possible modes of fibre buckling are considered, as shown in figure 7.15.

In the case of low fibre volume fractions it is feasible to imagine out-of-phase buckling of fibres which induces large tensile stresses in the matrix. More realistic behaviour is shown in figure 7.15(b) in which the fibres are shown to be buckling in phase with the same wavelengths. In this latter case the matrix is subjected to shear stresses.

This approach leads to the following two expressions for the compressive strength of the composite σ_{comp}.

The out-of-phase buckling gives

$$\sigma_x = 2V_f[V_f \quad E_m \quad E_f/3(1 - V_f)]^{1/2}. \tag{7.60}$$

The more realistic in-phase shear buckling is predicted to give

$$\sigma_x = G_m/(1 - V_f). \tag{7.61}$$

Neither equation is found to agree very well with experimental results and although the calculation based on shear buckling leads to much lower

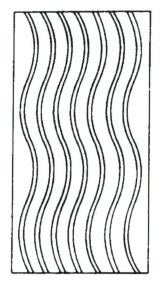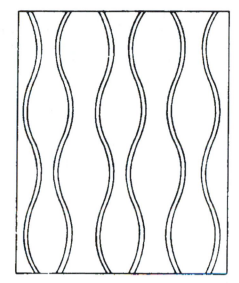

Figure 7.15. Two idealized buckling modes for fibres in a composite subjected to compressive loads: (a) in phase and (b) out of phase.

estimates of compressive strength than does the presumption of matrix tensile failure they are still much too high. The probable mechanisms which control compressive failure are illustrated in figure 7.16.

Buckling of the fibres will begin at a point where, locally, the fibres are not perfectly aligned parallel with the applied force. The displacement of the fibres requires the matrix to yield locally and it has been suggested that this is the important criterion. If matrix yielding initiates failure then

$$\sigma_{comp} = V_f \sigma_f + V_m \sigma_{my} \qquad (7.62)$$

where σ_{my} is the yield tensile strength of the matrix and σ_f is the stress supported when the matrix starts to yield plastically: $\sigma_f = \sigma_{my} E_f / E_m$.

If the failure criterion is not controlled by matrix yielding but by fibre failure then two further cases can be considered.

If the fibres are elastic they will fail from the convex surface of the bend when the fibre tensile failure stress is reached. Then failure in compression of the composite will occur when

$$\sigma_{comp} = V_f \sigma_{fu} + V_m \sigma_{my} \qquad (7.63)$$

where σ_{fu} is the tensile strength of the fibre.

If the fibres yield plastically in compression the fibres deform plastically when the concave surface experiences the elastic stress limit in compression

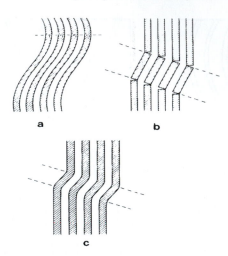

Figure 7.16. Probable failure processes in composites loaded in compression parallel to the fibre direction.

of the fibre:

$$\sigma_{comp} = V_f \sigma_{fy} + V_m \sigma_{my}. \tag{7.64}$$

In the case where matrix softening occurs due to heating or water uptake the compressive failure mode may be seen to change from the shear mode to the buckling mode.

7.7 Short fibre composites

Many composite structures consist of short fibres embedded in a matrix. This type of composite generally possesses much lower mechanical properties than continuous well-arranged fibre composites because of poor fibre alignment and lower fibre volume fractions. Short fibre reinforced composites are of considerable interest, however, as they can be press moulded or produced by injection moulding with much higher production speeds than is possible with continuous fibres.

The analysis detailed in section 7.1.2 shows how a rigid fibre embedded in a deformable matrix with good bonding between the two will experience a load transfer from the matrix. The stress supported by the fibre builds up from zero at the ends and, if the fibre is long enough, reaches the stress applied to the composite over the central section of the fibre. The average stress on the fibre is therefore less than the maximum supported stress because of the stress build up at the ends and is given by equation (7.18). The effective modulus of the short fibre is given by dividing the average

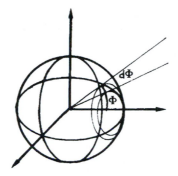

Figure 7.17. For an isotropic three-dimensional distribution of short fibres the distribution of fibres in a given direction is the ratio of the surface of the fibres in that direction to that of the sphere.

stress by the composite strain so that, using the law of mixtures, equation (7.24) gives a Young's modulus of

$$E_c = E_f\left\{1 - \frac{\tanh(ns)}{ns}\right\}V_f + E_m(1 - V_f).\qquad(7.65)$$

Equation (7.65) gives the Young's modulus of a short fibre reinforced composites in which all the fibres are aligned parallel to the applied force. This can occur in some parts of injection moulded structures but is not always the case, so the orientation efficiency factors η calculated by Krenchel and discussed in section 7.5 must be introduced into the equation so that it becomes

$$E_c = \eta E_f\left\{1 - \frac{\tanh(ns)}{ns}\right\}V_f + E_m(1 - V_f).\qquad(7.66)$$

In the case of random in-plane orientation of the fibres the value of η is 3/8, as calculated in section 7.5. If the fibres are uniformly arranged in three dimensions, Krenchel showed that the efficiency factor can be calculated by considering a sphere as shown in figure 7.17.

As with the two-dimensional cases considered in section 7.5, the fibres make an angle ϕ with respect to the X axis along which the load is applied. They are considered to pass through the centre of the sphere. The proportion a_ϕ is equal to the ratio between the area of the zone of the sphere (ϕ to $\phi + \delta\phi$) and the area of the hemisphere

$$a_\phi = \sin\phi\,d\phi.$$

Reinforcement comes from the total effect of fibres passing through all the elemental rings of thickness $\delta\phi$ from ϕ to $\pi/2$ so that

$$\eta = \int_0^{\pi/2} \sin\phi\cos^4\phi\,d\phi = -\left[\frac{\cos^5\phi}{5}\right]_0^{\pi/2} = \frac{1}{5}.\qquad(7.67)$$

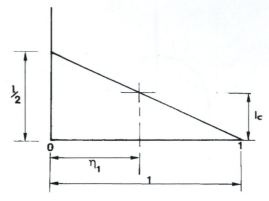

Figure 7.18. Distribution of lengths of short fibre projecting over the section of the sphere will vary in a uniform manner from 0 to $l/2$.

A further reduction in reinforcement efficiency occurs in the case of short fibres, as not all fibres crossing any given cross section can attain their maximum stress value. Some fibres projecting through the arbitrarily chosen cross section of a unidirectional short fibre composite will be too short, at least on one side of the section, to attain their breaking stress in the section. If failure occurred in the chosen section these fibres would pull out. The load transfer length which we are considering is given by equation (7.27) so that

$$l_c = \frac{\sigma_f r_0}{2\tau_y}$$

where σ_f is the breaking strain of the fibre and r_0 its radius, τ_y is the elastic yield stress in shear of the matrix. If all the short fibres have the same length l we can now consider those fibres which are intersected by the arbitrary section. The lengths of fibres projecting over the section considered will vary in a uniform manner from 0 to $l/2$, so arranging the projections in order of length would give the straight slope shown in figure 7.18. If the abscissa in figure 7.18 is given unit length then an additional efficiency factor η_l can be read directly from the graph as from geometrical factors:

$$l_c/(l/2) = 1 - \eta_l/1$$

so that

$$\eta_l = 1 - 2l_c/l. \tag{7.68}$$

7.8   Failure of a composite

7.8.1   The ACK model

The simplest form of a fibre reinforced composite is one in which all the fibres are aligned parallel to one another and the composite is loaded in

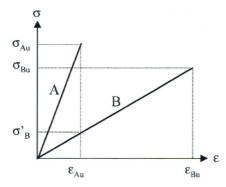

Figure 7.19. Elastic properties of the two components, A being less deformable than B. σ_{Au} and σ_{Bu} are the failure stresses of phases A and B, σ_B' is the stress supported by phase B at the failure strain of phase A, ε_{Au} and ε_{Bu} are the failure strain of phases A and B.

the direction of the fibres. In this case both the fibres and the matrix experience the same deformation ε and the applied load is shared by the fibres and the matrix in proportions which depend on their Young's moduli and their respective volume fractions. In the general case where the fibres and matrix are considered as simply two distinct phases, A and B, the stress σ_c in the composite can be written as

$$\sigma_c = \varepsilon E_A \cdot V_A + \varepsilon E_B \cdot V_B = \sigma_A \cdot V_A + \sigma_B \cdot V_B \qquad (7.69)$$

where E_A, V_A and E_B, V_B are the moduli and volume fraction of the two phases.

The model of composite behaviour, which is named 'ACK' after the initials of its authors, J Aveston, G A Cooper and A Kelly, treats the behaviour of such a general composite. One component in the composite can be expected to break before the other and therefore we will consider that a crack is formed in the phase which fails. The stress supported by the composite at the failure of the phase A is given by

$$\sigma_c = \sigma_{Au} \cdot V_A + \sigma_B' \cdot V_B \qquad (7.70)$$

where σ_{Au} is the failure stress of phase A and σ_B' is the stress supported by phase B at the failure strain of phase A as represented in figure 7.19.

In the crack plane, as all the load is only supported by the remaining component, the load from the broken component A is transferred to the phase B, whereas at a sufficient distance from the crack plane Cox's model allows us to see that the stress distribution across the composite remains unchanged if the interface between the two phases stays intact, as seen in figure 7.20.

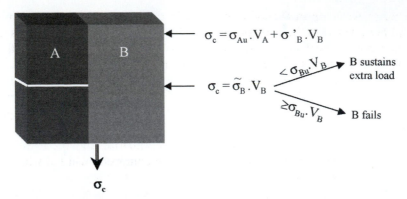

Figure 7.20. A first crack is formed in the less deformable phase A. In the crack plane the load is only supported by the phase B. At a long distance from the crack this load is shared between A and B.

The stress supported by the phase B in the crack plane, $\tilde{\sigma}_B$, is now given by

$$\sigma_c = \tilde{\sigma}_B \cdot V_B = \sigma_{Au} \cdot V_A + \sigma'_B \cdot V_B. \tag{7.71}$$

When this happens there are two possible results depending on the respective values of $\tilde{\sigma}_B$ and σ_{Bu}. Either the remaining component can support the additional load, which is possible if

$$\tilde{\sigma}_B \cdot V_B < \sigma_{Bu} \cdot V_B$$

that is if

$$\sigma_{Au} \cdot V_A + \sigma'_B \cdot V_B < \sigma_{Bu} \cdot V_B \tag{7.72}$$

or it will in turn immediately break if

$$\sigma_{Au} \cdot V_A + \sigma'_B \cdot V_B \geq \sigma_{Bu} \cdot V_B. \tag{7.73}$$

Two possible cases can be considered for a unidirectional composite depending on the failure strain of the two components.

7.8.2   First failure = fibre failure

In the case in which the failure strain of each fibre, ε_{fu}, is less than that of the matrix, ε_{mu}, and, assuming that all the fibres fail at the same stress, the matrix will fail as soon as the fibres break if

$$\sigma'_m V_m + \sigma_{fu} V_f \geq \sigma_{mu} V_m \tag{7.74}$$

where σ'_m is the stress supported by the matrix at ε_{fu}. However, if

$$\sigma'_m V_m + \sigma_{fu} V_f < \sigma_{mu} V_m \tag{7.75}$$

the matrix will support the additional load. In this case and if there is an interfacial bond between the two components, load transfer will occur across the break through local intact shear of the matrix and the effect of the fibre break will be isolated, so permitting it to be reloaded at a distance from the break. The result is that, when the stress applied on the composite reaches $\sigma_c = \sigma_{fu} V_f + \sigma'_m V_m$, the fibres will suffer multiple failures without any increase in composite stress up to the point when the fibre will be broken into lengths between l_c and $2l_c$ as we have seen above in Cox's model and represented in figure 7.8. At this point the composite can be reloaded and the additional load will only be supported by the matrix. The composite will fail when the stress supported by the matrix will reach the matrix failure stress:

$$\sigma_{cu} = \sigma_{mu} V_m. \tag{7.76}$$

This analysis shows that for an unidirectional composite containing fibres which are less deformable than the matrix, the failure stress of the composite σ_{cu} can be predicted by

$$\sigma_{cu} = \max(\sigma_{mu} V_m, \sigma'_m V_m + \sigma_{fu} V_f). \tag{7.77}$$

This is valid in the case where both constituents exhibit elastic behaviour, in the range of deformations imposed, and assuming perfect interfacial bonds.

The two functions $\sigma_{mu} V_m$ and $\sigma'_m V_m + \sigma_{fu} V_f$ can be plotted as a function of V_f $(= 1 - V_m)$ as shown in figure 7.21. This graph shows the regions for which composite failure is controlled, in turn, by the failure of the matrix and by the failure of the fibres. The transition between these two failure modes is determined by the volume fraction of fibres for which the two expressions have the same value. This threshold fibre volume fraction V_{th} is given by

$$V_{th} = \frac{\sigma_{mu} - \sigma'_m}{\sigma_{fu} + \sigma_{mu} - \sigma'_m} = \frac{E_m[(\varepsilon_{mu}/\varepsilon_{fu}) - 1]}{E_f + E_m[(\varepsilon_{mu}/\varepsilon_{fu}) - 1]}. \tag{7.78}$$

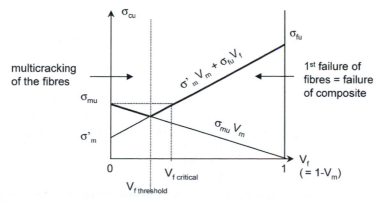

Figure 7.21. Failure stress and fracture mode of a composite composed of less deformable fibres, as a function of fibre volume fraction.

For a small volume fraction of fibres, it can be seen that the strength of the composite is lower than that of the unreinforced matrix. The introduction of a small amount of the fibres allows multicracking of these fibres and the failure is hence governed by the failure of the matrix. However, as the volume fraction of the matrix supporting the load is reduced by the introduction of fibres, the ultimate load on the composite, which is only supported by the remaining matrix, is less than that with an unreinforced matrix of the same section. The ultimate strength of such a composite, $\sigma_{mu} V_m$, is identical to that which would be obtained with a block of matrix of the same dimension in which fibres are replaced by holes, in the absence of any stress concentrations. For the fibres to increase the strength of the composite, that is for σ_c to be greater than σ_{mu}, their volume fraction must be greater than $V_{critical}$ given by

$$V_{critical} = \frac{\sigma_{mu} - \sigma'_m}{\sigma_{fu} - \sigma'_m} = \frac{E_m[(\varepsilon_{mu}/\varepsilon_{fu}) - 1]}{E_f - E_m}. \qquad (7.79)$$

In most of the structural composites for which the role of the fibre is to bring stiffness to the matrix, E_f is much greater than E_m, so that V_{th} and $V_{critical}$ are very similar and are of a few percent. For example, the threshold and critical volume fractions for an epoxy matrix ($E_m = 5.3$ GPa, $\varepsilon_{mu} = 0.02$) reinforced by carbon fibres (T300 $E_f = 240$ GPa, $\varepsilon_{mu} = 0.013$) are $V_{th} = 0.0118$ and $V_{critical} = 0.0122$.

Application of Cox's model

In the cases considered above using the ACK model it is important to recall the imposed limits to the model which include the supposition that all the fibres break at the same load and in the same plane. This is not the case in reality as reinforcements fail over a range of stresses and strains. Their failure points are unlikely to be in the same plane, so when a fibre breaks the effect of its failure is isolated within the section of the composite. Fibres in composites can fail more than once even with high volume fractions, as the shear of the matrix around a fibre break can effectively isolate it by reloading the fibre over a suitable load transfer length. This means that at a distance along the fibre from the point of failure the fibre supports the same load as it did before the failure occurred. After the first failure of a fibre, subsequent failure will take place at points along its length which are progressively stronger, so an exact analysis should take into account the statistical nature of fibre failure treated in chapter 2. Nevertheless the form of the curve shown in figure 7.21, which reveals that the composite strength passes through a minimum as the fibre volume fraction increases, can be clearly demonstrated experimentally. The calculation of the curve is most often carried out using the average failure strengths of the fibres and the matrix.

A specimen having the minimum V_f, that is containing only one fibre, can be used to determine l_c and τ as given by Cox's model from equation (7.27), by the measurement of the broken lengths of fibres after a tensile test. This is known as a multifragmentation test.

7.8.3 First failure = matrix failure

If $\varepsilon_{mu} < \varepsilon_{fu}$ it is the matrix which will break first and a single failure of the matrix will occur when the fibres in the crack plane cannot support the additional load:

$$\sigma_{mu} V_m + \sigma'_f V_f > \sigma_{fu} V_f. \tag{7.80}$$

The matrix will break into multiple parts when

$$\sigma_{mu} V_m + \sigma'_f V_f \leq \sigma_{fu} V_f \tag{7.81}$$

where σ'_f is the stress supported by the fibre at ε_{mu}.

The threshold volume fraction of fibre between the two failure modes, V_{th}, can be obtained by plotting $\sigma_{fu} V_f$ and $\sigma_{mu} V_m + \sigma'_f V_f$ as a function of $V_f (= 1 - V_m)$ using an analogous approach to that developed previously for a composite where $\varepsilon_{mu} > \varepsilon_{fu}$. In this previous case, however, the fibres were introduced to bring stiffness to the composite, and the Young's modulus of the fibres obviously had to be larger than that of the matrix. In the case that we are now considering, where $\varepsilon_{mu} < \varepsilon_{fu}$, the fibres are introduced to increase the toughness of a brittle matrix, by several processes such as by fibre debonding crack bridging and fibre pullout. This can be achieved with fibres which are less or more compliant than the matrix if their strain to failure is greater than that of the matrix. If the stiffness of the matrix is greater than that of the fibres then $\sigma_{mu} > \sigma'_f$, the slope of $\sigma_{mu} V_m + \sigma'_f V_f$ is negative, and this induces a critical volume fraction of fibres from which the failure stress of the composite σ_{cu} is larger than that of the unreinforced matrix σ_{mu}, as can be seen from figure 7.22(a). If the stiffness of the matrix is lower than that of the fibres then $\sigma_{mu} < \sigma'_f$, the slope of $\sigma_{mu} V_m + \sigma'_f V_f$ is positive, and σ_{cu} is always greater than σ_{mu}, as can be seen from figure 7.22(b).

The values of V_{th} and $V_{critical}$ are given by the following equations:

$$V_{th} = \frac{\sigma_{mu}}{\sigma_{fu} - \sigma'_f + \sigma_{mu}} = \frac{E_m}{E_f[(\varepsilon_{fu}/\varepsilon_{mu}) - 1] + E_m} \tag{7.82}$$

$$V_{critical} = \frac{\sigma_{mu}}{\sigma_{fu}} = \frac{E_m \cdot \varepsilon_{mu}}{E_f \cdot \varepsilon_{fu}} \quad \text{if } E_f < E_m. \tag{7.83}$$

For a silicon carbide matrix ($E_m = 350\,\text{GPa}$, $\varepsilon_{mu} = 0.00035$) reinforced by the first generation of SiC-based fibres with $E_f < E_m$ (Nicalon NL-200: $E_f = 180\,\text{GPa}$, $\varepsilon_{fu} = 0.011$), $V_{th} = 0.060$ and $V_{critical} = 0.062$. If this matrix is reinforced by near stoichiometric silicon carbide fibres with $E_f > E_m$ (Nicalon type S: $E_f = 386\,\text{GPa}$, $\varepsilon_{fu} = 0.0078$) $V_{th} = 0.041$.

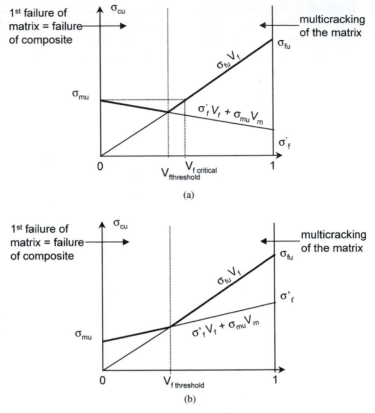

Figure 7.22. Failure stress and fracture mode of a composite composed of a less deformable matrix, as a function of fibre volume fraction. (a) The matrix stiffness is greater than that of the fibre, (b) the matrix stiffness is less than that of the fibre.

Critical length

The reinforcement of a brittle matrix by more deformable fibres requires that a propagating matrix crack does not pass through the fibres but is stopped or deflected by the fibre–matrix interfaces. This is only possible if these interfaces are sufficiently weak to allow fibre debonding. In the case when $V_f > V_{th}$, the debonded fibres can bridge a matrix crack without failing, and stress transfer from the fibre to the matrix allows the matrix to be progressively reloaded up to the stress level that an unbroken fibre would support. The distance necessary for the matrix to fully recover this stress level is again called the critical length.

To calculate the rate of stress transfer and the critical length, we will consider that this stress transfer occurs by interfacial friction between the debonded fibres and the matrix at a constant shear stress τ and will neglect

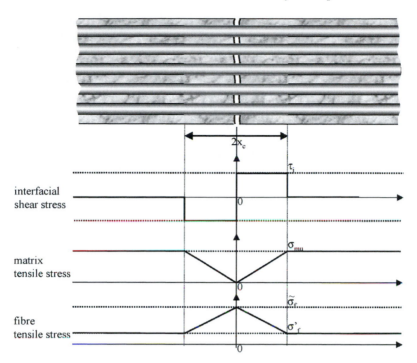

Figure 7.23. Tensile and shear stresses around a matrix crack.

any elastic stress transfer. The rate of stress transfer between matrix and fibre is determined by the maximum stress τ that the interface can sustain without failure in shear of the interface or of the matrix.

In an analogous manner to that used to derive equation (7.8), we can write that in each element of the fibre the shear forces at the fibre–matrix interface are balanced by the tensile forces in the fibre, so that

$$2\pi r \tau_i \, dx + \pi r^2 \, d\sigma_f = 0. \tag{7.84}$$

This equation can be integrated using the following boundary limits, ensuring the maximum rate of stress transfer at $x = 0$, $\sigma_f = \tilde{\sigma}_f$ and, at $x = x_c$, $\sigma_f = \sigma_f'$ with the notation introduced in the beginning of this section and presented in figure 7.23.

$$\int_{\tilde{\sigma}_f}^{\sigma_f'} d\sigma_f = -2 \frac{\tau_i}{r} \int_0^{x_c} dx. \tag{7.85}$$

This leads to

$$\tilde{\sigma}_f - \sigma_f' = 2 \frac{\tau_i}{r} x_c. \tag{7.86}$$

The stress $\tilde{\sigma}_f$ supported by each of the fibres in the crack plane is given by

$$\tilde{\sigma}_f V_f = \sigma_f' V_f + \sigma_{mu} V_m. \tag{7.87}$$

Combining equations (7.86) and (7.87) gives the length necessary for the matrix to fully recover the load which is also, with the hypothesis of this model, the length of decohesion of the fibre–matrix interface:

$$x_c = \frac{\sigma_{mu}}{\tau_i} \cdot \frac{V_m}{V_f} \cdot \frac{r}{2}. \tag{7.88}$$

Equation (7.88) can be compared with equation (7.27) giving the load transfer length in the more usual case where the fibres break before the matrix. It can be seen that both lengths depend on the fibre diameter, interfacial shear stress and tensile failure stress of the less deformable component. When the matrix cracks first, the load transfer depends in addition on the ratio of the volume fraction of the matrix to that of the fibres.

With an identical process to that described for fibre fragmentation when the stress applied on the composite reaches $\sigma_c = E_c \varepsilon_{mu} = \sigma_f' V_f + \sigma_{mu} V_m$, the matrix will suffer multiple failures without any increase in composite stress up to the point when the matrix will be broken into lengths between x_c and $2x_c$. This supposes that the matrix has a single failure strain.

Multicracking and composite strain calculation

If the fibres continue to support the load after the first matrix cracks, the fibres are subjected to higher loads around the matrix cracks and deform more even though the total load on the composite remains unchanged. The average increase in fibre strain inside a block of length $2x_c$ can be calculated knowing that at x_c from the crack the matrix again fully supports the load. The increase in stress on the fibres therefore varies from a maximum of $(\sigma_{mu} V_m)/V_f$ at the crack to zero at x_c, whereas the increase in strain varies $(\sigma_{mu} V_m)/E_f V_f$ at the crack to 0 at x_c, as seen in figure 7.24, and is given by

$$\frac{1}{2} \frac{\sigma_{mu} V_m}{E_f V_f} = \frac{\varepsilon_{mu}}{2} \frac{E_m V_m}{E_f V_f} = \frac{1}{2} \varepsilon_{mu} \alpha \tag{7.89}$$

where α is a dimensionless constant defined as

$$\alpha = \frac{E_m V_m}{E_f V_f}. \tag{7.90}$$

In the shortest block, of length x_c, the increase in stress is limited to a length of $x_c/2$ so that only half of this stress is transferred to the matrix. The increase in stress on the fibres therefore varies from a maximum of $\sigma_{mu} V_m/V_f$ at the crack to $(\sigma_{mu} V_m/V_f)/2$ at $x_c/2$. The average increase in composite strain over x_c is therefore given by

$$\frac{1}{2}\left(\frac{\sigma_{mu} V_m}{E_f V_f} + \frac{1}{2} \frac{\sigma_{mu} V_m}{E_f V_f} \right) = \frac{3}{4} \varepsilon_{mu} \alpha. \tag{7.91}$$

tensile stress inside the fibre

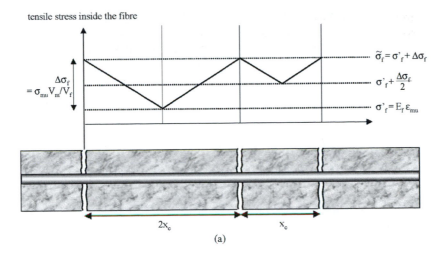

$$\tilde{\sigma}_f = \sigma'_f + \Delta\sigma_f$$

$$\sigma'_f + \frac{\Delta\sigma_f}{2}$$

$$\sigma'_f = E_f\varepsilon_{mu}$$

$$\Delta\sigma_f = \sigma_{mu}V_m/V_f$$

$$2x_c \qquad x_c$$

(a)

tensile strain inside the fibre

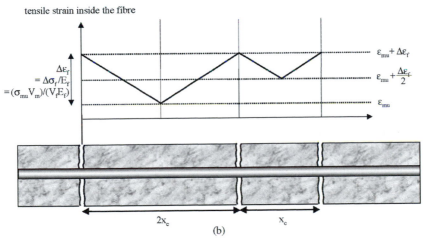

$$\varepsilon_{mu} + \Delta\varepsilon_f$$

$$\varepsilon_{mu} + \frac{\Delta\varepsilon_f}{2}$$

$$\varepsilon_{mu}$$

$$\Delta\varepsilon_f = \Delta\sigma_f/E_f = (\sigma_{mu}V_m)/(V_fE_f)$$

$$2x_c \qquad x_c$$

(b)

Figure 7.24. (a) Tensile stress and (b) tensile strain inside a fibre included in blocks of length $2x_c$ and x_c.

At the end of the repeated cracking all the block have a length lying between x_c and $2x_c$ so that the total strain of the composite is

$$\varepsilon_{mu}(1 + \alpha/2) < \varepsilon_c < \varepsilon_{mu}(1 + 3\alpha/4) \qquad (7.92)$$

whereas the stress remains equal to $\sigma_c = \varepsilon_{mu}E_c$, as seen in figure 7.25.

Fibre loading and ultimate composite stress

At this point the composite can be reloaded and the additional load is only supported by the fibres which will slide through the matrix after debonding.

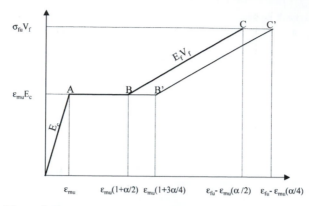

Figure 7.25. Theoretical stress–strain curve of a composite with more deformable fibres.

The slope of the stress–strain curve becomes $E_f V_f$. The composite will fail when the stress supported by the fibres will reach the fibre failure stress:

$$\sigma_{cu} = \sigma_{fu} V_f. \qquad (7.93)$$

The additional strain of the composite from the end of the multifissuration process up to the failure of the composite is given by

$$\frac{\sigma_{fu} V_f - \varepsilon_{mu} E_c}{E_f V_f} \qquad (7.94)$$

with $\sigma_{fu} = E_f \varepsilon_{fu}$ and E_c the Young's modulus of the composite before any matrix crack which is given by the rule of mixture $E_c = E_f V_f + E_m V_m$ so that the additional strain is

$$\varepsilon_{fu} - \varepsilon_{mu}(1 + \alpha). \qquad (7.95)$$

The failure strain of the composite is then obtained by adding equation (7.95) to equation (7.92)

$$(\varepsilon_{fu} - \alpha \varepsilon_{mu}/2) < \varepsilon_{cu} < (\varepsilon_{fu} - \alpha \varepsilon_{mu}/4). \qquad (7.96)$$

The entire stress–strain curve derived from this analysis is presented in figure 7.25. It can be seen that if the failure stress of the composite is $\sigma_{fu} V_f$ the failure strain is less than ε_{fu} as a part of the stress is still supported by the blocks of matrix.

Real stress–strain curve

The ideal stress–strain curve of a composite for which the fibres are more deformable than the matrix is given in figure 7.25. The horizontal line A–B supposes that the matrix has a single failure strain. This is rarely the case, so in reality the first failure of the matrix occurs from the largest or sharpest flaw, then a small increase in the composite stress is necessary to produce

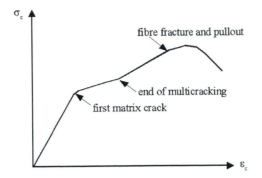

σ_c

fibre fracture and pullout

end of multicracking

first matrix crack

ε_c

Figure 7.26. Real stress–strain curve a SiC–SiC composite.

further cracks so that the line A–B has a small positive slope. The fibre fractures also occur for a range of stress and finally fibre pullout allows an increase of the strain with a decrease of the stress so that a real stress–strain curve is closer to that presented in figure 7.26. The length of the region between the first and last matrix crack is proportional to $\varepsilon_{mu}[(E_m V_m)/(E_f V_f)]$. Multicracking in vitroceramics reinforced by SiC fibres gives rise to a smaller strain than SiC/SiC composites. For cement reinforced by a small volume fraction of glass fibres ($V_f = 0.04$) this region is negligible compared with the strain in the region of fibre loading as $\varepsilon_{fu} > \varepsilon_{mu}$.

This type of behaviour can be expected and is desirable in ceramic matrix composites. The multiple cracking absorbs energy so that, although neither the fibre nor the matrix can deform plastically, the repeated cracking of ceramic matrix composite provides a mechanism which produces pseudo-ductility. Overall toughness is therefore enhanced. A word of warning, however, is perhaps necessary. The processes described above must be understood when speaking of increased toughness and it cannot be assumed that composite is any more capable of retaining its original pristine condition after a mechanical or thermal shock than the bulk ceramic. The enhanced toughness comes from the presence of the fibres which inhibit crack growth. Toughness in these systems is therefore associated with damage accumulation and the ability of the composite to support this damage. This has to be considered in the use of the composite. If a porcelain cup is dropped it will smash into a thousand parts and is clearly no longer useful. However, reinforced with ceramic fibres it might retain its shape and appear usable. If, however, tea which is poured into it passes through the cracks it could lead to the user finding out in a wet and uncomfortable way that the cup is still not usable. Perhaps more to the point would be that the cracks that form in the matrix to provide enhanced toughness also provide channels for the environment to reach the fibres. This can have two undesirable effects. The reinforcements might be attacked and degraded by the environment and the bond between the matrix and the fibres might be changed. This latter effect is a major difficulty for CMCs as the enhanced toughness described above can only occur if the cracking of the matrix does not induce failure of the fibres. For this to be possible the bond between the matrix and the fibres should not be strong so that cracks are either stopped at the interface or deviated along the interface. The fibre pullout which occurs during failure of the composite is an important mechanism in determining overall failure toughness. A common technique for ensuring a sufficiently weak interface is to make the surface of the fibre carbon rich or by coating it with carbon. Oxygen diffusing at high temperature along cracks in the matrix will oxidize the carbon and in many systems create silica at the interface. The presence of silica will provide an excellent interfacial bond and cracks arriving at the interface will not be stopped but will pass through the fibre, cutting them and preventing fibre pullout. The result will be a dramatic fall in toughness and strength. The effect may vary with temperature so that at room temperature a CMC system may have good toughness but this may fall catastrophically at say 700 °C when the interface is replaced by silica, but toughness may be regained by heating to say 1300 °C when the silica starts to soften and the interface once more stops crack propagation.

7.9   The process of crack propagation

Cracks propagate in any material when enough energy is provided to the crack to separate the material and form fracture surfaces. Energy is concentrated into a particular zone by irregularities or small pre-existing cracks which create stress concentrations. This has been known since 1913 when Inglis showed that holes in a loaded material produced stress concentrations. This original work considered the effects of rivet holes in the hulls of steel ships but had much more general implications for the propagation of all cracks. If the hole is elliptical and the load applied parallel to its minor axis, the stress concentration is increased with respect to that produced with a circular hole. The longer and thinner the ellipse the greater the stress concentration produced, and this brings us to the effect of pre-existing or induced cracks in any medium. The stress field around a hole is shown in figure 7.27. In an isotropic medium the controlling component of stress is σ_{11}, or σ_y in X–Y coordinates. This is a maximum at the crack tip and the component at right angles σ_{22}, or σ_x, is zero. The consequence is that the crack propagates along its axis. The value of σ_x increases, however, as we look along the crack axis, and passes through a maximum. This maximum value of σ_x is not often of importance in isotropic materials as it is only one fifth of the maximum value of σ_y, in the elastic case. However, if the medium is anisotropic with a strength less than one fifth when loaded in

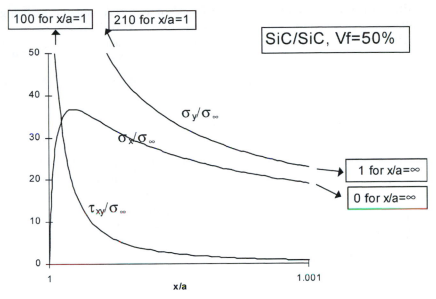

Figure 7.27. Variation of σ_x and σ_y ahead of an elliptic crack tip.

the X direction rather than in the Y direction the maximum value of σ_x becomes extremely important and a crack will be produced normal to the original crack and in the direction of the applied load. It is important to realize that this is not a phenomenon particular to composites but can be seen with any anisotropic material, such as highly drawn polymers. In CMCs, however, the interface between the fibre and the matrix provides a weak line which can be broken and crack propagation hindered. Cracks can still go around the fibres and continue to propagate, but their advancement requires the crack to open and this is impeded by the intact fibres behind the crack tip which bridge the crack. These processes of interfacial debonding and fibre pullout determine the toughness of the ceramic matrix composite.

7.9.1 Fracture mechanics

It can be seen from what has been discussed above that composite failure is not a simple process. Composite failure does not normally involve simple crack propagation from one side of the specimen to the other and slow crack growth is almost never seen. Nevertheless it is worthwhile examining fracture theory as applied to a continuum and which is known to describe failure in many metals. Linear elastic fracture mechanics finds its origins in the work of Inglis who showed that a nominal stress σ_0 applied to a body in a direction normal to the axis of a crack or notch considered as being elliptical is magnified and attains a value of

$$\sigma = \sigma_0 2(a/\rho)^{1/2} \tag{7.97}$$

in which a is half the crack length and ρ the radius of curvature of the crack tip. The ratio σ/σ_0 is called the stress concentration coefficient and explains why the presence of defects in a material weakens it so that its theoretical maximum strength is never attained. This is particularly evident with glass fibres, as is discussed in chapter 2.

In an ideally elastic case when the local stress reaches the value σ_c, necessary to separate the atomic structure, the crack will propagate. Griffith argued that crack propagation can only occur when the rate of energy (u) released by the material is equal to or greater than the energy required to create the new fracture surfaces so that for crack growth

$$\frac{\partial}{\partial a}(u - \nu) \geq 0 \tag{7.98}$$

where ν is the surface energy.

Griffith went on to show that

$$\sigma_c = \sqrt{E\nu/a}. \tag{7.99}$$

The factor ν is usually increased by plastic deformation or other processes around the crack. In order to have maximum energy absorbed during fracture, that is to have greatest toughness, $\sqrt{\sigma_c^2 a}$ should clearly be as great as possible and this factor is known, rather confusingly as the stress intensity factor K. For a specimen of finite size it is necessary to introduce a correcting factor Y so that

$$K = Y\sigma(a)^{1/2} \tag{7.100}$$

when the value of K exceeds a limit denoted by the critical stress intensity factor K_c, or simply the fracture toughness, and which depends on the material properties, then the crack becomes unstable and propagates. Irwin was also able to relate the rate of energy release from the body (G) to the factor and show that

$$K_c^2 = E^* G_c \tag{7.101}$$

where $E^* = E$ in the plane stress situation and $E/(1 - \nu^2)$ in plane strain, where ν is Poisson's ratio.

Dugdale and Barenblatt, in considering the effect of plastic deformation at the root of the notch, added a theoretical extension to the crack of length equivalent to the length of the plastic zone. This extension was considered to experience a closing stress which in reality accounted for the limiting of the maximum stress given by equation (7.97) by plastic deformation. In composite materials this closing stress can really exist and is due to unbroken fibres bridging the crack tip and ahead of the crack. The tensile stresses in these fibres reduce the local stresses in the region of the crack tip.

7.9.2 Physical processes involved in composite failure

The fracture toughness of a composite material depends on the many processes which can absorb energy during failure such as:

1. matrix cracking
2. fibre fracture
3. failure of the fibre–matrix interface
4. elastic energy released after the interface failure
5. fibre pullout from the matrix
6. shear failure of fibres not aligned at right angles to the crack
7. interlaminar delamination.

Different composite materials exhibit different modes of failure, and whereas some are amenable to an analysis based on fracture mechanics it is not so in all cases. The reason for this also explains why fibre reinforced composites have superior fatigue properties than do metals. The original work of Inglis was concerned with isotropic material containing holes and it was clear that the stress concentration described above was the determining factor in crack development. Further work, notably by Cook and Gordon, considered the whole stress field around a crack. These workers found that if a stress was applied at right angles to the crack axis then the stress concentration at the crack tip produced a maximum stress σ_y at right angles to the crack and that at the point the stress σ_x along the crack axis was zero. However, as mentioned earlier, it was found that σ_x increased from zero and passed through a maximum just ahead of the crack tip, as shown in figure 7.27. The ratio of $\sigma_{y_{max}}/\sigma_{x_{max}}$ was 5 for isotropic materials. Extending this argument to anisotropic materials it can be seen that, if the difference in strengths in the two orthogonal directions x and y is great, a crack at right angles to the initial crack will be created. This is indeed seen in highly drawn polymers and fibres where fibrillation rather than brittle failure occurs under tensile loading. The layers of a composite laminate are highly anisotropic and the fibre–matrix interface provides a usually weak discontinuity which can separate and blunt the developing crack. Composite materials therefore have their inbuilt crack growth inhibitors. Figure 7.28 shows how the stress in the direction of the crack can cause cracking parallel to the direction of applied load. If the fibre–matrix interface is too strong the fracture toughness of the composite is seen to fall, since a crack can cross the barrier with little difficulty.

The effects of the stress component, $\sigma_{x_{max}}$, are illustrated in figure 7.29. A crack normal to the original crack is created so that the original runs into it

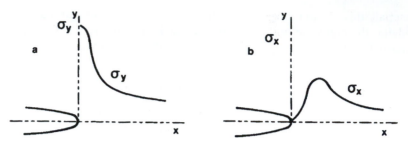

Figure 7.28. The components of the stress concentrations ahead of a crack show maximum values. The stress component tending to open the crack along its axis is a maximum at the crack tip but the component tending to open the material normal to the crack plane is ahead of the crack tip. In an anisotropic material or one which contains lines of weakness such as fibre–matrix interfaces this can lead to crack blunting.

and is blunted. In effect the radius of the crack tip is greatly increased so reducing the stress concentration, as described by equation (7.97).

7.9.3   Failure in a short fibre brittle matrix composite

We are accustomed to think that composites are designed so that greater benefit can be made of the fibre properties, but this is not always so. A class of composites exists for which the matrix is the most important component. For example, cement reinforced by randomly distributed fibres is a type of composite material which predates by centuries composite material, such as glass fibre reinforced resin. Cement is a brittle material which has very little tensile strength, no ductility and which cracks easily. It is also a suitable material to which the theories of fracture mechanics can be applied. Fibres are added to cement in order improve its properties and in particular to inhibit crack propagation. The fibre volume fraction typically used is of the order of 6%. So from figure 7.9 it can be seen that little increase in strength can be expected.

If a compact tension specimen shape is used, as shown in figure 7.30, it is possible to observe slow crack growth in this brittle matrix composite and study the stages of the load-displacement curve shown in figure 7.31. If a

Figure 7.29. The effects of the component of stress normal to the crack plane can be to create another crack, or a debond in a composite, which blunts the original crack.

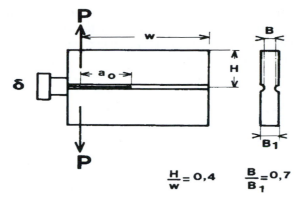

Figure 7.30. A compact tension specimen used to study crack growth in brittle materials.

transducer is placed in contact with the composite during the test it is possible to detect the vibrations produced by the internal failure processes. This technique is known as acoustic emission and has revealed that there are four stages to crack propagation in fibre reinforced cement. Acoustic emission activity starts at the point A in the linear elastic region of the load-displacement curve and shows that at this point a zone of microcracks is created ahead of the notch. At the point of non-linearity (B) the zone of microcracks has attained a certain size and the crack begin to propagate slowly. During this period of stable propagation the zone of microcracking increases in size until the point C is reached, after which it remain constant. During the propagation of the crack it is possible to observe the pullout of fibres from the cement matrix across the principal crack.

The straining of fibres in the zone of microcracking in front of the crack and the pullout of fibres which bridge it act as energy-absorbing mechanisms and increases the toughness and the resistance of the material to crack propagation. These two mechanisms supplement each other at the beginning of slow crack growth, but after the point C it is the increasing crack length

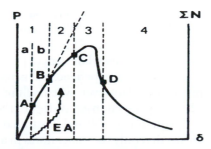

Figure 7.31. Load elongation curve of a C-T fibre reinforced cement specimen together with the emission acoustic recorded during loading.

over which fibre bridging occurs which accounts for a contributing increase in resistance. At the point D the opening of the crack corresponds to a maximum length of fibre-bridged crack, giving a maximum resistance to propagation. From this point the crack resistance of the material remains constant and as the specimen is tested in an increasing displacement mode it quickly fails.

The presence of the fibres are an inhibiting factor (K_r) as they act to prevent microcracks and the major crack from opening. We can therefore write for the stress intensity factor (K_R) necessary to cause propagation

$$K_R = K_0 + K_r \qquad (7.102)$$

where (K_0) is the stress intensity factor of the matrix in the presence of the fibres.

7.9.4 Failure of unidirectional composites

The behaviour and an approach to understanding the failure of unidirectional continuous fibre reinforced composites has been discussed in section 7.8.1. However, the ACK model makes assumptions which are not fully justified, such as each type of component having a unique failure stress. We have seen in chapter 2 that a statistical approach has to be adopted in characterizing the behaviour of fibres as they show considerable dispersion in their failure stresses. We have also seen in section 7.1 that the bond between fibres and the matrix leads to hear of the more deformable matrix a transfer of load from the matrix to the fibres. This means that even for short fibres, at a distance from the fibres' ends, they support the load that a continuous fibre would support. This means that when a continuous fibre breaks in a composite its fracture is isolated due to the effect of load transfer, just as is described in Cox's model. An approach to describing the accumulation of damage in a composite which takes into account these mechanisms should reflect better the behaviour of the composite.

Let us consider the difference between a bundle of fibres such as we might find in a simple fibre tow and the same bundle but impregnated with a matrix. If both specimens are progressively loaded the scatter in fibre strengths will mean that fibres will eventually begin to break. The failure of a fibre in the tow will mean that when a fibre fails it will cease participating in supporting the applied load over its whole length [see figure 7.32(a)]. This means that the other intact fibres will experience an increase in the loads they support over their entire lengths. However, when a fibre breaks in a composite the break is isolated in a short section equivalent to the critical load transfer length, as described by equation (7.27) and illustrated in figure 7.32(b). In the simplest case this means that only the intact fibres in a section of the composite of width $2l_c$ will experience an increase in load and elsewhere, outside of this section, the fibres remain unaffected by the break. If several fibre

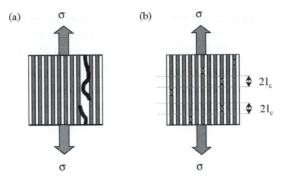

Figure 7.32. Part (a) illustrates a bundle of fibres with one broken. In this case the fibre ceases to carry load over its whole length. Part (b) shows a similar bundle but impregnated with a matrix. In this case the load transfer from the matrix to the fibre isolates the breaks and allows the fibres to break more than once.

breaks occur in the same section there is an increased probability of further breaks occurring in the section (see figure 7.33). In reality the effect of the fibre break is even more limited as it is only those fibres immediately neighbouring the break that are affected and which experience a local increase in stress. The breaks therefore have an insignificant effect on the overall performance of the composite. No change in stiffness can be detected nor can any change in overall strain be measured, but the breaks can be detected by the use of a technique called acoustic emission in which a piezoelectric transducer is used to pick up the vibrations caused by the fibre breaks.

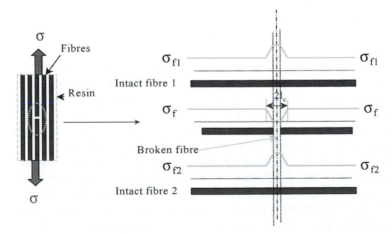

Figure 7.33. The intact fibres immediately neighbouring a broken fibre experience an increase of stress which is transferred to them by the shear of the matrix. This concentration, however, is limited to a section of the composite equal to the critical length $2l_c$.

The failure of a unidirectional fibre reinforced composite which takes into account the effects of the scatter in fibre strengths was first investigated by Rosen who likened the behaviour of the composite to that of a chain with each link being the length of $2l_c$. This is explained in section 7.1.3. In this way the model takes into account the contribution of the matrix in defining the critical length but afterwards ignores it. As a fibre breaks, the effect is limited to the link in which the break is situated. The failure stress of a unidirectional composite loaded in the direction of the fibres can therefore be seen to be due to the accumulation of fibre breaks which are initially isolated from one another. The rate of fibre failure is governed by the statistics of fibre strength, as described in chapter 2. However, eventually fibre breaks begin to interfere with one another as the density of failures increases and the composite fails.

7.9.5   Delamination of composite laminates

Crossplied composites are made up of stacks of either unidirectional fibre composite layers or stacks of woven layers. In both cases the layers are bonded together by resin. This means that the layers can be separated by shear forces, if they are great enough. The failure essentially involves crack propagation through an isotropic medium, the resin, in which case it is appropriate to apply fracture mechanics to this type of failure and the analysis reduces to that of the failure of an isotropic continuum. In order to improve interlaminar shear failure stress, some composites are made with fibres placed in the direction normal to the lamina. This can be achieved by needle punching or by special weaving techniques.

7.10   Conclusions

The mechanisms governing composite behaviour are very different from those which are encountered with traditional materials. The two-phase nature of composites means that a full understanding of their behaviour must involve the fibre reinforcements, the matrix and their mutual inter-action at the fibre–matrix interface. The quality of the bond between fibre and matrix is fundamental in determining ultimate properties. In most composites the matrix is much less rigid than the fibres and its deformation in shear allows the fundamental mechanism of load transfer to the fibres to take place. In some other composites the matrix is the more brittle com-ponent, but load transfer still determines the failure process. The nature of fibre reinforcement implies that at the level of the fibres the material is heterogeneous and anisotropic, and this has to be understood so as to benefit fully from these materials and in order to understand the failure processes which are involved.

7.11   Revision exercises

1. Explain how load can be transferred to short elastic fibre embedded in a softer elastic matrix when a load is applied parallel to the fibre. Who developed the model to explain this behaviour? What are the assumptions in this model? Why do the values of the shear forces induced in the matrix around the ends of the fibre have opposite signs?

2. In the case of a short elastic fibre embedded in a perfectly plastic matrix, write the balance of forces equation for the tensile stresses in the fibre and the shear forces in the matrix. How does this allow us to calculate the critical length of the reinforcement?

3. Explain how a single long fibre embedded in a matrix can be broken several times if the specimen is loaded in the direction of the fibre, without breaking the matrix. Why does this process of multicracking cease when the lengths of broken fibre reach a minimum size?

4. If a single carbon fibre embedded in a resin is broken into short lengths ranging from 70 μm to 140 μm what is the critical length for load transfer?

5. Calculate the interfacial shear strength of the carbon fibre system considered in question 4, given that the strength of the fibre can be taken as 3.5 GPa. (Answer 175 MPa.)

6. Develop the law of mixtures for a unidirectional composite loaded parallel to the fibres. If the fibre volume fraction is 0.6, the Young's modulus of the carbon fibres is 240 GPa and the Young's modulus of the epoxy resin is 4GPa, show that the fibres control the ultimate rigidity of the composite.

7. Calculate the Young's modulus of the composite described in question 4, at right angles to the fibres, assuming a perfect interfacial bond. (Answer 9.75 GPa.)

8. If the strain to failure of carbon fibres is taken as 1.5% and that of the epoxy resin 3%, the strength of the fibres is 3 GPa and that of the matrix 50 MPa, explain how the law of mixtures give the strength of the composite, σ_c, by the equation $\sigma_c = 3000 \times 0.6 + 25 \times 0.4$.

9. Calculate the Young's modulus of a glass fibre reinforced unidirectional composite loaded parallel to the fibres if the fibre volume fraction is 0.5. Also calculate the strength of the composite loaded parallel to the fibres. Take the Young's modulus of the glass fibres to be 70 GPa and their average strength 3.5 GPa, the Young's modulus of the resin to be 2 GPa and its strength 50 MPa. (Answers 36 GPa and 1.775 GPa.)

10. A single glass fibre, having a diameter of 14 μm and the properties specified in question 8, is embedded in a resin. The specimen is loaded parallel to the fibre, which is found to break into short pieces having lengths varying from 810 μm to twice this length. Calculate the interfacial shear strength.

11. If a unidirectional glass fibre composite is loaded parallel to the fibres it is found to fail at 884 MPa but at 45° with respect to the fibre direction it fails at 30 MPa. Explain this and calculate the shear failure stress of the composite. What would be the strength of the composite if loaded at 30° with respect to the fibre direction?

12. If the composite considered in Question 11 is loaded at right angles to the fibre direction it fails at 20 MPa. What is the process of failure? Calculate the strength of the composite when it is loaded at 60° with respect to the fibres.

13. What solution to overcome the anisotropy of wood has been adapted to make less anisotropic composite laminates. Explain how a quasi-isotropic laminate can be made.

14. According to the approach developed by Krenchel the reinforcing efficiency of layers containing fibres at right angles to the loading direction is 0, and 1 for layers in which the fibres are parallel to the applied load is 1. Using the formula for the efficiency of a crossplied laminate, calculate the efficiency of a laminate composed of two layers at 0°, two at $\pi/2$, and one each at $\pi/4$ and $-\pi/4$. (Answer 5/12.)

15. Discuss the ACK model. Under which conditions can multicracking of fibres occur in a resin matrix composite? Explain the role of the matrix and of the fibres in a ceramic matrix composite. What is the role of the interface in both of these composites?

16. Taking the values from question 6 for the characteristics of a unidirectional carbon fibre reinforced composite loaded in the fibre direction, calculate the threshold fibre volume fraction below which multicracking of the fibres will occur, and also calculate the fibre volume fraction at which the composite strength equals that of the unreinforced matrix material. Consider the ACK model. (Answers 0.826%, 1.2%.)

17. In the case of a brittle ceramic matrix with a Young's modulus of 400 GPa reinforced with ceramic fibres having a Young's modulus of approximately 200 GPa and assuming that the failure strains of the matrix and fibres are respectively 0.05% and 1%, calculate the threshold fibre volume fraction for multicracking of the matrix. Now calculate the critical volume fraction when the strength of the composite equals that of the matrix. (Answers 9.5%, 10%.)

18. The ultimate strength of a reinforcing fibre calculated from first principles based on the atomic forces binding the matter together gives a failure stress of 18 GPa but it fails at a much lower stress of 3 GPa. A study by scanning electron microscopy reveals cracks of 1 μm in length. What is the radius of curvature of the crack tip?

19. Describe a mechanism inherent in composite materials which means that fatigue crack propagation normal to the reinforcing fibres is practically impossible. Discuss the role of crack bridging in the failure of a brittle matrix composite.

20. What are the differences between the failure of a bundle of unimpregnated fibres loaded in the direction of the fibres and the same bundle embedded in a matrix. How does failure ultimately occur?

References

Cook J and Gordon J E 1964 A mechanism for the control of crack propagation in all brittle systems *Proc. Roy. Soc.* **A282** 508–520

Cox H L 1952 The elasticity and strength of paper and other fibrous materials *Br. J. Appl. Phys.* **3** 72–79

Inglis C 1913 Stress in a plate due to the presence of cracks and sharp corners *Trans. Inst. Naval Architecture* **55** 219–230

Griffith A A 1920 The phenomena of rupture and flow in solids *Phil. Trans. Roy Soc.* **A221** 163–197

Hull D and Clyne T W 1996 *An Introduction to Composite Materials* 2nd edition (Cambridge: Cambridge University Press)

Kelly A 1986 *Strong Solids* 3rd edition (Oxford: Oxford University Press and Macmillan NH)

Kelly A 1994 *Concise Encyclopedia of Composite Materials* (Oxford: Pergamon Elsevier)

Kelly A and Zweben C 2000 *Comprehension Composite Materials* Vol 1–6 (Oxford: Pergamon Elsevier)

Krenchel H 1964 *Fibre Reinforcement—Theoretical and Practical Investigations of the Eelasticity and Strength of Fibre-reinforced Materials* (Copenhagen: Akademisk Forlag)

Piggott 1980 *Load Bearing Fibre Composites* (Oxford: Pergamon Press)

Weibull W 1951 A statistical distribution function of wide applicability *J. Appl. Mech.* **18** 293–305

Chapter 8

Theory and analysis of laminated composites

8.1 Basic lamina and laminated structures

This chapter is devoted to laminated structures, composed by the stacking of layers which have different properties. These can be of many types—for example, continuous or short fibres in a matrix, honeycomb structures with composite skins or particles embedded in a matrix. In the case of long fibres, the fibre arrangement can vary to include unidirectional, bidirectional, woven, discontinuous or randomly distributed. In each case the ply is the basic entity and its mechanical properties are dependent on the reinforcements and their orientations. Unidirectional fibre plies are rigid and strong in the fibre direction but in the transverse direction properties are weaker. Poor fibre–matrix adhesion considerably decreases the transverse properties.

A laminate is a collection of laminae stacked to achieve the desired stiffness and thickness in the final composite. The sequence of various orientations in the plies of a fibre-reinforced composite laminate is called the lamination scheme or stacking sequence. In most applications, composite laminates are primarily two-dimensional with much greater dimensions in the plane of the laminate than in the thickness. Often laminates are used in applications that require axial and bending strengths. Therefore, flat composite are treated as plate elements. Further, in some applications, for example in naval and aeronautical applications, such as the Airbus A380, primary structures are made of laminated composites. To support high stresses these structures have to be sufficiently thick with thicknesses of 10 mm or more. In such cases the structure can no longer be considered flat and theories have to account for stresses through the thickness. These out-of-plane stresses are very important because they can cause delamination between plies, and they cannot be ignored.

The following analyses of composite plates will begin with a description of two-dimensional theories:

276

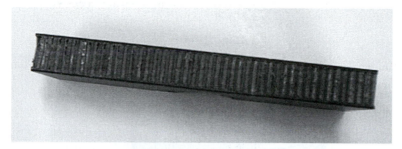

Figure 8.1. Honeycomb sandwich structure.

- Shear deformation laminate theories which do not neglect out-of-plane stresses will be considered.
- Classical laminate theory, which can correctly describe the behaviour of thin structures will be studied.

In the case of thick laminate structures (more than 10 mm), we are going to propose a homogenization method based on three-dimensional elasticity theory. Figure 8.1 illustrates one of these thick composites. This is a honeycomb sandwich structure with carbon skins. We are going to study these two approaches using the simplest theories.

8.2 First-order two-dimensional analysis

8.2.1 Kinematically possible displacements and deformations

We are going to limit our analysis to small deformations. Our purpose is to model the behaviour of a three-dimensional laminated plate as a two-dimensional plate. This implies, as has already been mentioned, a relatively thin thickness compared with other in-plane dimensions. Let us consider a laminated plate composed of n plies stacked in the x_3 direction and the volume of which is $v = \lfloor -h/2, h/2 \rfloor \times \Omega$. Ω represents the surface of the plate and h its thickness. $\partial\Omega$ is the boundary of Ω. The x_3 axis is taken to be positive upward from the midplane. The kth layer is located between the points $x_3 = x_3^k$ and $x_3 = x_3^{k+1}$ in the thickness direction. High-order laminate theories are based on the following displacement field.

$$u_1(x_1, x_2, x_3) = u_1(x_1, x_2, 0) + x_3\phi_1(x_1, x_2) + x_3^2\varphi_1(x_1, x_2) + \cdots$$

$$u_2(x_1, x_2, x_3) = u_2(x_1, x_2, 0) + x_3\phi_2(x_1, x_2) + x_3^2\varphi_2(x_1, x_2) + \cdots \quad (8.1)$$

$$u_3(x_1, x_2, x_3) = w(x_1, x_2, 0) + x_3\phi_3(x_1, x_2) + \cdots$$

where u_1 and u_2 need a supplementary order to account for the solid rotation of the normal to the midplane when bending or rotation are induced. Further

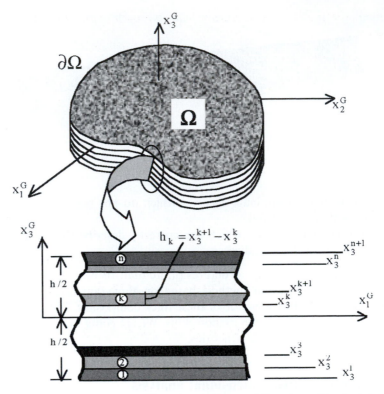

Figure 8.2. Coordinate system and layer numbering used for a typical laminated plate.

we shall note $u_1^0(x_1, x_2) = u_1(x_1, x_2, 0)$ and $u_2^0(x_1, x_2) = u_2(x_1, x_2, 0)$, the displacement of a material point in the $x_3 = 0$ plane and $w(x_1, x_2, 0) = w(x_1, x_2)$ which is the deflection of the plate that is supposed to be constant through the thickness.

The differences between the different theories is due to the limitations of the displacement field. The first order theory, known as the Mindlin theory, requires the displacement u_1, u_2, u_3 to be such that:

$$u_1(x_1, x_2, x_3) = u_1^0(x_1, x_2) + x_3\phi_1(x_1, x_2)$$

$$u_2(x_1, x_2, x_3) = u_2^0(x_1, x_2) + x_3\phi_2(x_1, x_2) \qquad (8.2)$$

$$u_3(x_1, x_2, x_3) = w(x_1, x_2)$$

where u_1^0, u_2^0, w, ϕ_1 and ϕ_2 are unknown functions which have to be determined. ϕ_1 and ϕ_2 respectively represent the rotations of the transverse normal about the x_1 and x_2 axes as

$$\frac{\partial u_1}{\partial x_3} = \phi_1, \qquad \frac{\partial u_2}{\partial x_3} = \phi_2. \qquad (8.3)$$

By differentiating the displacement field, we obtain the following deformations:

$$\varepsilon_1 = \frac{\partial u_1}{\partial x_1} = \frac{\partial u_1^0}{\partial x_1} + x_3 \frac{\partial \phi_1}{\partial x_1} = \varepsilon_1^0 + x_3 \kappa_1$$

$$\varepsilon_2 = \frac{\partial u_2}{\partial x_2} = \frac{\partial u_2^0}{\partial x_2} + x_3 \frac{\partial \phi_2}{\partial x_2} = \varepsilon_2^0 + x_3 \kappa_2$$

$$\varepsilon_3 = \frac{\partial u_3}{\partial x_3} = 0$$

$$\varepsilon_4 = \frac{\partial u_3}{\partial x_2} + \frac{\partial u_2}{\partial x_3} = \frac{\partial w}{\partial x_2} + \phi_2$$

$$\varepsilon_5 = \frac{\partial u_3}{\partial x_1} + \frac{\partial u_1}{\partial x_3} = \frac{\partial w}{\partial x_1} + \phi_1$$

$$\varepsilon_6 = \frac{\partial u_1}{\partial x_2} + \frac{\partial u_2}{\partial x_1} = \frac{\partial u_1^0}{\partial x_2} + x_3 \frac{\partial \phi_1}{\partial x_2} + \frac{\partial u_2^0}{\partial x_1} + x_3 \frac{\partial \phi_2}{\partial x_1} = \varepsilon_6^0 + x_3 \kappa_6$$

(8.4)

where ε_1^0, ε_2^0, ε_6^0 are the membrane strains and κ_1, κ_2, κ_6 are the flexural (bending) strains, known as curvatures. ε_4 and ε_5 are the transversal shear strains.

8.2.2 Resultant force and resultant moments

Resultant efforts are the quantities (N_1, N_2, N_6) called the in-plane force resultants, and (M_1, M_2, M_6) which are called the moment resultants. Q_4 and Q_5 denote the transverse force resultants. We are going to define these quantities.

At the kth ply level, the behaviour in the global reference coordinates of the structure is expressed with relation to section 5.14. After rotation of angle θ about the x_3 axis, we obtain

$$\{\sigma^G\}_k = [C(\theta)]_k \{\varepsilon^G\}_k.$$

The stacking sequence is made of parallel plies having the same state of strain. In this condition, we can define the resultant forces and resultant moments by a summation through the thickness.

The in-plane resultant force is defined as

$$N(x_1, x_2) = \begin{Bmatrix} N_1 \\ N_2 \\ N_6 \end{Bmatrix} = \int_{-h/2}^{h/2} \begin{Bmatrix} \sigma_1^G \\ \sigma_2^G \\ \sigma_6^G \end{Bmatrix} dx_3 \qquad (8.5)$$

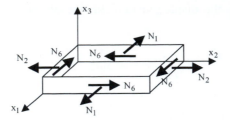

Figure 8.3. Resultant in-plane forces.

or

$$N(x_1, x_2) = \left\{ \begin{array}{c} N_1 \\ N_2 \\ N_6 \end{array} \right\} = \sum_{k=1}^{n} \int_{x_3^k}^{x_3^{k+1}} \left\{ \begin{array}{c} \sigma_1^G \\ \sigma_2^G \\ \sigma_6^G \end{array} \right\}_k \mathrm{d}x_3 \qquad (8.6)$$

where N_1, N_2, N_6 are resultants per unit length of normal and shear stresses in the x_1, x_2 plane. The integrated unit of these stresses is Pa m or N m^{-1} if we consider these resultants as forces per unit of thickness. These forces are those shown in figure 8.3.

Bending and torsion resultant moments are defined as

$$M(x_1, x_2) = \left\{ \begin{array}{c} M_1 \\ M_2 \\ M_6 \end{array} \right\} = \int_{-h/2}^{h/2} \left\{ \begin{array}{c} \sigma_1^G \\ \sigma_2^G \\ \sigma_6^G \end{array} \right\} x_3 \, \mathrm{d}x_3 \qquad (8.7)$$

or

$$M(x_1, x_2) = \left\{ \begin{array}{c} M_1 \\ M_2 \\ M_6 \end{array} \right\} = \sum_{k=1}^{n} \int_{x_3^k}^{x_3^{k+1}} \left\{ \begin{array}{c} \sigma_1^G \\ \sigma_2^G \\ \sigma_6^G \end{array} \right\}_k x_3 \, \mathrm{d}x_3 \qquad (8.8)$$

where M_1, M_2, M_6 are resultant moments per unit length in bending and in torsion (figure 8.4).

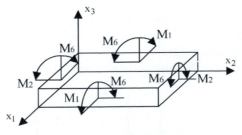

Figure 8.4. Bending and torsion resultant moments.

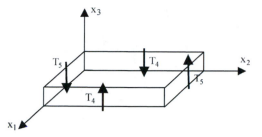

Figure 8.5. Resultant transverse shear forces.

Resultant transverse shear forces are integrated over the thickness to give

$$T(x_1, x_2) = \left\{ \begin{array}{c} T_4 \\ T_5 \end{array} \right\} = \sum_{k=1}^{n} \int_{x_3^k}^{x_3^{k+1}} \left\{ \begin{array}{c} \sigma_4^G \\ \sigma_5^G \end{array} \right\}_k dx_3 \qquad (8.9)$$

and summarized in figure 8.5.

8.2.3 Equations of motion

We are going to present equations of motion by applying the principle of virtual displacement of an elementary volume of the laminated plate. The variation of the potential energy for any kinematically displacement field is given by

$$\delta\pi(u) = \int_v \sigma \, \delta\varepsilon \, dv - \int_{S_f} F^d \, \delta u \, ds - \int_v f^d \, \delta u \, dv. \qquad (8.10)$$

The two last integrals represent the work of external forces over the surface and in the volume. In the case of laminated plates with first order displacement field (8.2) and strain state (8.4), we obtain the strain energy as being

$$\int_v \sigma \, \delta\varepsilon \, dv = \int_v \left[\sigma_1 \delta\left(\frac{\partial u_1^0}{\partial x_1} + x_3 \frac{\partial \phi_1}{\partial x_1} \right) + \sigma_2 \delta\left(\frac{\partial u_2^0}{\partial x_2} + x_3 \frac{\partial \phi_2}{\partial x_2} \right) \right.$$

$$+ \sigma_6 \delta\left(\frac{\partial u_1^0}{\partial x_2} + \frac{\partial u_2^0}{\partial x_1} + x_3 \frac{\partial \phi_1}{\partial x_2} + x_3 \frac{\partial \phi_2}{\partial x_1} \right)$$

$$\left. + \sigma_4 \delta\left(\frac{\partial w}{\partial x_2} + \phi_2 \right) + \sigma_5 \delta\left(\frac{\partial w}{\partial x_1} + \phi_1 \right) \right] dv \qquad (8.11)$$

because

$$\frac{\partial}{\partial x_j}(\sigma_{ij}\delta u_i) = \frac{\partial \sigma_{ij}}{\partial x_j} \delta u_i + \sigma_{ij} \frac{\partial(\delta u_i)}{\partial x_j}.$$

The former relations become ($\alpha, \beta = 1, 2$):

$$\int_v \sigma \delta \varepsilon \, dv = - \int_v \frac{\partial \sigma_{\alpha\beta}}{\partial x_\beta} \delta u_\alpha^0 \, dv - \int_v x_3 \frac{\partial \sigma_{\alpha\beta}}{\partial x_\beta} \delta \phi_\alpha \, dv - \int_v \frac{\partial \sigma_{\alpha 3}}{\partial x_\alpha} \delta w \, dv$$

$$+ \int_v \sigma_{\alpha 3} \delta \phi_\alpha \, dv + \int_{\partial v} \sigma_{\alpha\beta} \delta u_\alpha^0 n_\beta \, ds + \int_{\partial v} x_3 \sigma_{\alpha\beta} \delta \phi_\alpha n_\beta \, ds$$

$$+ \int_{\partial v} \sigma_{\alpha 3} \delta w n_\alpha \, ds.$$

If we split the integrals in the volume into an in-plane integral and an integration in the thickness direction, the integration in the thickness direction reveals the resultant forces and moment introduced in section 8.2.2.

$$- \int_\Omega \frac{\partial N_{\alpha\beta}}{\partial x_\beta} \delta u_\alpha^0 \, dx_1 \, dx_2 - \int_\Omega \frac{\partial M_{\alpha\beta}}{\partial x_\beta} \delta \phi_\alpha \, dx_1 \, dx_2 - \int_\Omega \frac{\partial T_\alpha}{\partial x_\alpha} \delta w \, dx_1 \, dx_2$$

$$+ \int_\Omega T_\alpha \delta \phi_\alpha \, dx_1 \, dx_2 + \int_{\partial \Omega} N_{\alpha\beta} \delta u_\alpha^0 n_\beta \, dl + \int_{\partial \Omega} M_{\alpha\beta} \delta \phi_\alpha n_\beta \, dl$$

$$+ \int_{\partial \Omega} T_\alpha \delta w n_\alpha \, dl$$

where l is the curvilinear abscissa of the boundary contour $\partial \Omega$ and $\vec{n}$ is the normal vector of this contour. By gathering the terms which are linked to the same displacement, we obtain

$$\int_v \sigma \delta \varepsilon \, dv = - \int_\Omega \left[\frac{\partial N_{\alpha\beta}}{\partial x_\beta} \delta u_\alpha^0 + \left(\frac{\partial M_{\alpha\beta}}{\partial x_\beta} - T_\alpha \right) \delta \phi_\alpha + \frac{\partial T_\alpha}{\partial x_\alpha} \delta w \right] dx_1 \, dx_2$$

$$+ \int_{\partial \Omega} (N_{\alpha\beta} \delta u_\alpha^0 n_\beta + M_{\alpha\beta} \delta \phi_\alpha n_\beta + T_\alpha n_\alpha \delta w) \, dl.$$

The kinematically admissible field is the solution when $\delta \pi = 0$. Then, the internal work is in equilibrium with the external work (assuming no dynamic work). The external work forces can be broken down into body forces and boundary forces. In the case of plates, body forces are surface forces comprising

- uniformly distributed extensional forces $p_\alpha(x_1, x_2)$
- uniformly distributed transverse forces $p_3(x_1, x_2)$
- uniformly distributed moments $m_\alpha(x_1, x_2)$.

Boundary forces are defined as

- extensional forces F_α
- transverse forces F_3
- bending and torsion moments C_α.

External work generated is then given by

$$\delta W_e = \int_\Omega (p_\alpha \delta u_\alpha^0 + m_\alpha \delta \phi_\alpha + p_3 \delta w)\, dx_1\, dx_2$$

$$+ \int_{\partial\Omega} (F_\alpha \delta u_\alpha^0 + C_\alpha \delta \phi_\alpha + F_3 \delta w)\, dl.$$

The maximum or minimum of the potential energy (derivative equal zero) gives the equations of equilibrium for the generalized forces:

$$\frac{\partial N_{\alpha\beta}}{\partial x_\beta} + p_\alpha = 0, \qquad \frac{\partial M_{\alpha\beta}}{\partial x_\beta} - T_\alpha + m_\alpha = 0, \qquad \frac{\partial T_\alpha}{\partial x_\alpha} + p_3 = 0$$

and the boundary conditions of the contour of the plate:

$$N_{\alpha\beta} n_\beta = F_\alpha, \qquad M_{\alpha\beta} n_\beta = C_\alpha, \qquad T_\alpha n_\alpha = F_3.$$

8.2.4 Laminate constitutive equations

Mindlin's theory assumes that the normal stress σ_3 is negligible in the thickness direction compared with other stress values. If this is so the constitutive law referred to the global coordinate (x_1, x_2, x_3) is given by relation (5.31):

$$\begin{Bmatrix} \sigma_1 \\ \sigma_2 \\ 0 \\ \sigma_4 \\ \sigma_5 \\ \sigma_6 \end{Bmatrix} = \begin{bmatrix} C_{11}(\theta) & C_{12}(\theta) & C_{13}(\theta) & 0 & 0 & C_{16}(\theta) \\ C_{12}(\theta) & C_{22}(\theta) & C_{23}(\theta) & 0 & 0 & C_{26}(\theta) \\ C_{13}(\theta) & C_{23}(\theta) & C_{33}(\theta) & 0 & 0 & C_{36}(\theta) \\ 0 & 0 & 0 & C_{44}(\theta) & C_{45}(\theta) & 0 \\ 0 & 0 & 0 & C_{45}(\theta) & C_{55}(\theta) & 0 \\ C_{16}(\theta) & C_{26}(\theta) & C_{36}(\theta) & 0 & 0 & C_{66}(\theta) \end{bmatrix} \begin{Bmatrix} \varepsilon_1 \\ \varepsilon_2 \\ \varepsilon_3 \\ \varepsilon_4 \\ \varepsilon_5 \\ \varepsilon_6 \end{Bmatrix}.$$

$$(8.12)$$

This constitutive law describes the behaviour of the ply and depends on the orientation of the global coordinates with respect to the material coordinate systems. In the following we are not going to repeat the angle θ but we are going to relate the kth layer to the k indices. Then combining the in-plane stresses and strains and out-of-plane stresses and strains, the relation (8.12) can be written in the global coordinate system as

$$\begin{Bmatrix} \sigma_1 \\ \sigma_2 \\ \sigma_6 \\ -- \\ \sigma_4 \\ \sigma_5 \end{Bmatrix} = \begin{bmatrix} Q_{11} & Q_{12} & Q_{16} & 0 & 0 \\ Q_{12} & Q_{22} & Q_{26} & 0 & 0 \\ Q_{16} & Q_{26} & Q_{66} & 0 & 0 \\ -- & -- & -- & -- & -- \\ 0 & 0 & 0 & C_{44} & C_{45} \\ 0 & 0 & 0 & C_{45} & C_{55} \end{bmatrix}_k \begin{Bmatrix} \varepsilon_1 \\ \varepsilon_2 \\ \varepsilon_6 \\ -- \\ \varepsilon_4 \\ \varepsilon_5 \end{Bmatrix} \qquad (8.13)$$

with

$$Q_{\alpha\beta} = C_{\alpha\beta} - \frac{C_{\alpha3}C_{\beta3}}{C_{33}} \qquad (\alpha, \beta = 1, 2, 6)$$

and

$$\varepsilon_3 = -\frac{1}{C_{33}}(C_{13}\varepsilon_1 + C_{23}\varepsilon_2 + C_{36}\varepsilon_6). \qquad (8.14)$$

The change of the behaviour, from layer to layer, leads to the change of stress in different materials and orientated layers. The extension stresses of the kth layer can be expressed as

$$\begin{Bmatrix} \sigma_1 \\ \sigma_2 \\ \sigma_6 \end{Bmatrix}_k = \begin{bmatrix} Q_{11} & Q_{12} & Q_{16} \\ Q_{12} & Q_{22} & Q_{26} \\ Q_{16} & Q_{26} & Q_{66} \end{bmatrix}_k \begin{Bmatrix} \varepsilon_1^0 \\ \varepsilon_2^0 \\ \varepsilon_6^0 \end{Bmatrix} + x_3 \begin{bmatrix} Q_{11} & Q_{12} & Q_{16} \\ Q_{12} & Q_{22} & Q_{26} \\ Q_{16} & Q_{26} & Q_{66} \end{bmatrix}_k \begin{Bmatrix} \kappa_1 \\ \kappa_2 \\ \kappa_6 \end{Bmatrix}.$$

$$(8.15)$$

The transversal shear stresses of the kth layer can be expressed as

$$\begin{Bmatrix} \sigma_4 \\ \sigma_5 \end{Bmatrix}_k = \begin{bmatrix} C_{44} & C_{45} \\ C_{45} & C_{55} \end{bmatrix}_k \begin{Bmatrix} \varepsilon_4 \\ \varepsilon_5 \end{Bmatrix}. \qquad (8.16)$$

We obtain the global constitutive law for the laminated plate by integration in the thickness direction. In this way we obtain the constitutive equations in terms of generalized forces and strains. According to equations introduced in section 8.2.2, we obtain the extensional force resultants:

$$N(x_1, x_2) = \sum_{k=1}^{n} \int_{x_3^k}^{x_3^{k+1}} [Q_k \varepsilon^0(x_1, x_2) + x_3 Q_k \kappa(x_1, x_2)] \, dx_3.$$

If we suppose the ply behaviour to be homogeneous, we obtain

$$N(x_1, x_2) = \sum_{k=1}^{n} \left[Q_k \varepsilon^0(x_1, x_2) \int_{x_3^k}^{x_3^{k+1}} dx_3 \right] + \sum_{k=1}^{n} \left[Q_k \kappa(x_1, x_2) \int_{x_3^k}^{x_3^{k+1}} x_3 \, dx_3 \right]$$

or, by summation of each ply,

$$N(x_1, x_2) = \left[\sum_{k=1}^{n} (x_3^{k+1} - x_3^k) Q_k \right] \varepsilon^0(x_1, x_2)$$

$$+ \left[\frac{1}{2} \sum_{k=1}^{n} [(x_3^{k+1})^2 - (x_3^k)^2] Q_k \right] \kappa(x_1, x_2)$$

which can be written globally as

$$N(x_1, x_2) = A\varepsilon^0(x_1, x_2) + B\kappa(x_1, x_2). \qquad (8.17)$$

The extensional stiffness matrix $[A]$ and the bending-extensional coupling stiffness matrix $[B]$ are defined in terms of the lamina stiffnesses Q_k:

$$A = \sum_{k=1}^{n}(x_3^{k+1} - x_3^k)Q_k, \qquad B = \sum_{k=1}^{n}\frac{1}{2}[(x_3^{k+1})^2 - (x_3^k)^2]Q_k \qquad (8.18)$$

or

$$A_{\alpha\beta} = \sum_{k=1}^{n}(x_3^{k+1} - x_3^k)(Q_{\alpha\beta})_k$$

$$B_{\alpha\beta} = \frac{1}{2}\sum_{k=1}^{n}[(x_3^{k+1})^2 - (x_3^k)^2](Q_{\alpha\beta})_k, \qquad \alpha, \beta = 1, 2, 6.$$

The moment resultants are expressed with relation (8.8).

The introduction of stresses leads to

$$M(x_1, x_2) = \sum_{k=1}^{n}\int_{x_3^k}^{x_3^{k+1}}[x_3 Q_k \varepsilon^0(x_1, x_2) + x_3^2 Q_k(x_1, x_2)]\,dx_3$$

or by summation over all the plies:

$$M(x_1, x_2) = \left[\frac{1}{2}\sum_{k=1}^{n}[(x_3^{k+1})^2 - (x_3^k)^2]Q_k\right]\varepsilon^0(x_1, x_2)$$

$$+ \left[\frac{1}{3}\sum_{k=1}^{n}[(x_3^{k+1})^3 - (x_3^k)^3]Q_k\right]\kappa(x_1, x_2).$$

Therefore

$$M(x_1, x_2) = B\varepsilon^0(x_1, x_2) + D\kappa(x_1, x_2). \qquad (8.19)$$

The matrix $[D]$ characterizes the bending stiffness which is defined in terms of the Q_k lamina as

$$D = \frac{1}{3}\sum_{k=1}^{n}[(x_3^{k+1})^3 - (x_3^k)^3]Q_k \quad \text{or} \quad D_{\alpha\beta} = \frac{1}{3}\sum_{k=1}^{n}[(x_3^{k+1})^3 - (x_3^k)^3](Q_{\alpha\beta})_k. \qquad (8.20)$$

Note that the Q_k and therefore the A, B and D are, in general, functions of position (x_1, x_2).

Transverse force resultants are defined by using (8.9):

$$T(x_1, x_2) = \begin{Bmatrix} T_4 \\ T_5 \end{Bmatrix} = \sum_{k=1}^{n}\int_{x_3^k}^{x_3^{k+1}}\begin{Bmatrix} \sigma_4 \\ \sigma_5 \end{Bmatrix}_k dx_3$$

where

$$\begin{Bmatrix} T_4 \\ T_5 \end{Bmatrix} = \begin{bmatrix} H_{44} & H_{45} \\ H_{45} & H_{55} \end{bmatrix}\begin{Bmatrix} \varepsilon_4 \\ \varepsilon_5 \end{Bmatrix}. \qquad (8.21)$$

H_{44}, H_{45} and H_{55} are the transverse shear stiffnesses:

$$H_{\gamma\delta} = \sum_{k=1}^{n} (x_3^{k+1} - x_3^k)(C_{\gamma\delta})_k, \qquad \gamma, \delta = 4, 5. \qquad (8.22)$$

The force resultants are related to the membrane and flexural strains by

$$\begin{Bmatrix} N_1 \\ N_2 \\ N_6 \\ M_1 \\ M_2 \\ M_6 \\ T_4 \\ T_5 \end{Bmatrix} = \begin{bmatrix} A_{11} & A_{12} & A_{16} & B_{11} & B_{12} & B_{16} & 0 & 0 \\ A_{21} & A_{22} & A_{26} & B_{21} & B_{22} & B_{26} & 0 & 0 \\ A_{61} & A_{62} & A_{66} & B_{61} & B_{62} & B_{66} & 0 & 0 \\ B_{11} & B_{12} & B_{16} & D_{11} & D_{12} & D_{16} & 0 & 0 \\ B_{21} & B_{22} & B_{26} & D_{21} & D_{22} & D_{26} & 0 & 0 \\ B_{61} & B_{62} & B_{66} & D_{61} & D_{62} & D_{66} & 0 & 0 \\ 0 & 0 & 0 & 0 & 0 & 0 & H_{44} & H_{45} \\ 0 & 0 & 0 & 0 & 0 & 0 & H_{54} & H_{55} \end{bmatrix} \begin{Bmatrix} \varepsilon_1^0 \\ \varepsilon_2^0 \\ \varepsilon_6^0 \\ \kappa_1 \\ \kappa_2 \\ \kappa_6 \\ \varepsilon_4 \\ \varepsilon_5 \end{Bmatrix}. \qquad (8.23)$$

The coupling coefficients B_{ij} decrease in magnitude (hence the effect of coupling decreases) with the increase in the number of layers (for the same total thickness of the plate) for antisymmetric laminates, and are zero for symmetric laminates.

Relation (8.23) can be written in a concise way such that

$$\begin{Bmatrix} N \\ M \\ T \end{Bmatrix} = \begin{bmatrix} A & B & 0 \\ B & D & 0 \\ 0 & 0 & H \end{bmatrix} \begin{Bmatrix} \varepsilon^0 \\ \kappa \\ \varepsilon^T \end{Bmatrix}.$$

8.2.5   Classical laminate theory

Classical laminate theory uses the displacement field of Mindlin (8.2) with one additional hypothesis, which is that a normal line to the midplane remains normal to this plane after deformation of the plate. It follows that the transverse shear strains $\varepsilon_4, \varepsilon_5$ are zero. According to (8.4), we obtain

$$\phi_1 = -\frac{\partial w}{\partial x_1}, \qquad \phi_2 = -\frac{\partial w}{\partial x_2}.$$

This hypothesis is similar to a Kirchhoff hypothesis applied to the case of thin plates, which requires the displacements (u_1, u_2, u_3) to be such that

$$u_1 = u_1^0 - x_3 \frac{\partial w}{\partial x_1}, \qquad u_2 = u_2^0 - x_3 \frac{\partial w}{\partial x_2}, \qquad u_3 = w \qquad (8.24)$$

where $u_\alpha^0(x_1, x_2)$, $\alpha = 1, 2$ are the extensional displacements of a material point on the midplane $(x_1, x_2, 0)$ and w the deflection always constant through the thickness. Note that once the midplane displacements are known, the displacements of any arbitrary point in the three-dimensional continuum can be determined using equation (8.24).

The displacement field allows the strain fields to be obtained so that

$$\varepsilon_1 = \frac{\partial u_1^0}{\partial x_1} - x_3 \frac{\partial^2 w}{\partial x_1^2}$$

$$\varepsilon_2 = \frac{\partial u_2^0}{\partial x_2} - x_3 \frac{\partial^2 w}{\partial x_3^2}$$

$$\varepsilon_6 = \left(\frac{\partial u_1^0}{\partial x_2} + \frac{\partial u_2^0}{\partial x_1} \right) - 2x_3 \frac{\partial^2 w}{\partial x_1 \partial x_2}$$

$$\varepsilon_3 = \varepsilon_4 = \varepsilon_5 = 0.$$

(8.25)

The non-zero strains reduces to $\varepsilon_1, \varepsilon_2, \varepsilon_6$. The extension strains, which are only functions of the in-plane displacements u_1^0 and u_2^0, are

$$\left\{ \begin{array}{c} \varepsilon_1^0 \\ \varepsilon_2^0 \\ \varepsilon_6^0 \end{array} \right\} = \left\{ \begin{array}{c} \dfrac{\partial u_1^0}{\partial x_1} \\[2mm] \dfrac{\partial u_2^0}{\partial x_2} \\[2mm] \dfrac{\partial u_1^0}{\partial x_2} + \dfrac{\partial u_2^0}{\partial x_1} \end{array} \right\}$$

(8.26)

and the bending and torsion strains are

$$\left\{ \begin{array}{c} \varepsilon_1^f \\ \varepsilon_2^f \\ \varepsilon_6^f \end{array} \right\} = \left\{ \begin{array}{c} -x_3 \dfrac{\partial^2 w}{\partial x_1^2} \\[2mm] -x_3 \dfrac{\partial^2 w}{\partial x_2^2} \\[2mm] -2x_3 \dfrac{\partial^2 w}{\partial x_1 \partial x_2} \end{array} \right\}.$$

(8.27)

If we introduce curvatures,

$$\left\{ \begin{array}{c} \kappa_1 \\ \kappa_2 \\ \kappa_6 \end{array} \right\} = \left\{ \begin{array}{c} -\dfrac{\partial^2 w}{\partial x_1^2} \\[2mm] -\dfrac{\partial^2 w}{\partial x_2^2} \\[2mm] -2 \dfrac{\partial^2 w}{\partial x_1 \partial x_2} \end{array} \right\}$$

(8.28)

the strain field given by classical plate theory is obtained:

$$\left\{ \begin{array}{c} \varepsilon_1 \\ \varepsilon_2 \\ \varepsilon_6 \end{array} \right\} = \left\{ \begin{array}{c} \varepsilon_1^0 \\ \varepsilon_2^0 \\ \varepsilon_6^0 \end{array} \right\} + x_3 \left\{ \begin{array}{c} \kappa_1 \\ \kappa_2 \\ \kappa_6 \end{array} \right\}.$$

(8.29)

8.2.6.   Constitutive behaviour in the laminate

In the case of classical laminate theory hypotheses the constitutive behaviour reduces to

$$
\left\{ \begin{array}{c} \sigma_1 \\ \sigma_2 \\ \sigma_6 \end{array} \right\}_k = \begin{bmatrix} Q_{11} & Q_{12} & Q_{16} \\ Q_{12} & Q_{22} & Q_{26} \\ Q_{16} & Q_{26} & Q_{66} \end{bmatrix}_k \left\{ \begin{array}{c} \varepsilon_1 \\ \varepsilon_2 \\ \varepsilon_6 \end{array} \right\}.
$$

If we now introduce membrane strains and flexural (bending strains), we find again relation (8.15) without (8.16) because the shear strains are zero:

$$
\left\{ \begin{array}{c} \sigma_1 \\ \sigma_2 \\ \sigma_6 \end{array} \right\}_k = \begin{bmatrix} Q_{11} & Q_{12} & Q_{16} \\ Q_{12} & Q_{22} & Q_{26} \\ Q_{16} & Q_{26} & Q_{66} \end{bmatrix}_k \left\{ \begin{array}{c} \varepsilon_1^0 \\ \varepsilon_2^0 \\ \varepsilon_6^0 \end{array} \right\} + x_3 \begin{bmatrix} Q_{11} & Q_{12} & Q_{16} \\ Q_{12} & Q_{22} & Q_{26} \\ Q_{16} & Q_{26} & Q_{66} \end{bmatrix}_k \left\{ \begin{array}{c} \kappa_1 \\ \kappa_2 \\ \kappa_6 \end{array} \right\}.
$$

$$(8.30)$$

After integration through the thickness, we find relation (8.23) without the transverse force resultants:

$$
\left\{ \begin{array}{c} N_1 \\ N_2 \\ N_6 \\ M_1 \\ M_2 \\ M_6 \end{array} \right\} = \begin{bmatrix} A_{11} & A_{12} & A_{16} & B_{11} & B_{12} & B_{16} \\ A_{21} & A_{22} & A_{26} & B_{21} & B_{22} & B_{26} \\ A_{61} & A_{62} & A_{66} & B_{61} & B_{62} & B_{66} \\ B_{11} & B_{12} & B_{16} & D_{11} & D_{12} & D_{16} \\ B_{21} & B_{22} & B_{26} & D_{21} & D_{22} & D_{26} \\ B_{61} & B_{62} & B_{66} & D_{61} & D_{62} & D_{66} \end{bmatrix} \left\{ \begin{array}{c} \varepsilon_1^0 \\ \varepsilon_2^0 \\ \varepsilon_6^0 \\ \kappa_1 \\ \kappa_2 \\ \kappa_6 \end{array} \right\}
$$

$$(8.31)$$

where $[A]$, $[B]$, $[D]$ are identical, are those in relations (8.18) and (8.20).

8.3   Three-dimensional approach

The proposed method for a three-dimensional thick structure is to isolate, when possible, a stacking sequence, the repetition through the thickness of which can describe the complete laminate. This elementary staking sequence can be used as a periodic elementary cell representative of the whole laminated structure. It is then possible to use the 'average method' for a periodic medium described in chapter 6 to evaluate the equivalent homogenized medium of the laminated structure. Let us call Y, this elementary cell, h_Y its height, composed of n i-plies with the height of each of them, h_i. If s represents the number of cells through the thickness and h the total height of the laminate, we obtain

$$
h = s h_Y
$$

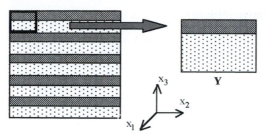

Figure 8.6. Thick laminate and its elementary cell.

Then the problem P_Y is described by equation (6.79) as

$$P_Y \begin{cases} \operatorname{div} \sigma = 0 \\ \sigma = C\varepsilon(u) \\ \sigma, \varepsilon \quad Y \text{ periodic} \\ \langle \varepsilon(u) \rangle = E \end{cases}$$

and we have to evaluate the stiffness matrix as

$$\langle \sigma \rangle = C^{\mathrm{hom}} E.$$

Each ply of the laminate is supposed to be homogeneous and monoclinic as defined in chapter 5 by relation (5.8)

$$\begin{Bmatrix} \sigma_1 \\ \sigma_2 \\ \sigma_3 \\ \sigma_4 \\ \sigma_5 \\ \sigma_6 \end{Bmatrix} = \begin{Bmatrix} C_{11} & C_{12} & C_{13} & 0 & 0 & C_{16} \\ C_{12} & C_{22} & C_{23} & 0 & 0 & C_{26} \\ C_{13} & C_{23} & C_{33} & 0 & 0 & C_{36} \\ 0 & 0 & 0 & C_{44} & C_{45} & 0 \\ 0 & 0 & 0 & C_{45} & C_{55} & 0 \\ C_{16} & C_{26} & C_{36} & 0 & 0 & C_{66} \end{Bmatrix} \begin{Bmatrix} \varepsilon_1 \\ \varepsilon_2 \\ \varepsilon_3 \\ \varepsilon_4 \\ \varepsilon_5 \\ \varepsilon_6 \end{Bmatrix}.$$

If the plate is large enough, its behaviour does not depend on the position of the cell in the plane (x_1, x_2). Thus the behaviour does not then depend on the in-plane position and we can choose a unit length cell in these directions. Consequently, all the variables depend only on x_3, as do the stresses and strains:

$$\sigma = \sigma(x_3), \qquad \varepsilon = \varepsilon(x_3). \qquad (8.32)$$

Obeying these conditions, the equilibrium equations give $\partial \sigma_{ij}/\partial x_j = 0$ $(i, j = 1, 2, 3)$

$$\sigma_5 = C^{\mathrm{te}}, \qquad \sigma_4 = C^{\mathrm{te}}, \qquad \sigma_3 = C^{\mathrm{te}}.$$

The compatibility equations give the relations

$$\frac{\mathrm{d}^2 \varepsilon_1}{\mathrm{d}x_3^2} = 0, \qquad \frac{\mathrm{d}^2 \varepsilon_2}{\mathrm{d}x_3^2} = 0, \qquad \frac{\mathrm{d}^2 \varepsilon_6}{\mathrm{d}x_3^2} = 0.$$

After integrating we obtain

$$\varepsilon_1 = \alpha_1 x_3 + \beta_1, \qquad \varepsilon_2 = \alpha_2 x_3 + \beta_2, \qquad \varepsilon_6 = \alpha_6 x_3 + \beta_6.$$

Periodicity conditions oblige $\alpha_1, \alpha_2, \alpha_6$, to be zero and so $\varepsilon_1, \varepsilon_2, \varepsilon_6$ are constants.

We can separate the strains and stresses which are constants from those depending on x_3 with a relation which is more explicit:

$$\vec{b}(x_3) = L(x_3)\vec{a} \tag{8.33}$$

with $\vec{a} = (\varepsilon_1, \varepsilon_2, \varepsilon_6, \sigma_5, \sigma_4, \sigma_3)$ and $\vec{b} = (\varepsilon_5, \varepsilon_4, \varepsilon_3, \sigma_1, \sigma_2, \sigma_6)$. Again using the stiffness matrix which characterizes the monoclinic laminate, we obtain

$$\sigma_1 = C_{11}\varepsilon_1 + C_{12}\varepsilon_2 + C_{13}\varepsilon_3 + C_{16}\varepsilon_6$$

$$\sigma_2 = C_{12}\varepsilon_1 + C_{22}\varepsilon_2 + C_{23}\varepsilon_3 + C_{26}\varepsilon_6$$

$$\sigma_3 = C_{13}\varepsilon_1 + C_{23}\varepsilon_2 + C_{33}\varepsilon_3 + C_{36}\varepsilon_6$$

$$\sigma_4 = C_{44}\varepsilon_4 + C_{45}\varepsilon_5$$

$$\sigma_5 = C_{45}\varepsilon_4 + C_{55}\varepsilon_5$$

$$\sigma_6 = C_{16}\varepsilon_1 + C_{26}\varepsilon_2 + C_{36}\varepsilon_3 + C_{66}\varepsilon_6.$$

The third relation gives

$$\varepsilon_3 = -\frac{C_{13}}{C_{33}}\varepsilon_1 - \frac{C_{23}}{C_{33}}\varepsilon_2 - \frac{C_{36}}{C_{33}}\varepsilon_6 + \frac{1}{C_{33}}\sigma_3. \tag{8.34}$$

From equation (8.34), we can now express $\sigma_1, \sigma_2, \sigma_6$ in a different manner such that

$$\sigma_1 = \left(C_{11} - \frac{C_{13}^2}{C_{33}}\right)\varepsilon_1 + \left(C_{12} - \frac{C_{13}C_{23}}{C_{33}}\right)\varepsilon_2 + \left(C_{16} - \frac{C_{13}C_{36}}{C_{33}}\right)\varepsilon_6 + \frac{C_{13}}{C_{33}}\sigma_3$$

$$\sigma_2 = \left(C_{12} - \frac{C_{13}C_{23}}{C_{33}}\right)\varepsilon_1 + \left(C_{22} - \frac{C_{23}^2}{C_{33}}\right)\varepsilon_2 + \left(C_{26} - \frac{C_{23}C_{36}}{C_{33}}\right)\varepsilon_6 + \frac{C_{23}}{C_{33}}\sigma_3$$

$$\sigma_6 = \left(C_{16} - \frac{C_{13}C_{36}}{C_{33}}\right)\varepsilon_1 + \left(C_{26} - \frac{C_{23}C_{36}}{C_{33}}\right)\varepsilon_2 + \left(C_{66} - \frac{C_{36}^2}{C_{33}}\right)\varepsilon_6 + \frac{C_{36}}{C_{33}}\sigma_3.$$

$$\tag{8.35}$$

By inverting the fourth and the fifth relations, we obtain

$$\varepsilon_4 = \frac{1}{C_{44}C_{55} - C_{45}^2}(C_{55}\sigma_4 - C_{45}\sigma_5)$$

$$\tag{8.36}$$

$$\varepsilon_5 = \frac{1}{C_{44}C_{55} - C_{45}^2}(-C_{45}\sigma_4 + C_{44}\sigma_5).$$

Equations (8.34), (8.35) and (8.36) can be written in a similar manner to expression (8.33) which gives

$$\vec{b}(x_3) = \begin{bmatrix} 0 & 0 & 0 & L_{14} & L_{15} & 0 \\ 0 & 0 & 0 & L_{24} & L_{25} & 0 \\ L_{31} & L_{32} & L_{33} & 0 & 0 & L_{36} \\ L_{41} & L_{42} & L_{43} & 0 & 0 & L_{46} \\ L_{51} & L_{52} & L_{53} & 0 & 0 & L_{56} \\ L_{61} & L_{62} & L_{63} & 0 & 0 & L_{66} \end{bmatrix} \vec{a}.$$

We can then calculate $\langle L(x_3) \rangle$ as the average of the former relation (remember that $\vec{a}$ is a constant):

$$\langle \vec{b}(x_3) \rangle = \langle L(x_3) \rangle \vec{a}.$$

Then

$$\begin{Bmatrix} \langle e_5 \rangle \\ \langle e_4 \rangle \\ \langle e_3 \rangle \\ \langle \sigma_1 \rangle \\ \langle \sigma_2 \rangle \\ \langle \sigma_6 \rangle \end{Bmatrix} = \begin{bmatrix} 0 & 0 & 0 & \langle L_{14} \rangle & \langle L_{15} \rangle & 0 \\ 0 & 0 & 0 & \langle L_{24} \rangle & \langle L_{25} \rangle & 0 \\ \langle L_{31} \rangle & \langle L_{32} \rangle & \langle L_{33} \rangle & 0 & 0 & \langle L_{36} \rangle \\ \langle L_{41} \rangle & \langle L_{42} \rangle & \langle L_{43} \rangle & 0 & 0 & \langle L_{46} \rangle \\ \langle L_{51} \rangle & \langle L_{52} \rangle & \langle L_{53} \rangle & 0 & 0 & \langle L_{56} \rangle \\ \langle L_{61} \rangle & \langle L_{62} \rangle & \langle L_{63} \rangle & 0 & 0 & \langle L_{66} \rangle \end{bmatrix} \begin{Bmatrix} \varepsilon_1 \\ \varepsilon_2 \\ \varepsilon_6 \\ \sigma_5 \\ \sigma_4 \\ \sigma_3 \end{Bmatrix}$$

with

$$L_{14} = \frac{C_{44}}{C_{44}C_{55} - C_{45}^2}, \qquad L_{15} = -\frac{C_{45}}{C_{44}C_{55} - C_{45}^2}$$

$$L_{24} = -\frac{C_{45}}{C_{44}C_{55} - C_{45}^2}, \qquad L_{25} = \frac{C_{55}}{C_{44}C_{55} - C_{45}^2}$$

$$L_{31} = -\frac{C_{13}}{C_{33}}, \qquad L_{32} = -\frac{C_{23}}{C_{33}}, \qquad L_{33} = -\frac{C_{36}}{C_{33}}, \qquad L_{36} = \frac{1}{C_{33}}$$

$$L_{41} = C_{11} - \frac{C_{13}^2}{C_{33}}, \qquad L_{42} = C_{12} - \frac{C_{13}C_{23}}{C_{33}}, \qquad L_{43} = C_{16} - \frac{C_{13}C_{36}}{C_{33}}$$

$$L_{46} = \frac{C_{13}}{C_{33}}, \qquad L_{51} = C_{12} - \frac{C_{13}C_{23}}{C_{33}}, \qquad L_{52} = C_{22} - \frac{C_{23}^2}{C_{33}}$$

$$L_{53} = C_{26} - \frac{C_{23}C_{36}}{C_{33}}, \qquad L_{56} = \frac{C_{23}}{C_{33}}, \qquad L_{61} = C_{16} - \frac{C_{13}C_{36}}{C_{33}}$$

$$L_{62} = C_{26} - \frac{C_{23}C_{36}}{C_{33}}, \qquad L_{63} = C_{66} - \frac{C_{36}^2}{C_{33}}, \qquad L_{66} = \frac{C_{36}}{C_{33}}.$$

We have now to make explicit $\langle \sigma \rangle = C^{\text{hom}} \langle \varepsilon \rangle$ to obtain the homogenized behaviour of the laminate.

If we now write the homogenized relations

$$\langle \varepsilon_5 \rangle = \langle L_{14} \rangle \langle \sigma_5 \rangle + \langle L_{15} \rangle \langle \sigma_4 \rangle$$

$$\langle \varepsilon_4 \rangle = \langle L_{24} \rangle \langle \sigma_5 \rangle + \langle L_{25} \rangle \langle \sigma_4 \rangle$$

we obtain

$$\langle \sigma_5 \rangle = \frac{1}{\langle L_{14} \rangle \langle L_{25} \rangle - \langle L_{24} \rangle \langle L_{15} \rangle} (\langle L_{25} \rangle \langle \varepsilon_5 \rangle - \langle L_{15} \rangle \langle \varepsilon_4 \rangle)$$

$$\langle \sigma_4 \rangle = \frac{1}{\langle L_{14} \rangle \langle L_{25} \rangle - \langle L_{24} \rangle \langle L_{15} \rangle} (-\langle L_{24} \rangle \langle \varepsilon_5 \rangle + \langle L_{14} \rangle \langle \varepsilon_4 \rangle).$$

Similarly,

$$\langle \varepsilon_3 \rangle = \langle L_{31} \rangle \langle \varepsilon_1 \rangle + \langle L_{32} \rangle \langle \varepsilon_2 \rangle + \langle L_{33} \rangle \langle \varepsilon_6 \rangle + \langle L_{36} \rangle \langle \sigma_3 \rangle$$

and so

$$\langle \sigma_3 \rangle = -\frac{\langle L_{31} \rangle}{\langle L_{36} \rangle} \langle \varepsilon_1 \rangle - \frac{\langle L_{32} \rangle}{\langle L_{36} \rangle} \langle \varepsilon_2 \rangle - \frac{\langle L_{33} \rangle}{\langle L_{36} \rangle} \langle \varepsilon_6 \rangle + \frac{1}{\langle L_{36} \rangle} \langle \varepsilon_3 \rangle.$$

Similarly,

$$\langle \sigma_1 \rangle = \langle L_{41} \rangle \langle \varepsilon_1 \rangle + \langle L_{42} \rangle \langle \varepsilon_2 \rangle + \langle L_{43} \rangle \langle \varepsilon_6 \rangle + \langle L_{46} \rangle \langle \sigma_3 \rangle$$

then

$$\langle \sigma_1 \rangle = \left(\langle L_{41} \rangle - \frac{\langle L_{46} \rangle}{\langle L_{36} \rangle} \langle L_{31} \rangle \right) \langle \varepsilon_1 \rangle + \left(\langle L_{42} \rangle - \frac{\langle L_{46} \rangle}{\langle L_{36} \rangle} \langle L_{32} \rangle \right) \langle \varepsilon_2 \rangle$$

$$+ \frac{\langle L_{46} \rangle}{\langle L_{36} \rangle} \langle \varepsilon_3 \rangle + \left(\langle L_{43} \rangle - \frac{\langle L_{46} \rangle}{\langle L_{36} \rangle} \langle L_{33} \rangle \right) \langle \varepsilon_6 \rangle.$$

Also we can write

$$\langle \sigma_2 \rangle = \langle L_{51} \rangle \langle \varepsilon_1 \rangle + \langle L_{52} \rangle \langle \varepsilon_2 \rangle + \langle L_{53} \rangle \langle \varepsilon_6 \rangle + \langle L_{56} \rangle \langle \sigma_3 \rangle$$

giving

$$\langle \sigma_2 \rangle = \left(\langle L_{51} \rangle - \frac{\langle L_{56} \rangle}{\langle L_{36} \rangle} \langle L_{31} \rangle \right) \langle \varepsilon_1 \rangle + \left(\langle L_{52} \rangle - \frac{\langle L_{56} \rangle}{\langle L_{36} \rangle} \langle L_{32} \rangle \right) \langle \varepsilon_2 \rangle$$

$$+ \frac{\langle L_{56} \rangle}{\langle L_{36} \rangle} \langle \varepsilon_3 \rangle + \left(\langle L_{53} \rangle - \frac{\langle L_{56} \rangle}{\langle L_{36} \rangle} \langle L_{33} \rangle \right) \langle \varepsilon_6 \rangle.$$

Further, and with the same approach,

$$\langle \sigma_6 \rangle = \langle L_{61} \rangle \langle \varepsilon_1 \rangle + \langle L_{62} \rangle \langle \varepsilon_2 \rangle + \langle L_{63} \rangle \langle \varepsilon_6 \rangle + \langle L_{66} \rangle \langle \sigma_3 \rangle$$

so that

$$\langle \sigma_6 \rangle = \left(\langle L_{61} \rangle - \frac{\langle L_{66} \rangle}{\langle L_{36} \rangle} \langle L_{31} \rangle \right) \langle \varepsilon_1 \rangle + \left(\langle L_{62} \rangle - \frac{\langle L_{66} \rangle}{\langle L_{36} \rangle} \langle L_{32} \rangle \right) \langle \varepsilon_2 \rangle$$
$$+ \frac{\langle L_{66} \rangle}{\langle L_{36} \rangle} \langle \varepsilon_3 \rangle + \left(\langle L_{63} \rangle - \frac{\langle L_{66} \rangle}{\langle L_{36} \rangle} \langle L_{33} \rangle \right) \langle \varepsilon_6 \rangle.$$

We therefore obtain all the terms of the homogenized stiffness matrix C^{hom} which are:

$$C_{11}^{\text{hom}} = \left\langle C_{11} - \frac{C_{13}^2}{C_{33}} \right\rangle + \frac{\left\langle \dfrac{C_{13}}{C_{33}} \right\rangle^2}{\left\langle \dfrac{1}{C_{33}} \right\rangle}$$

$$C_{12}^{\text{hom}} = C_{21}^{\text{hom}} = \left\langle C_{12} - \frac{C_{13} C_{23}}{C_{33}} \right\rangle + \frac{\left\langle \dfrac{C_{13}}{C_{33}} \right\rangle \left\langle \dfrac{C_{23}}{C_{33}} \right\rangle}{\left\langle \dfrac{1}{C_{33}} \right\rangle}$$

$$C_{13}^{\text{hom}} = C_{31}^{\text{hom}} = \frac{\left\langle \dfrac{C_{13}}{C_{33}} \right\rangle}{\left\langle \dfrac{1}{C_{33}} \right\rangle}$$

$$C_{16}^{\text{hom}} = C_{61}^{\text{hom}} = \left\langle C_{16} - \frac{C_{13} C_{63}}{C_{33}} \right\rangle + \frac{\left\langle \dfrac{C_{63}}{C_{33}} \right\rangle \left\langle \dfrac{C_{13}}{C_{33}} \right\rangle}{\left\langle \dfrac{1}{C_{33}} \right\rangle}$$

$$C_{22}^{\text{hom}} = \left\langle C_{22} - \frac{C_{23}^2}{C_{33}} \right\rangle + \frac{\left\langle \dfrac{C_{23}}{C_{33}} \right\rangle^2}{\left\langle \dfrac{1}{C_{33}} \right\rangle}$$

$$C_{23}^{\text{hom}} = C_{32}^{\text{hom}} = \frac{\left\langle \dfrac{C_{23}}{C_{33}} \right\rangle}{\left\langle \dfrac{1}{C_{33}} \right\rangle}; \qquad C_{33}^{\text{hom}} = \frac{1}{\left\langle \dfrac{1}{C_{33}} \right\rangle}$$

$$C_{26}^{\text{hom}} = C_{62}^{\text{hom}} = \left\langle C_{26} - \frac{C_{23} C_{63}}{C_{33}} \right\rangle + \frac{\left\langle \dfrac{C_{63}}{C_{33}} \right\rangle \left\langle \dfrac{C_{23}}{C_{33}} \right\rangle}{\left\langle \dfrac{1}{C_{33}} \right\rangle}$$

$$C_{36}^{\text{hom}} = C_{63}^{\text{hom}} = \frac{\left\langle \dfrac{C_{36}}{C_{33}} \right\rangle}{\left\langle \dfrac{1}{C_{33}} \right\rangle}$$

$$C_{44}^{\text{hom}} = \frac{\left\langle \dfrac{C_{44}}{C_{44}C_{55} - C_{45}^2} \right\rangle}{\left\langle \dfrac{C_{44}}{C_{44}C_{55} - C_{45}^2} \right\rangle \left\langle \dfrac{C_{55}}{C_{44}C_{55} - C_{45}^2} \right\rangle - \left\langle \dfrac{C_{45}}{C_{44}C_{55} - C_{45}^2} \right\rangle^2}$$

$$C_{45}^{\text{hom}} = \frac{\left\langle \dfrac{C_{45}}{C_{44}C_{55} - C_{45}^2} \right\rangle}{\left\langle \dfrac{C_{44}}{C_{44}C_{55} - C_{45}^2} \right\rangle \left\langle \dfrac{C_{55}}{C_{44}C_{55} - C_{45}^2} \right\rangle - \left\langle \dfrac{C_{45}}{C_{44}C_{55} - C_{45}^2} \right\rangle^2}$$

$$C_{55}^{\text{hom}} = \frac{\left\langle \dfrac{C_{55}}{C_{44}C_{55} - C_{45}^2} \right\rangle}{\left\langle \dfrac{C_{44}}{C_{44}C_{55} - C_{45}^2} \right\rangle \left\langle \dfrac{C_{55}}{C_{44}C_{55} - C_{45}^2} \right\rangle - \left\langle \dfrac{C_{45}}{C_{44}C_{55} - C_{45}^2} \right\rangle^2}$$

$$C_{66}^{\text{hom}} = \left\langle C_{66} - \frac{C_{63}^2}{C_{33}} \right\rangle + \frac{\left\langle \dfrac{C_{63}}{C_{33}} \right\rangle^2}{\left\langle \dfrac{1}{C_{33}} \right\rangle}.$$

8.4   Conclusions

To conclude this chapter, the purpose of which was to show the ease with which laminate theories can be used, it is also necessary to understand the limits of these theories. The structures are always considered to be plane plates. With the increase in use of composite materials, such a geometrical limitation is too restrictive. Manufactures have to increasingly design thick structures which have to be able to carry large loads, as in the case of primary structures. In the aeronautical field, composite parts in the future Airbus A380 will represent more than 22% of the overall weight of the airframe, without the engines and landing gear. The Boeing 7E7 is planned to have 50% of its weight as composite material. In both examples the composites are used for important primary structures having considerable thicknesses. The continual increase in composite use in successive generations of aircraft results from the accumulating knowledge and experience acquired in the applications of composite materials to demanding applications. The consequence is that metallic load-bearing structures are being replaced by advanced composites. This increase is mirrored in other areas such as naval structures so that large minesweepers are now routinely made with thick composite hulls, with advantages over traditional materials which are not limited to mechanical performance but which also include a lowering in maintenance costs.

We have limited this chapter to considerations of first-order kinematic fields. Even more complicated fields, which often need finite element computation, do not adequately describe the edges of the structures, parts which contain holes, bonded and other assemblies, and other structures. As a consequence the use of three-dimensional finite element calculations is gaining increasing acceptance as it proves to be more accurate for high gradients zones where the stress state is completely three-dimensional and for which classical theory is not appropriate. Three-dimensional calculations have the advantage of making no hypotheses and they lead to good calculated approximations of the structures considered. This way of calculating the response of composite materials to applied stresses is being facilitated by the ever-increasing computational power available with modern computers which permits complex calculations in shorter and shorter times and at low cost.

8.5 Revision exercises

1. If the ply material properties are

$$E_1 = 139\,300 \text{ MPa in the fibre direction}$$

$$E_2 = 19\,100 \text{ MPa transversaly to the fibres}$$

$$\nu_{12} = 0.36 \text{ Poisson's ratio}$$

$$G = 4340 \text{ MPa shear modulus}$$

give the in-plane stiffness matrix in the principle coordinate system of the ply (the plane of the ply is 1-2).

Answer: In the principal reference frame system, the orthotropic behaviour can be written as

$$\begin{Bmatrix} \sigma_1^k \\ \sigma_2^k \\ \sigma_6^k \end{Bmatrix} = \begin{bmatrix} 142.2 & 7.01 & 0 \\ 7.01 & 19.5 & 0 \\ 0 & 0 & 4.34 \end{bmatrix} \begin{Bmatrix} \varepsilon_1^k \\ \varepsilon_2^k \\ \varepsilon_6^k \end{Bmatrix}$$

where σ is in GPa and ε is non-dimensional.

2. If the ply properties are those of the former exercice, give in the case of $[0, 90, \pm 45]_s$ the in-plane stiffnesses of each ply referenced in the global coordinate system.

Answer: If the stresses evaluated in the principal frame system are marked k and k' in the global coordinate system of the laminated structure, we obtain k and k' for the studied laminated scheme:

Orientation of plies	k/k'
0 ply	1 8
90 ply	2 7
+45 ply	3 6
−45 ply	4 5

In the 0° plies the in-plane behaviour can be written

$$\begin{Bmatrix} \sigma_1^{1'} \\ \sigma_2^{1'} \\ \sigma_6^{1'} \end{Bmatrix} = \begin{bmatrix} Q_{11}^{1'} & Q_{12}^{1'} & 0 \\ Q_{12}^{1'} & Q_{22}^{1'} & 0 \\ 0 & 0 & Q_{66}^{1'} \end{bmatrix} \begin{Bmatrix} \varepsilon_1^{1'} \\ \varepsilon_2^{1'} \\ \varepsilon_6^{1'} \end{Bmatrix} = \begin{bmatrix} 142.2 & 7.01 & 0 \\ 7.01 & 19.5 & 0 \\ 0 & 0 & 4.34 \end{bmatrix} \begin{Bmatrix} \varepsilon_1^{1'} \\ \varepsilon_2^{1'} \\ \varepsilon_6^{1'} \end{Bmatrix}$$

$$\begin{Bmatrix} \sigma_1^{8'} \\ \sigma_2^{8'} \\ \sigma_6^{8'} \end{Bmatrix} = \begin{bmatrix} Q_{11}^{8'} & Q_{12}^{8'} & 0 \\ Q_{12}^{8'} & Q_{22}^{8'} & 0 \\ 0 & 0 & Q_{66}^{8'} \end{bmatrix} \begin{Bmatrix} \varepsilon_1^{8'} \\ \varepsilon_2^{8'} \\ \varepsilon_6^{8'} \end{Bmatrix} = \begin{bmatrix} 142.2 & 7.01 & 0 \\ 7.01 & 19.5 & 0 \\ 0 & 0 & 4.34 \end{bmatrix} \begin{Bmatrix} \varepsilon_1^{8'} \\ \varepsilon_2^{8'} \\ \varepsilon_6^{8'} \end{Bmatrix}$$

In the 90° plies the in-plane behaviour can be written

$$\begin{Bmatrix} \sigma_1^{2'} \\ \sigma_2^{2'} \\ \sigma_6^{2'} \end{Bmatrix} = \begin{bmatrix} Q_{11}^{2'} & Q_{12}^{2'} & 0 \\ Q_{12}^{2'} & Q_{22}^{2'} & 0 \\ 0 & 0 & Q_{66}^{2'} \end{bmatrix} \begin{Bmatrix} \varepsilon_1^{2'} \\ \varepsilon_2^{2'} \\ \varepsilon_6^{2'} \end{Bmatrix} = \begin{bmatrix} 19.5 & 7.01 & 0 \\ 7.01 & 142.2 & 0 \\ 0 & 0 & 4.34 \end{bmatrix} \begin{Bmatrix} \varepsilon_1^{2'} \\ \varepsilon_2^{2'} \\ \varepsilon_6^{2'} \end{Bmatrix}$$

$$\begin{Bmatrix} \sigma_1^{7'} \\ \sigma_2^{7'} \\ \sigma_6^{7'} \end{Bmatrix} = \begin{bmatrix} Q_{11}^{7'} & Q_{12}^{7'} & 0 \\ Q_{12}^{7'} & Q_{22}^{7'} & 0 \\ 0 & 0 & Q_{66}^{7'} \end{bmatrix} \begin{Bmatrix} \varepsilon_1^{7'} \\ \varepsilon_2^{7'} \\ \varepsilon_6^{7'} \end{Bmatrix} = \begin{bmatrix} 19.5 & 7.01 & 0 \\ 7.01 & 142.2 & 0 \\ 0 & 0 & 4.34 \end{bmatrix} \begin{Bmatrix} \varepsilon_1^{7'} \\ \varepsilon_2^{7'} \\ \varepsilon_6^{7'} \end{Bmatrix}$$

In the −45° plies the in-plane behaviour can be written

$$\begin{Bmatrix} \sigma_1^{3'} \\ \sigma_2^{3'} \\ \sigma_6^{3'} \end{Bmatrix} = \begin{bmatrix} Q_{11}^{3'} & Q_{12}^{3'} & Q_{16}^{3'} \\ Q_{12}^{3'} & Q_{22}^{3'} & Q_{26}^{3'} \\ Q_{16}^{3'} & Q_{26}^{3'} & Q_{66}^{3'} \end{bmatrix} \begin{Bmatrix} \varepsilon_1^{3'} \\ \varepsilon_2^{3'} \\ \varepsilon_6^{3'} \end{Bmatrix} = \begin{bmatrix} 48.3 & 39.6 & 30.6 \\ 39.6 & 48.3 & 30.6 \\ 30.6 & 30.6 & 37 \end{bmatrix} \begin{Bmatrix} \varepsilon_1^{3'} \\ \varepsilon_2^{3'} \\ \varepsilon_6^{3'} \end{Bmatrix}$$

$$\begin{Bmatrix} \sigma_1^{6'} \\ \sigma_2^{6'} \\ \sigma_6^{6'} \end{Bmatrix} = \begin{bmatrix} Q_{11}^{6'} & Q_{12}^{6'} & Q_{16}^{6'} \\ Q_{12}^{6'} & Q_{22}^{6'} & Q_{26}^{6'} \\ Q_{16}^{6'} & Q_{26}^{6'} & Q_{66}^{6'} \end{bmatrix} \begin{Bmatrix} \varepsilon_1^{3'} \\ \varepsilon_2^{3'} \\ \varepsilon_6^{3'} \end{Bmatrix} = \begin{bmatrix} 48.3 & 39.6 & 30.6 \\ 39.6 & 48.3 & 30.6 \\ 30.6 & 30.6 & 37 \end{bmatrix} \begin{Bmatrix} \varepsilon_1^{6'} \\ \varepsilon_2^{6'} \\ \varepsilon_6^{6'} \end{Bmatrix}$$

In the 90° plies the in-plane behaviour can be written

$$
\left\{ \begin{array}{c} \sigma_1^{4'} \\ \sigma_2^{4'} \\ \sigma_6^{4'} \end{array} \right\} = \left[\begin{array}{ccc} Q_{11}^{4'} & Q_{12}^{4'} & Q_{16}^{4'} \\ Q_{12}^{4'} & Q_{22}^{4'} & Q_{26}^{4'} \\ Q_{16}^{4'} & Q_{26}^{4'} & Q_{66}^{4'} \end{array} \right] \left\{ \begin{array}{c} \varepsilon_1^{4'} \\ \varepsilon_2^{4'} \\ \varepsilon_6^{4'} \end{array} \right\} = \left[\begin{array}{ccc} 48.3 & 39.6 & -30.6 \\ 39.6 & 48.3 & -30.6 \\ -30.6 & -30.6 & 37 \end{array} \right] \left\{ \begin{array}{c} \varepsilon_1^{4'} \\ \varepsilon_2^{4'} \\ \varepsilon_6^{4'} \end{array} \right\}
$$

$$
\left\{ \begin{array}{c} \sigma_1^{5'} \\ \sigma_2^{5'} \\ \sigma_6^{5'} \end{array} \right\} = \left[\begin{array}{ccc} Q_{11}^{5'} & Q_{12}^{5'} & Q_{16}^{5'} \\ Q_{12}^{5'} & Q_{22}^{5'} & Q_{26}^{5'} \\ Q_{16}^{5'} & Q_{26}^{5'} & Q_{66}^{5'} \end{array} \right] \left\{ \begin{array}{c} \varepsilon_1^{5'} \\ \varepsilon_2^{5'} \\ \varepsilon_6^{5'} \end{array} \right\} = \left[\begin{array}{ccc} 48.3 & 39.6 & -30.6 \\ 39.6 & 48.3 & -30.6 \\ -30.6 & -30.6 & 37 \end{array} \right] \left\{ \begin{array}{c} \varepsilon_1^{5'} \\ \varepsilon_2^{5'} \\ \varepsilon_6^{5'} \end{array} \right\}
$$

Chapter 9

Failure criteria for composites

9.1   Introduction

In this chapter, some failure models, which are widely used in designing composite material laminates, will be described. These models are generally developed based on the concept of strength. After a description of the most traditional criteria, we shall underline the limitations inherent in these models.

9.2   The ply strength constants

In the laminate theories discussed in chapter 8, the basic ply is homogenized which means that the details of the microstructure are not considered. Both the two- and three-dimensional laminate theory models are then used to calculate the stresses in each ply. Based on the material strength concept, the ply, which is anisotropic, is then assumed to possess a set of strength constants. In this way a comparison with the stress state of the ply can be used to predict failure. These constants can be determined either by a model based on the mechanics of the processes involved and developed at the fibre–matrix level or by testing the ply material in different directions to reveal the anisotropy of the mechanical properties. The first approach, similar in terms of strength to models described in chapter 6, is rarely used. These models based on the strength of the components are found to be poor in estimating ply strength. For this reason, ply strength constants are measured directly by testing the ply material.

Within the same notation adopted in chapter 5, the quantities which are described in the principal reference coordinate system of the ply are annotated as (L). Further the same quantities transformed in the global coordinate system of the laminate are shown as (G). Figure 9.1 describes these two systems with (1-2) in the plane of the laminate and x_1 parallel to fibre axis. Axis x_3 is perpendicular to the plane 1-2.

Within the frame of the two-dimensional laminate plate theory, each ply is in plane stress. Five independent ply failure modes are assumed: tensile and

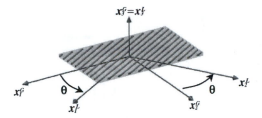

Figure 9.1. Local and global coordinate systems.

compressive failures in the fibre direction, tensile and compressive failures transverse to the fibre direction and in-plane shear failure. Thus, five strength constants have to be determined by testing the basic ply of the laminate composite structure.

1. X_t tensile breaking stress in the fibre direction.
2. X_c compressive breaking stress in the fibre direction.
3. Y_t tensile breaking stress transverse to the fibre direction.
4. Y_c compressive breaking stress transverse to the fibre direction.
5. S in plane shear breaking stress.

Experimental determination of these constants is sometimes difficult. Results often exhibit wide scatter and it is not always possible to determine these constants independently. The state of stress is not always homogeneous depending on the loading mode, the size and the geometry of the test specimen. Analysis of the state of stress has to be carefully made to be sure that the correct stress and the corresponding strength are determined. An example is the shear test where a mixture of all stresses cannot be avoided.

  Despite these difficulties, the above five ply constants are routinely determined and widely used in failure criteria for composite materials.

9.3   Maximum stress criterion and strain criterion

Let σ_1, σ_2 and σ_6 be the in-plane stresses at a point in the ply. The material at this point and also the ply itself are supposed to fail if any one of these stresses reaches or exceeds the limiting value. Thus the maximum stress criterion entails a set of five independent conditions:

$$
\begin{aligned}
\sigma_1^L &\leq X_t && \text{if } \sigma_1 \text{ is tensile} \\
|\sigma_1^L| &\leq X_c && \text{if } \sigma_1 \text{ is compressive} \\
\sigma_2^L &\leq Y_t && \text{if } \sigma_2 \text{ is tensile} \qquad (9.1) \\
|\sigma_2^L| &\leq Y_c && \text{if } \sigma_2 \text{ is compressive} \\
|\sigma_6^L| &\leq S.
\end{aligned}
$$

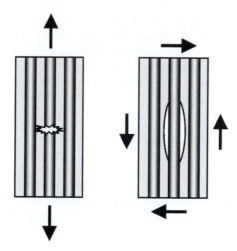

Figure 9.2. (a) Fibre failure, (b) interfacial debonding.

This criterion can also provide information about the failure mode of the ply. Namely, if $\sigma_1^L \leq X_t$ is satisfied alone, then the ply fails in tension [figure 9.2(a)]. Or if $|\sigma_6^L| \leq S$ is satisfied alone, then the ply fails by in-plane shear [figure 9.2(b)]. In this sense, failure of the ply is independently caused by one of the three in-plane stresses, whichever is predominant.

A variation of the maximum stress criterion can be obtained by using the ply strains ε_1^L, ε_2^L, ε_6^L:

$$\varepsilon_1^L \leq \frac{X}{E_L}, \qquad \varepsilon_2^L \leq \frac{Y}{E_T}, \qquad \varepsilon_6^L \leq \frac{S}{G}. \tag{9.2}$$

The corresponding limiting strains are found in the same tests with the limiting stresses. This criterion is known as the maximum strain criterion. These maximum stresses or maximum strains criteria lead to similar information in the elastic domain if the Poisson's ratios are low, but can be different if the ply material displays non-linearity before failure. Conceptually these criteria are identical. Each component of stress or strain has its own criterion and is not affected by the other components. That is to say that there is no interaction between them.

9.4 Quadratic criterion in stress space or in strain space

9.4.1 Generality

These criteria are often qualified 'interaction criteria' because they assume that the ply material at a point can fail even if none of the five conditions in relations (9.1) are satisfied, especially when σ_1^L, σ_2^L and σ_6^L are of the same order of magnitude. The premise of this assumption was postulated in the stress interaction effect in the yield criterion of von Mises which is based on the stored distorsional energy. After a short reminder of this criterion, we are going to analyse the main quadratic criteria which are used in the design and sizing of composite material structures.

9.4.2   The Von Mises criterion

Isotropy requires that the boundary of the domain of the criterion be invariant under a change of axes. Therefore, a criterion is to be only expressed with invariant quantities. The Von Mises criterion corresponds to the isotropic-hardening state of metals. As metals generally exhibit plastic incompressibility and yield-independence with respect to hydrostatic stress, it is sufficient to use the deviatoric stress tensor defined by

$$s = \sigma - \tfrac{1}{3}[\mathrm{tr}(\sigma)]I$$

where I is the unit second-order tensor and $[\mathrm{tr}(\sigma)]$ is the trace of the stress tensor. The Von Mises criterion is expressed as a function of the deviatoric stress tensor by the threshold function $f(s_{\mathrm{II}}, \sigma_s) = 0$.

s_{II} corresponds to the second invariant of the deviatoric stress tensor: $s_{\mathrm{II}} = \tfrac{1}{2}\mathrm{tr}(s^2)$.

σ_s characterizes the yield stress in simple tension. The Von Mises criterion postulates that the threshold is reached when s_{II}, which characterizes a three-dimensional state of stress, is equivalent to the pure tensile one-dimensional state, the threshold of which is σ_s.

As in the one-dimensional case, a pure tensile stress state and the deviatoric stress are expressed by

$$\sigma = \langle \sigma_s, 0, 0, 0, 0, 0 \rangle \qquad \text{and} \qquad s = \langle \tfrac{2}{3}\sigma_s, -\tfrac{1}{3}\sigma_s, -\tfrac{1}{3}\sigma_s, 0, 0, 0 \rangle$$

The Von Mises criterion is therefore written as

$$f(s_{\mathrm{II}}, \sigma_s) = s_{\mathrm{II}} - \tfrac{1}{3}\sigma_s^2 = 0. \tag{9.3}$$

If we develop this expression for a three-dimensional state of stress, we obtain a convex envelope in stress space:

$$\tfrac{1}{2}[(\sigma_1^L - \sigma_2^L)^2 + (\sigma_2^L - \sigma_3^L)^2 + (\sigma_3^L - \sigma_1^L)^2 + 6(\sigma_4^{L^2} + \sigma_5^{L^2} + \sigma_6^{L^2})] - \sigma_s^2 = 0. \tag{9.4}$$

This criterion has been established for isotropic materials. The criteria we are now going to present are extensions of the Von Mises criterion to materials with anisotropic properties.

9.4.3 The Hill criterion

The extension of the Von Mises criterion to anisotropic materials was first presented by Hill [1]. In this case the convex envelope which characterizes the threshold in the principal reference frame of the material is written as

$$F(\sigma_2^L - \sigma_3^L)^2 + G(\sigma_3^L - \sigma_1^L)^2 + H(\sigma_1^L - \sigma_2^L)^2$$

$$+ 2L\sigma_4^{L^2} + 2M\sigma_5^{L^2} + 2N\sigma_6^{L^2} = 1. \tag{9.5}$$

We can develop this expression and obtain the following:

$$(G+H)\sigma_1^{L^2} + (F+H)\sigma_2^{L^2} + (F+G)\sigma_3^{L^2} - 2H\sigma_1^L\sigma_2^L - 2G\sigma_1^L\sigma_3^L$$

$$- 2F\sigma_2^L\sigma_3^L + 2L\sigma_4^{L^2} + 2M\sigma_5^{L^2} + 2N\sigma_6^{L^2} = 1. \tag{9.6}$$

The parameters F, G, H, L, M and N are constants characteristic of the material and depend on the failure stresses X, Y and S of the ply material.

In the case of a tensile (or compressive) test in the fibre direction 1, the Hill criterion reduces to

$$G + H = \frac{1}{X^2} \tag{9.7}$$

where X is the tensile (or compressive) failure stress in the fibre direction. Similarly, we obtain

$$F + H = \frac{1}{Y^2} \tag{9.8}$$

$$F + G = \frac{1}{Z^2} \tag{9.9}$$

where Y and Z are the tensile (or compressive) failure stresses in the transverse directions (2 and 3). In the case of in-plane shear test, the Hill criterion reduces to

$$2N = \frac{1}{S_6^2} \tag{9.10}$$

where S_6 is the shear in-plane failure stress. This gives

$$2M = \frac{1}{S_5^2} \tag{9.11}$$

$$2L = \frac{1}{S_4^2} \tag{9.12}$$

where S_5 and S_4 are shear failure stresses, respectively in planes (1-3) and (2-3). The use of expressions from (9.7) to (9.12) allows the failure constants, F, G, L, M, N to be determined and leads to another form of the Hill criterion which is

$$\left(\frac{\sigma_1^L}{X}\right)^2 + \left(\frac{\sigma_2^L}{Y}\right)^2 + \left(\frac{\sigma_3^L}{Z}\right)^2 - \left(\frac{1}{X^2} + \frac{1}{Y^2} - \frac{1}{Z^2}\right)\sigma_1^L\sigma_2^L$$

$$- \left(\frac{1}{X^2} + \frac{1}{Z^2} - \frac{1}{Y^2}\right)\sigma_1^L\sigma_3^L - \left(\frac{1}{Y^2} + \frac{1}{Z^2} - \frac{1}{X^2}\right)\sigma_2^L\sigma_3^L$$

$$+ \left(\frac{\sigma_4^L}{S_4}\right)^2 + \left(\frac{\sigma_5^L}{S_5}\right)^2 + \left(\frac{\sigma_6^L}{S_6}\right)^2 = 1. \tag{9.13}$$

Let us point out that the Hill criterion is not able to consider possible differences between tensile and compressive behaviours.

In the case of a plane stress behaviour in the (1-2) plane, $\sigma_3^L = \sigma_4^L = \sigma_5^L = 0$, the Hill criterion can be simplified and reduces to

$$\left(\frac{\sigma_1^L}{X}\right)^2 + \left(\frac{\sigma_2^L}{Y}\right)^2 - \left(\frac{1}{X^2} + \frac{1}{Y^2} - \frac{1}{Z^2}\right)\sigma_1^L\sigma_2^L + \left(\frac{\sigma_6^L}{S_6}\right)^2 = 1. \tag{9.14}$$

9.4.4   The Tsai–Hill criterion

The former criterion has been simplified by Azzi and Tsai [2] for composite laminates. In this case $Z = Y$ and for the ply under in-plane stresses, (9.14) reduces to the so-called Tsai–Hill criterion. It is expressed as

$$\left(\frac{\sigma_1^L}{X}\right)^2 - \left(\frac{\sigma_1^L}{X}\right)\left(\frac{\sigma_2^L}{Y}\right) + \left(\frac{\sigma_2^L}{Y}\right)^2 + \left(\frac{\sigma_6^L}{S}\right)^2 = 1 \tag{9.15}$$

where $X = X_t$ if σ_1^L is tensile, $X = X_c$ if σ_1^L is compressive, $Y = Y_t$ if σ_2^L is tensile, and $Y = Y_c$ if σ_2^L is compressive.

9.4.5   The Hoffman criterion

A possible generalization of the Hill criterion, which takes into account possible differences between tensile and compressive behaviours has been proposed by Hoffman [3]. This criterion predicts the failure of ply material when the following condition is obeyed:

$$C_1(\sigma_2^L - \sigma_3^L)^2 + C_2(\sigma_3^L - \sigma_1^L)^2 + C_3(\sigma_1^L - \sigma_2^L)^2$$

$$+ C_4\sigma_1^L + C_5\sigma_2^L + C_6\sigma_3^L + C_7\sigma_4^{L^2} + C_8\sigma_5^{L^2} + C_9\sigma_6^{L^2} = 1 \tag{9.16}$$

where C_1, C_2, C_3, C_4, C_5, C_6, C_7, C_8, C_9 are characteristic constants depending on the failure stresses of the material such that

$$C_1 = \frac{1}{2}\left[\frac{1}{Y_t Y_c} + \frac{1}{Z_t Z_c} - \frac{1}{X_t X_c} \right]$$

$$C_2 = \frac{1}{2}\left[\frac{1}{Z_t Z_c} + \frac{1}{X_t X_c} - \frac{1}{Y_t Y_c} \right]$$

$$C_3 = \frac{1}{2}\left[\frac{1}{X_t X_c} + \frac{1}{Y_t Y_c} - \frac{1}{Z_t Z_c} \right]$$

$$C_4 = \frac{1}{X_t} - \frac{1}{X_c}, \qquad C_7 = \frac{1}{S_4^2}$$

$$C_5 = \frac{1}{Y_t} - \frac{1}{Y_c}, \qquad C_8 = \frac{1}{S_5^2}$$

$$C_6 = \frac{1}{Z_t} - \frac{1}{Z_c}, \qquad C_9 = \frac{1}{S_6^2}.$$

(9.17)

In the case of an in-plane stress state in the (1-2) plane, the Hoffman criterion reduces to

$$\frac{\sigma_1^{L2}}{X_t X_c} + \frac{\sigma_2^{L2}}{Y_t Y_c} - \frac{\sigma_1^L \sigma_2^L}{X_t X_c} + \frac{X_c - X_t}{X_t X_c}\sigma_1^L + \frac{Y_c - Y_t}{Y_t Y_c}\sigma_2^L + \frac{\sigma_6^{L2}}{S_6^2} = 1. \qquad (9.18)$$

We can see that the formulation of these criteria is based on a threshold concept, the evaluation of which is difficult in the case of composite materials. In fact, even when they are ductile, damage modes are coupled and different. The type of threshold depends on the type of damage which occurs.

In the case of the former criteria, the only coupling which is possible is between the longitudinal and transversal stresses σ_1^L and σ_2^L. Depending on the loading, we can imagine other interactions such as shear with transversal or longitudinal directions. That explain new criteria proposal with all possible interactions.

9.4.6   The Tsai–Wu criterion

In anisotropic materials, the energies involved in the distortion and dilatation of the material are not usually separable. It must be understood that the Tsai–Hill criterion for yielding does not introduce coupling with shear deformations. The Tsai–Wu criterion is more a mathematical than a physical concept. Variations of the Tsai–Hill proposal have led to criteria of a purely mathematical nature being developed, without any attempt to associate them to specific failure modes. The general form for the Tsai–Wu criterion is a potential function expandable into a power series truncated after the quadratic terms. As a function of stress, this criterion can be written $(i,j = 1,2,6)$

$$F_{ij}\sigma_i^L\sigma_j^L + F_i\sigma_i^L = 1 \tag{9.19}$$

or in strain alone

$$G_{ij}\varepsilon_i^L\varepsilon_j^L + G_i\varepsilon_j^L = 1. \tag{9.20}$$

These criteria can be applied to plies under two- or three-dimensional stresses and are invariant with respect to the reference coordinate frame used. The constants F_i and F_{ij} or G_i and G_{ij} have to be determined from tests on the ply material and can be dependent on the strength constants introduced in section 9.2. Because of material symmetry in the unidirectional ply, which is assumed to be either transversely isotropic or orthotropic, there are three independent constants in F_i (similarly G_i) and nine in F_{ij} (similarly G_{ij}) if the ply is under three-dimensional stresses. Their number reduces to two in F_i and four in F_{ij} if the ply is under two-dimensional in-plane stresses. In the two-dimensional stress case, the constants F_{ij} and F_i can be related to the ply strength constants X_t, X_c, Y_t, Y_c and S obtained from testing the ply material.

With the assumption of a two-dimensional stress state, the expanded criterion can be written as

$$F_{11}\sigma_1^{L^2} + 2F_{12}\sigma_1^L\sigma_2^L + F_{22}\sigma_2^{L^2} + F_{66}\sigma_6^{L^2} + 2F_{16}\sigma_1^L\sigma_6^L + 2F_{26}\sigma_2^L\sigma_6^L$$
$$+ F_1\sigma_1^L + F_2\sigma_2^L + F_6\sigma_6^L = 1. \tag{9.21}$$

Referenced to the principal framework of the ply and by virtue of material symmetry, the yield surface has not to be affected by the sign of the shear. In this way the following shear terms disappear:

$$F_{16}\sigma_1^L\sigma_6^L, \qquad F_{26}\sigma_2^L\sigma_6^L, \qquad F_6\sigma_6. \tag{9.22}$$

Nevertheless, changing the sign for tensile or compressive loading modifies the failure of material. This leads to the general form

$$F_{11}\sigma_1^{L^2} + 2F_{12}\sigma_1^L\sigma_2^L + F_{22}\sigma_2^{L^2} + F_{66}\sigma_6^{L^2} + F_1\sigma_1^L + F_2\sigma_2^L = 1. \tag{9.23}$$

Thus, we have to determine six constants in the two-dimensional case. This can be achieved by simple tests.

Tensile–compressive loading in the fibre direction

If $\sigma_1^L = X_t$, we obtain

$$F_{11}X_t^2 + F_1X_t = 1. \tag{9.24}$$

Even if $\sigma_1^L = -X_c$

$$F_{11}X_c^2 - F_1X_c = 1. \tag{9.25}$$

We then obtain two equations for two unknown constants leading to the following solutions:

$$F_{11} = \frac{1}{X_t X_c} \tag{9.26}$$

$$F_1 = \frac{1}{X_t} - \frac{1}{X_c}. \tag{9.27}$$

Tensile–compressive loading in the transverse direction

In a similar way, we obtain

$$F_{22} = \frac{1}{Y_t Y_c} \tag{9.28}$$

$$F_2 = \frac{1}{Y_t} - \frac{1}{Y_c}. \tag{9.29}$$

In-plane shearing

From in-plane shearing tests, we obtain directly $F_{66} = 1/S^2$.

With the above tests, we are able to obtain five of the six necessary coefficients to completely write the criterion. The remaining constant F_{12} is related to the coupling between the transversal and longitudinal components. Biaxial tests would be necessary, with both normal stress components to be non-zero, but these tests are often difficult to perform as well as to analyse. A homogeneous area is not always guaranteed and is sometimes very small. The method which is generally used consists in determining the lower and upper bounds of this constant based on geometric considerations. The equation of this quadratic function describes a curve which can go from an ellipse to parallel lines, and to hyperbola depending on the value of the interaction term we are looking for. The criterion that indicates which branch of the quadratic curve it belongs to, is based on the value of the following discriminant.

$$\text{Discriminant} = F_{11}F_{22} - F_{12}^2 \begin{cases} > 0 & \text{for ellipse} \\ = 0 & \text{for parallel lines} \\ < 0 & \text{for hyperbola.} \end{cases} \tag{9.30}$$

In order to avoid a calculation leading to infinite strength, we have to ensure that the failure criterion represents a closed curve in stress space. Consequently, this discriminant is constrained by the value shown for the ellipse in the above equation.

If we introduce a dimensionless or normalized interaction term, then

$$F_{12}^* = F_{12}/\sqrt{F_{11}F_{22}}. \tag{9.31}$$

The range of values of the discriminant in equation (9.30) can now be expressed by the range of values of the normalized interaction term:

$$-1 < F_{12}^* < 1 \qquad \text{for ellipse}$$

$$F_{12}^* = \pm 1 \qquad \text{for parallel lines} \tag{9.32}$$

$$F_{12}^* < -1 \qquad \text{and} \quad F_{12}^* > 1 \quad \text{for hyperbola.}$$

We can rearrange equation (9.23) in terms of the dimensionless parameters

$$x = \sqrt{F_{11}}\sigma_1, \qquad y = \sqrt{F_{22}}\sigma_2, \qquad z = \sqrt{F_{66}}\sigma_6 \tag{9.33}$$

and

$$F_1^* = \frac{F_1}{\sqrt{F_{11}}}, \qquad F_2^* = \frac{F_2}{\sqrt{F_{22}}}.$$

Equation (9.23) becomes

$$x^2 + 2F_{12}^* xy + y^2 + z^2 + F_1^* x + F_2^* y = 1. \tag{9.34}$$

This equation represents a family of ellipses. In the $z = 0$ plane, by analogy with the Von Mises criterion, we obtain

$$x^2 - xy + y^2 = 1 \tag{9.35}$$

as generally we take $F_{12}^* = -1/2$. The linear terms will determine the displacements of the centre. Further, the failure curve in stress space will depend on the choice of this value (Tsai [4]).

The stress failure criterion in equation (9.23) for unidirectional composites can be expressed in strain space. This is often more convenient than in stress-space because strain distributions across the thickness of a laminate is generally assumed to be constant or at most a linear function of the x_3 axis.

In order to resolve equation (9.20) from (9.19), we only need to substitute the stress components by strain components using the in-plane constitutive law:

$$\left\{ \begin{array}{c} \sigma_1^L \\ \sigma_2^L \\ \sigma_6^L \end{array} \right\} = \begin{bmatrix} Q_{11} & Q_{12} & 0 \\ Q_{12} & Q_{22} & 0 \\ 0 & 0 & Q_{66} \end{bmatrix} \left\{ \begin{array}{c} \varepsilon_1^L \\ \varepsilon_2^L \\ \varepsilon_6^L \end{array} \right\}. \tag{9.36}$$

Thus

$$F_{ij}Q_{ik}Q_{jf}\varepsilon_k^L \varepsilon_f^L + F_i Q_{ij}\varepsilon_j^L = 1. \tag{9.37}$$

If we define

$$G_{kf} = F_{ij}Q_{ik}Q_{jf}, \qquad G_j = F_i Q_{ij} \tag{9.38}$$

so that the failure criterion in strain space is

$$G_{kj}\varepsilon_k^L\varepsilon_j^L + G_k\varepsilon_k^L = 1. \tag{9.39}$$

If we expand the equation and invoke symmetry, we obtain

$$G_{11}\varepsilon_1^{L^2} + 2G_{12}\varepsilon_1^L\varepsilon_2^L + G_{22}\varepsilon_2^{L^2} + G_{66}\varepsilon_6^{L^2} + G_1\varepsilon_1^L + G_2\varepsilon_2^L = 1 \tag{9.40}$$

where

$$\begin{aligned}
G_{11} &= F_{11}Q_{11}^2 + 2F_{12}Q_{11}Q_{12} + F_{22}Q_{12}^2 \\
G_{22} &= F_{11}Q_{12}^2 + 2F_{12}Q_{12}Q_{22} + F_{22}Q_{22}^2 \\
G_{12} &= F_{11}Q_{11}Q_{12} + F_{12}[Q_{11}Q_{22} + Q_{12}^2] + F_{22}Q_{12}Q_{22} \\
G_{66} &= F_{66}Q_{66}^2 = [Q_{66}/S]^2 \\
G_1 &= F_1Q_{11} + F_2Q_{12} \\
G_2 &= F_1Q_{12} + F_2Q_{22}.
\end{aligned} \tag{9.41}$$

This equation is dimensionless, leading to the same values for the materials constants whatever the physical dimension system (SI or Imperial units).

Let us note that these criteria in stress space or in strain space are written in the principal reference frame of the ply. We can establish the off-axis criterion by using the transformation equations.

9.5 Transformation equations for failure criteria

The transformation equations are the same as the transformation of the constitutive laws (stiffness or compliance) from the local reference system to a global reference system turned through an angle θ (figure 9.1). We can derive this by substituting the stress transformation (or strain transformation) in the on-axis failure criterion (ply reference system) in equation (9.19). Thus, we obtain a quadratic criterion in the off-axis reference system ($m = \cos\theta$, $n = \sin\theta$):

$$\begin{Bmatrix} F_{11}^\theta \\ F_{22}^\theta \\ F_{12}^\theta \\ F_{66}^\theta \\ F_{16}^\theta \\ F_{26}^\theta \end{Bmatrix} = \begin{bmatrix} m^4 & n^4 & 2m^2n^2 & m^2n^2 \\ n^4 & m^4 & 2m^2n^2 & m^2n^2 \\ m^2n^2 & m^2n^2 & m^4+n^4 & -m^2n^2 \\ 4m^2n^2 & 4m^2n^2 & -8m^2n^2 & (m^2-n^2)^2 \\ 2m^3n & -2mn^3 & 2(mn^3-m^3n) & mn^3-m^3n \\ 2mn^3 & -2m^3n & 2(m^3n-mn^3) & m^3n-mn^3 \end{bmatrix} \begin{Bmatrix} F_{11} \\ F_{22} \\ F_{12} \\ F_{66} \end{Bmatrix}. \tag{9.42}$$

$$\begin{Bmatrix} F_1^{\theta} \\ F_2^{\theta} \\ F_6^{\theta} \end{Bmatrix} = \begin{bmatrix} m^2 & n^2 \\ n^2 & m^2 \\ 2mn & -2mn \end{bmatrix} \begin{Bmatrix} F_1 \\ F_2 \end{Bmatrix}.$$

In a similar way, in strain space we obtain

$$\begin{Bmatrix} G_{11}^{\theta} \\ G_{22}^{\theta} \\ G_{12}^{\theta} \\ G_{66}^{\theta} \\ G_{16}^{\theta} \\ G_{26}^{\theta} \end{Bmatrix} = \begin{bmatrix} m^4 & n^4 & 2m^2n^2 & 4m^2n^2 \\ n^4 & m^4 & 2m^2n^2 & 4m^2n^2 \\ m^2n^2 & m^2n^2 & m^4+n^4 & -4m^2n^2 \\ m^2n^2 & m^2n^2 & -2m^2n^2 & (m^2-n^2)^2 \\ m^3n & -mn^3 & mn^3-m^3n & 2(mn^3-m^3n) \\ mn^3 & -m^3n & m^3n-mn^3 & 2(m^3n-mn^3) \end{bmatrix} \begin{Bmatrix} G_{11} \\ G_{22} \\ G_{12} \\ G_{66} \end{Bmatrix} \quad (9.43)$$

$$\begin{Bmatrix} G_1^{\theta} \\ G_2^{\theta} \\ G_6^{\theta} \end{Bmatrix} = \begin{bmatrix} m^2 & n^2 \\ n^2 & m^2 \\ mn & -mn \end{bmatrix} \begin{Bmatrix} G_1 \\ G_2 \end{Bmatrix}.$$

9.5.1   Extension to the three-dimensional case

If the ply is under three-dimensional stresses, the criterion is still valid if the out-of-plane stresses are taken into account. Supplementary tests have to be performed to determine all the strength constants. In practice, criteria based on the strength concept are mostly limited to two-dimensional stress cases. The values of the constants depend on the type of test which has been performed and do not represent the *in situ* ply strength when the ply is an integral part of a laminate. This is due to the interaction between plies or some interply phenomena which happens during loading. This important limitation of these criteria leads to more accurate criteria being proposed for the designing of composite material structures which are able to predict composite material failure mechanisms.

9.6   Progressive ply failures in laminates

The various ply failure criteria discussed above are used in the failure analysis of laminates. Failure of the laminate may thus be a process in which the weaker ply fails first whilst the rest of the plies in the laminate still carry the load. The progressive ply failure model assumes that when one ply fails, it no longer can carry the stress in it and then its contribution to laminate strength should be removed from the laminate. A new iterative analysis of the laminate then has to be carried out to determine if any of the remaining plies will fail under the previously applied loads. If one ply does fail due to the redistribution of the load, the failed ply is again removed from consideration. If no ply fails after load redistribution, the applied load is progressively increased until another ply fails. This procedure is repeated until the last ply of the laminate fails.

Modification of this 'fail-or-no-fail' procedure have been proposed by considering that the failed ply, instead of failing in a brittle manner, has yielded in a non-linear sense such as showing perfect-plasticity behaviour or strain hardening plasticity. In this case, the concerned ply remains inside the laminate and continues to carry the stress as a yielded ply. We are not going to go into the details of such modifications because they always discard the effects of ply-to-ply interactions and the real physics of the materials. This is for various reasons. First, in the two-dimensional laminated plate, the out-of-plane stresses are not taken into account and only the in-plane stresses are homogenized at the ply level. This means that the ply interface has no physical meaning and failure in the ply interface cannot be predicted. Secondly, the microstructure is lost in the ply homogenization process as are local effects and interactions between constituents at the ply level and between neighbouring plies. The next chapter will propose more accurate criteria based on the microstructure of the material and the failure mechanisms occurring in composite materials.

9.7   Physical failure mechanisms in composite materials

According to the applied loading, composite material structures can develop three types of damage which are rarely observed in conventional homogeneous materials.

1. Ply cracking, generally called transverse cracking, which appears first in the plies which contain fibres which are off-axis compared with the applied load (figure 9.3).
2. Debonding or delamination between plies.
3. Fibre failure which occurs mainly near the point of complete composite failure.

Inside the laminate, each ply interacts with its neighbouring plies through all bonded surfaces which are the fibre–matrix interfaces and ply-to-ply interfaces. This is the essence of a composite material which is already composed as a structure. These bonded surfaces induce the previously mentioned failure modes which cannot be predicted by former failure criteria based on ply strength.

To illustrate this, we can use a famous example developed by Crossman and Wang [5]. Let us consider a carbon epoxy specimen with two types of layup. Both are loaded in the same manner: $[0/90n/0]$ and $[25/-25/90n/-25/25]$ with $n = 1, 2, 3$. When loaded under uniaxial tension

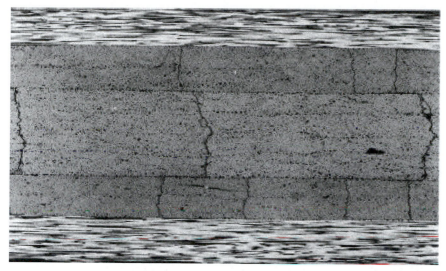

Figure 9.3. Typical multiple cracking pattern in a $[0, \pm55, 90]_s$ composite.

both types of composite developed in-plane stresses except close to the edge where the stress state is three-dimensional. The failure processes in both composites would be the same if any of the ply failure criteria discussed previously were used. However, if we carry out tests on these specimens, the failure mode developed in the first stacking is fundamentally different from that developed in the second. The failure process in the $[25/-25/90n/-25/25]$ composite is interply delamination which could never be predicted by the ply failure criteria. We are going to analyse these different failure modes in terms of their physical characteristics and then examine the consequences on the failure mechanisms.

9.7.1 Intralaminar cracking

Calculations for the $[0/90n/0]$ laminates show that the three-dimensional stresses close to free edges are not predominant compared with two-dimensional stresses in the interior of the laminate. The stresses in the outer $0°$ layers are $(\sigma_1)_0$ in the fibre direction, and $(\sigma_2)_0$ in the transverse direction. Both are tensile. In the middle layer of $n - 90°$ plies, the stresses are a compressive stress $(\sigma_1)_{90}$ along the fibres and a tensile transverse stress $(\sigma_2)_{90}$ perpendicular to the fibres. Microscopic analysis during testing shows that cracks parallel to the fibres appear in the $n - 90°$ plies during loading. The first crack occurs suddenly and across the whole width of $90°$ layer specimen. The crack, however remains in the $90°$ layer due to the constraining effect provided by the outside $0°$ plies. Hence, the laminate remains intact under the applied tension. This result is in conflict with ply failure criteria developed previously which predicts the failure of the laminate when the first ply failure occurs. The second disagreement concerns the stress $(\sigma_2)_{90}$ for the onset of the first transverse crack which depends on the thickness of the $n - 90°$ plies. In particular, the critical *in situ* stress $(\sigma_2)_{90}$ decreases as n increases. The asymptotic value of $(\sigma_2)_{90}$ with large n, however, tends to approach the tensile transverse strength of the ply, Y_t, tested with a 90° ply alone. As this value is used in ply failure criteria, that means that these criteria do not account for the influence of the 90° layer thickness concerning the first ply failure stress $(\sigma_2)_{90}$.

The ply thickness which controls the behaviour of intralaminar cracking is a consequence of the constraining effect of the stiff outer 0° plies on the possibility of deformation of the 90° layer through their bonded interfaces.

In addition, test results also show that the kinetics of transverse cracks also depend on the 90° ply thickness when increasing the loading on the laminate. The reasons are similar to those which have been previously evoked for the first crack. As the 90° ply thickness decreases the crack density increases.

The limiting case is when n becomes large. The 90° ply behaviour in the laminate tends to that of the 90° ply alone. In this sense, when n increases, the constraining effect of stiffener 0° plies reduces and the number of cracks decreases.

Once again, all these interaction effects are not taken into account by ply failure criteria which predict the first transverse crack at a much lower laminate loading and wrongly predict crack propagation kinetics and consequently a wrong crack density. The following chapter will propose more accurate models based on a more accurate description of physical mechanisms involved in the failure of composite materials.

9.7.2   Interlaminar failure

As in the $[0/90n/0]$ laminates discussed above, transverse cracking appears in the $n - 90°$ plies of $[25/-25/90n/-25/25]$ laminates when they are loaded under uniaxial tension. However, in this case, the three-dimensional stresses close to the laminate free edges are predominant compared with the two-dimensional in-plane stresses inside the laminate. In particular, a localized tensile interlaminar normal stress $(\sigma_3)_{90}$, which is zero in the interior of the laminate, exists in the thickness of the 90° layer close to free edge interface $-25/90$. This stress rises sharply in this vicinity as shown in figure 9.4.

Thus the first and main mode of degradation in such a laminate is an interply crack, known as free-edge delamination, which is mainly initiated by out-of-plane stresses close to the edge.

Even if a stress such as $(\sigma_3)_{90}$ does exist near the free edge interface in the $[0/90n/0]$ laminate, this stress is much smaller compared with in-plane

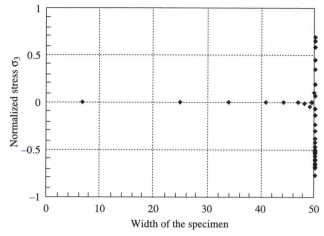

Figure 9.4. σ_3 stress along the $-25/90$ interface.

stresses leading to preponderant ply cracking failure. In general, free edge stresses develop in laminates which have neighbouring plies having markedly different alignments. In the $[0/90n/0]$ laminates, the outer $0°$ ply has a Poisson ratio, ν_{12}, of about 0.3 compared with about 0.1 for the middle $90°$ layer. In the $[25/-25/90n/-25/25]$ laminates, ν_{12} for the outer $\pm25°$ plies is about 0.9. This results in the large out-of-plane stresses at the free-edge interface.

9.8 Conclusions

The two failure modes previously described do not have the same consequences for the lifetime of the laminates. Delamination growth is more damaging for the laminates than transverse cracking. Delamination can grow in size under an increasing load and render the laminate structurally useless. If a delamination crack became unstable, this damage could provoke the failure of the structure. On the contrary, this is generally not the case with transverse cracking where neighbouring plies can constrain the effect of this damage and maintain the residual load-carrying capacity of the structure.

As a failure mode, delamination can occur near any stress risers in the laminate, such as holes, cut-outs etc., in addition to free edges. Even the formation of transverse cracks can induce localized delamination at the tips of transverse cracks. On the other hand, if a delamination crack occurs first, stress concentrations can develop at the crack tip area, which in turn can precipitate transverse cracking. Continuous formation of these competing damage processes means that the microstructure changes continuously when under load. This dependence on the loading-history of the laminate in service becomes crucial when long-term behaviour has to be analysed. As has already been pointed out, it is necessary to have a complete understanding of the processes occurring when the composite is under load, from the microstructure to the laminated structure. The following chapter will illustrate such a method which is based on the concept of damage mechanics.

9.9 Revision exercises

1. If the five lamina strength constants X_t, X_c, Y_t, Y_c and S are referred to as the principal material axes, draw the failure surface in the σ_1, σ_2 plane ($\sigma_6 = 0$) for a unidirectional lamina according to (a) the maximum stress criterion and (b) the Tsai–Wu criterion.
   Numerical values: $X_t = 2280$ MPa; $X_c = 335$ MPa; $Y_t = 57$ MPa; $Y_c = 158$ MPa; $S = 71$ MPa.
   Answer:

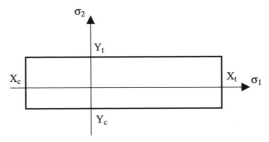

(a) maximum stress criterion

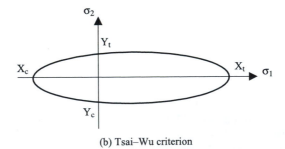

(b) Tsai–Wu criterion

References

[1] Hill R 1948 A theory of the yielding and plastic flaw of anisotropic metals *Proc. Royal Soc. A* **193** 281–297

[2] Azzi V O and Tsai S W 1965 Anisotropic strength of composites *Experimental Mechanics* **5** 283–288

[3] Hoffman O 1967 The brittle strength of orthotropic materials *J. Comp. Materials* **1** 200–206

[4] Tsai S W 1980 *Introduction to Composite Materials* (Technomic Publishing Company)

[5] Crossman F W and Wang A S D 1982 The dependence of transverse cracking and delamination on ply thickness in graphite-epoxy laminates *Damage in Composite Materials* ASTM STP **775** 118–139

Chapter 10

Damage in composites: from physical mechanisms to modelling

10.1 Failure criteria based on a multi-scale approach

The formulation of criteria based on a multi-scale approach, uses principally the concepts of damage mechanics. Two scale levels are considered in the analysis of heterogeneous materials. They are commonly used in the classical vocabulary of homogenization and are defined as:

- the 'macro' scale which analyses the structure as a completely homogeneous continuum;
- the 'micro' scale which takes into account all the heterogeneities or the constituents of the material.

Considering a laminated structure, the macro scale never examines the ply itself nor the arrangement of the plies. The whole structure is considered as a homogeneous continuum, the behaviour of which is described by an anisotropic constitutive law. This is in contrast to the micro scale which analyses phenomena at the level of the constituents. In the case of composite materials, the constituents are the fibres, the matrix and, if possible, the fibre–matrix interphase. Thus, in both cases, the ply is never considered as an entity. It has been seen that it is useful, in the case of composite materials, to introduce an intermediate scale between micro and macro, called the 'meso' scale, which considers the ply as the basic entity for the description of laminated structures. In this case it is the ply which is considered as homogeneous, the behaviour of which is supposed to be orthotropic or transversally isotropic in its principal reference frame. This scale is ideal for the description and the prediction of the failure of composite structures. Physical degradation generally appears inside the plies: this is the case of transverse cracking and fibre breakages, or between the plies, which is delamination. As far as the ply becomes the basic scale for the analysis of damage, its influence on a given laminate can be taken into account by using the following steps.

316

1. First, by carrying out appropriate tests, to define and identify parameters which are able to describe each physical mechanism and how it changes. As the degradation can be different from one part of the structure to another, these parameters must be a function of only local quantities. In this way these parameters can be used to describe the damage in each ply or interface which is included in the laminated structure.
2. The second step is to calculate the equivalent homogeneous behaviour of the damaged ply or cracked interface by using a homogenization method.
3. Finally, we have to study the evolution of damage when the ply, or the interface, is submitted to different loading conditions or paths. An analysis based on the thermodynamics of a continuum allows a law of the evolution of damage to be defined according to the first and second principles of thermodynamics. In this way the design and description of the damage of a complex structure becomes tractable and is really not so difficult. After an evaluation of strains or stresses in each ply of the laminated structure subjected to an external loading, it is easy to apply the proposed models to each ply and interface to appreciate their damage state. The evaluation of the global behaviour of the structure can then be made by using the following techniques:
   - the periodic homogenization in the case of a thick structure (as is described in chapter 8). Due to the periodic pattern through the thickness of the laminate, a three dimensional description of the state of stress or strain is needed;
   - classical laminate theory (as described in chapter 8), if plane stress can be assumed.

Two- or three-dimensional finite element numerical simulation allows the equivalent homogeneous behaviour of the cracked ply or interface to be described. Without restricting any generalities, we are going to develop the proposed method for the case of a two-dimensional composite application. It can be extended, without any difficulty, to a three-dimensional model. Thus we are going to limit our consideration to planar thin structures. In this case, it can be assumed that plane stress applies and that 'classical plate theory' can establish the link between the meso and the macro scale.

This chapter will describe the three main damage modes which can be encountered in composite material structures: progressive ply cracking, inter-ply delamination and fibre breakage. Ply cracking generally appears first. Without being really detrimental for the integrity of the structure, this damage directly affects the material properties of the ply over a large portion of its lifetime, and will accelerate other types of damage such as delamination and fibre breakage which are more detrimental to the structure. We are going to propose a deterministic approach to describe intra-ply transverse cracking damage and inter-ply delamination, followed by a probabilistic approach to estimate the accumulation of fibre failure. The proposed models lead to criteria which describe physical mechanisms and processes and allow the damage which is induced by these mechanisms to be predicted.

10.2 Intra-ply cracking or multiple transverse cracking

Tests, which are carried out on polymer matrix composites under load, show that the density of cracks which are induced in plies which are oriented at the largest angles, with respect to the loading direction, increases monotonously until a saturated state is reached which has a maximum number of cracks. This state is characterized by a relatively homogeneous cracking interspace. Two approaches can be proposed to analyse the mechanisms of this degradation.

1. Fracture mechanics, which mainly studies the onset and the propagation of a single crack. This restriction is not realistic due to the high densities of cracks but it is a good way for understanding the energy balances involved in the generation of such local phenomena.
2. Damage mechanics, the purpose of which is to decrease the stiffnesses of the elementary volume to take into account the effect of transverse cracking on the ply behaviour. These approaches use internal variables which describe the damage state of the material. These variables generally range between 0 and 1, to describe, respectively, the initial undamaged state and the degraded state of the elementary volume. In this family of models, we can mention simplified models based on 'shear lag analysis' [1], or numerical homogenization models using finite element computation.

We are going to pay much more attention to numerical homogenization which allows a more refined description of the physical mechanisms. The first step of such an analysis is a physical one, generally based on observations by microscopy.

10.2.1 Physical characterization of transverse cracking

When macroscopic damage consists only of transverse cracking, as it does in cross-ply laminates, cracks can be counted by recording the acoustic emissions associated with their formation, or alternatively by observing micrographs of the sides of specimens so as to count the cracks (figure 10.1) or by techniques such as ultrasonic C-SCAN analysis [figure 10.2(a)] or x-ray radiography [figure 10.2(b)].

   Thus the first crack and the subsequent crack accumulation can be continuously detected throughout tensile tests carried out up to the ultimate failure of the specimen. Such results will be used below as an experimental database to support the general assumptions of the numerical model.

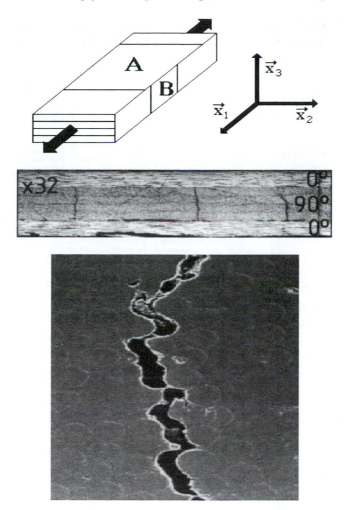

Figure 10.1. Transverse cracks in 90° plies (micrograph of the side B).

When the cumulative number of cracks is plotted as a function of the applied strain, curves similar to those shown in figure 10.3 are generated.

The curves represent the behaviours of four carbon/epoxy materials of the same $[0_2/90_4/0_2]$ lay-up, loaded to various levels. In spite of the large scatter from sample to sample, the curves are typical of the damage progression with cracks forming within a loading interval which depends on the material and the fracture strength of 0° plies. These curves illustrate, first, the onset of the cracking process depending on the material 'toughness', the sigmoidal type distribution and the final state at the end of the test with cracking apparently levelling off ('saturation').

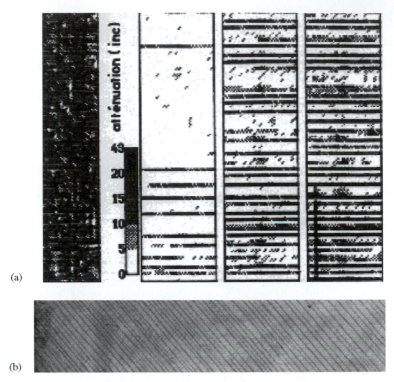

(a)

(b)

Figure 10.2. Detection of transverse cracking (zone A). (a) ultrasonic C-SCAN [2], (b) x-ray analysis.

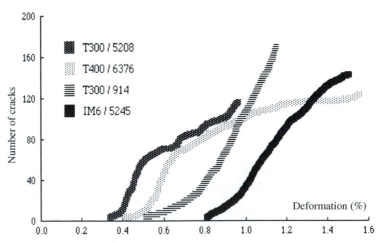

Figure 10.3. Number of cracks in the transverse ply of a $[0_2/90_4/0_2]$ laminate, versus the applied strain.

10.2.2   Crack geometry

Crack spacing

Typical distributions of the crack spacing along the specimen gauge length are shown in figure 10.4. The distributions refer to specimens that have been loaded close to final fracture at which the crack spacing are seen to be nearly uniform.

Crack length

Depending on the material and the ply thickness, a number of cracks do not span the whole width of the coupons. This is illustrated in figure 10.5 where

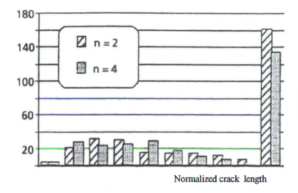

Figure 10.4. Typical crack spacing distribution [3].

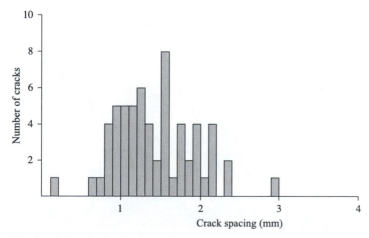

Figure 10.5. Crack length distributions [3].

the cracks in the last class are referred to as 'through-cracks'. Short cracks can occur during the tensile test, appearing at specimen edges, without any subsequent propagation. However, cracks starting from opposite edges can also be arrested somewhere within the width of the specimen owing to their interaction, which explains the relative number for intermediate crack length. An example is given in figure 10.5 for two 90° plies thicknesses, $n = 2$ and $n = 4$.

Saturation

As shown in figure 10.3, there is no clear point at which the saturation stage is reached. Generally we consider that saturation happens when it is no longer possible to create a new crack between two existing cracks. Due to fibre–matrix interface debonding, associated with each break, the critical load transfer length necessary to increase tensile load to the failure stress is not reached so that a new crack cannot be generated and saturation occurs. That explains the slowing of the cracking process and the quasi-plateau in the crack distribution curve. We shall consider the end of the curve as the saturation stage.

It should be noted that the same geometrical characteristics of transverse cracks can be found in textile composite lamina (figure 10.6).

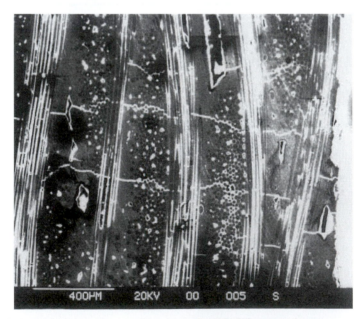

Figure 10.6. Typical transverse cracking in textile SiC–SiC composites.

10.2.3 Characteristic damage variable

Previous analyses which describe the evolution of the cracking process up to saturation of the crack density, show that this accumulation depends on the thickness of the cracked ply (figure 10.7). Further if we plot on the same graph, the decrease of stiffness of $[0_m, 90_n, 0_m]$ composites, not versus the crack density, d (figure 10.8), but versus a non-dimensional parameter which is the crack density normalized to the ply thickness, t, we obtain a kind of master curve. This characterizes a normalized evolution law for all of the experimental curves of density of cracks versus the applied stress and therefore is a master curve of stiffness reduction (figure 10.8).

In this way we can propose a characteristic damage variable, D, which is going to be used for a predictive model of ply degradation such that: $D = dt$.

The damage variable is characteristic of the degradation of the composite and we will use it to model the influence of this damage on the material behaviour.

The steps are as follows.

1. Homogenization calculations to evaluate the influence of damage on all the ply stiffnesses.
2. The thermodynamics of irreversible processes gives the global framework in which to formulate the evolution of damage as a function of local variables.

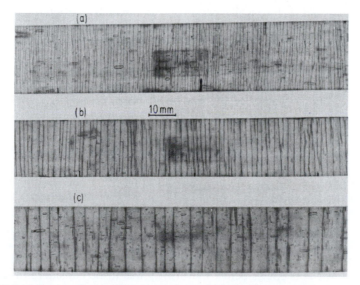

Figure 10.7. Influence of ply thickness on the multi-cracking process [4].

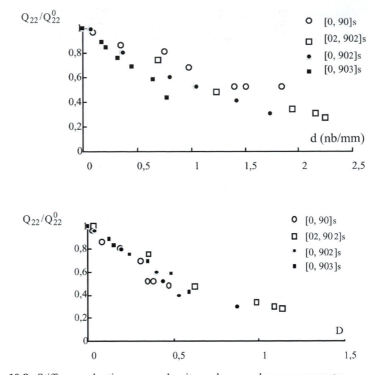

Figure 10.8. Stiffness reduction versus density and versus damage parameter.

3. Define a yield surface which can evolve and which characterizes not only
   the first value for the damage threshold but its changes too.

Considering the first step, we have to define the volume element characteristic
of the damaged ply. It is a portion of material containing a sufficient number
of defects to justify replacing it with an equivalent homogeneous material.
The relative homogeneity of the spacing between cracks, identified by the
observations, leads us, schematically, to consider the cracks to be uniformly
distributed, spaced apart by a constant length L, the appearance of a new
crack determining a fictitious rearrangement of the distribution of cracks
(figure 10.9).

It is clear that these hypotheses do not apply when the number of cracks
present in the ply is small. The model is therefore meaningful only when there
are a large number of cracks. Hence the concepts of Damage Mechanics can
be used. It should also be mentioned that whereas the homogeneous distribution of the cracks in the structure is realistic in the case of a flat specimen
under tension, this is no longer the case for a complex structure. In the
latter case the concept of defect uniformity has to be considered region by
region, assuming a relatively slow variation of damage between them.

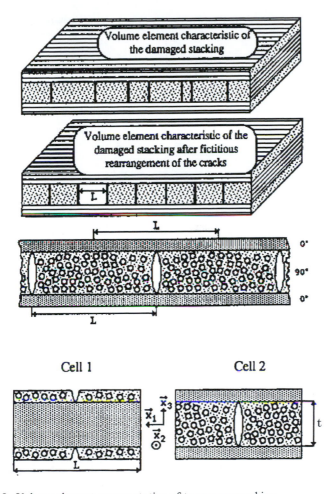

Figure 10.9. Volume element representative of transverse cracking.

The cracked ply cannot be considered other than included in a laminate. If it is not included in a laminate it will be completely fractured by the first crack, whereas in a laminated structure it continues to carry load away from the crack site. As in the case of $[0_m, 90_n, 0_m]$ composites the cracks are exclusively contained in the 90° plies and their distribution can evolve up to the saturated state. For this reason the elementary cell has been defined so as to take account of the stacking arrangement. The representative cell, the length of which is L, is defined as a portion of material containing a crack and the geometry of which is repetitive in the loading direction (figure 10.9). We take a unit length in the transverse in-plane direction (x_2 axis), which allows the problem of transverse cracking to be treated locally in a straight section of the structure. The results of a three-dimensional

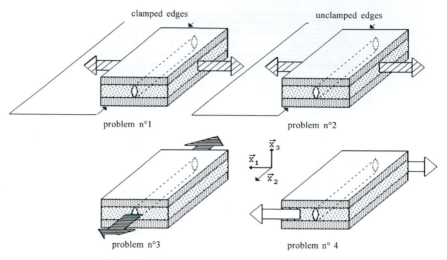

Figure 10.10. Scheme of different homogenization calculations.

finite element homogenization (figure 10.10) are used in order to obtain the effective homogenized behaviour of the cracked ply. These results show that transverse stresses can be neglected and that the plane stress assumption is realistic.

Then if we plot the in-plane stiffness, Q_{11} in the fibre direction, Q_{22} in transverse direction, Q_{12} the coupling between the transverse and fibre direction and Q_{66}, the shear modulus, a possible fit of the numerical result is

$$Q_{\gamma\delta}(D) = Q^0_{\gamma\delta} \exp(-k_{\gamma\delta}D)$$

where $Q^0_{\gamma\delta}$ represents the stiffness of the original material and the $k_{\gamma\delta}$ coefficients depend on the material (figure 10.11). Calculations with different adjacent plies show that the parameters $Q^0_{\gamma\delta}$ and $k_{\gamma\delta}$ are intrinsic to the cracked ply and so is the expression for $Q_{\gamma\delta}(D)$. Interaction with the neighbouring plies is made through the stress and strain fields which are of course dependent on the stacking and its evolution. However, for a given cracked state in a given ply of a laminate, the behaviour of the equivalent ply is unchanged.

This first step is similar to taking a photo of the material corresponding to a given damage state which gives, for this state, the residual stiffnesses of the material. The second step has to describe the development of this damage using the state variables, σ or ε, while the applied loading is changing during calculation. Always assuming plane stress, the thermodynamic potential (free energy) can be written as

$$\rho\psi(\varepsilon, D) = \tfrac{1}{2}Q_{\gamma\delta}(D)\varepsilon_\gamma\varepsilon_\delta, \qquad \gamma, \delta = 1, 2, 6$$

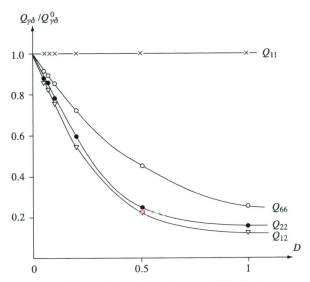

Figure 10.11. In-plane stiffness reductions (carbon epoxy) laminate.

where $Q_{\gamma\delta}(D)$ is the damageable elastic stiffness evaluated during the first step, and $\psi(\varepsilon, D)$ is the thermodynamic potential (chapter 5) which characterizes the state of the material function of the state variable ε and the internal variable D, which characterizes the damage state. The material damage process can be characterized by a yield surface $f(A, D)$, written as

$$f(A, D) = A - A^c(D) \geq 0$$

where A is the variable associated with damage expressed by $A = \rho \, \partial\psi/\partial D$, A^c is the damage threshold variable. The experimental curve, indicating the crack density versus the applied loading, allows the calculation of A^c from A. In this way, we find a polynomial equation similar to the interactive criteria described in chapter 9, but their justification is derived from a local analysis of the damage mechanisms. That leads to a better prediction of the studied damage stacking as shown hereafter. The coupling between all strains in the criterion, especially between transverse and shear strains, allows the evolution of cracks to be predicted, even if the plies are stressed in mode I, mode II or mixed mode. In addition, the dependence of $A^c(D)$ on the damage state through the variable D, includes the damage which has previously occurred in the considered ply. The criterion is then able to predict not only the first crack as interactive criteria, but also any further evolution of damage. This is a major difference compared with classical criteria. The proposed criterion is powerful as it is able to evolve so as to describe the changing state of damage. Damage will only occur if the two following conditions are satisfied.

1. The representative point of the damaged ply has to occur on the yield surface:

$$f(A, D) = 0.$$

2. This point does not leave the yield surface during the damage process and as the damage stage changes the evolution of this surface as respects the following consistency condition:

$$df(\varepsilon, D) = \frac{\partial f}{\partial \varepsilon} \, d\varepsilon + \frac{\partial f}{\partial D} \, dD = 0.$$

The increase of the damage parameter is then expressed as a function of the increase of the local strains in the ply:

$$dD = \frac{-\dfrac{\partial A}{\partial \varepsilon}}{\dfrac{\partial A}{\partial D} - \dfrac{\partial A^c}{\partial D}} \, d\varepsilon.$$

As the description of damage is at the ply level, any stacking can be analysed and damage can be predicted in all plies. Academic but instructive examples are given in figure 10.12. Transverse and longitudinal strains are recorded. These cases propose two different scenarios of damage evolution in each ply.

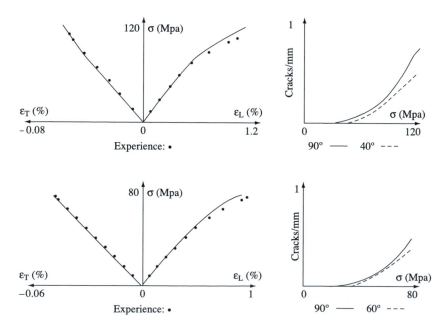

Figure 10.12. (a) Tensile test on a T300/914 laminate $(40_2, 90_2)_s$. (b) Tensile test on a T300/914 laminate $(60_2, 90_2)_s$.

10.3   Inter-ply cracking: delamination

Experiments carried out on layered composite materials, show that inter-facial delamination occurring at the free edge could be very detrimental for the safety of such structures. This phenomenon, commonly called the 'edge effect', is due to the combination of different factors:

- any one layer in the structure is basically orthotropic;
- adjacent layers have different directions of orthotropy.

The major consequence of this is the coupling, at the interface, of the in-plane and out-of-plane strain and stress fields close to the edge. This problem, which is inherent to laminated structures, has to be completely understood for the design of these structures. Because the edge effect has been considered by many analytical approaches which give a variety of solutions that can be found in the literature, an explanation of this mechanism will be proposed using a numerical approach.

10.3.1   Physical characterization of delamination

Experiments carried out on glass fibre epoxy matrix angle-ply composites, show that the possible failure modes can be very different, depending on the orientation of the plies (figure 10.13).

Observations, of specimens with different lay-ups, by optical and scanning electron microscopy as well as x-radiography, after tensile tests, allow the physical damage mechanisms which occur to be observed.

In the case of a $[\pm 30]_s$ lay-up, we observe an inter-ply delamination which initiates at the edge and propagates towards the centre of the specimen. This delamination, which is the consequence of an 'edge effect', is

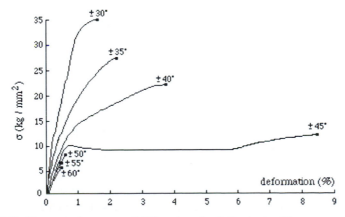

Figure 10.13. Stress–strain curves of different angle-ply laminates [5].

characterized by high out-of-plane shear stresses. It does not greatly affect the in-plane stiffness of the laminate or its elongation to failure and few transverse cracks are generated inside the plies. These observations explain the quasi-linearity of the loading curve up to the failure of the specimen.

It is remarkable to observe the change of behaviour between the $[\pm 40]_s$ and the $[\pm 45]_s$ laminates. The material behaviour changes from being quasi-brittle to ductile, exhibiting failure strains of about 7%. In this latter case, the first damage to appear is transverse cracking coupled with the non-linear behaviour of the matrix. Their combination results in the plateau in the loading curve, as shown in figure 10.13. Delamination appears during the last stage of the curve coupled with a realignment of fibres of about 2%, which explains the recovery of the material before it breaks.

If we now increase the angle θ, the failure mode completely changes. Damage is by intra-ply cracking which initiates and propagates in a mixed mode manner. Transverse cracking is the major damage process up to failure of the specimen.

It is interesting to point out that if failure mechanisms of $[\pm 30]_s$, $[\pm 35]_s$ and $[\pm 40]_s$ are different from those of $[\pm 50]_s$, $[\pm 55]_s$ and $[\pm 60]_s$, the stress–strain curves are not very different. The behaviour is mainly brittle. This is in contrast to the behaviour of the $[\pm 45]_s$ laminate, which is ductile.

These few experiments show the necessity of better appreciating the stress states of angle-ply laminates $[\pm \theta]_s$ with θ lower than 45°, the failure mode of these composites is by inter-ply delamination initiated by the 'edge effect'.

If we consider laminate theory, the generalized constitutive law sets relations between the generalized efforts (forces and couples) and in-plane strains and curvatures by

$$\left\{ \begin{matrix} N \\ M \end{matrix} \right\} = \left[\begin{matrix} A & B \\ B & D \end{matrix} \right] \left\{ \begin{matrix} \varepsilon^0 \\ \kappa \end{matrix} \right\}.$$

In the case of a symmetrical stacking sequence, the coupling rigidities between membrane and bending actions are zero. Forces are related to membrane strains and moments with curvatures. To illustrate the edge effect phenomenon, supposing that there is no bending, it can be seen that the constitutive law reduces to

$$\left\{ \begin{matrix} N_1 \\ N_2 \\ N_6 \end{matrix} \right\} = \left[\begin{matrix} A_{11} & A_{12} & A_{16} \\ A_{12} & A_{22} & A_{26} \\ A_{16} & A_{26} & A_{66} \end{matrix} \right] \left\{ \begin{matrix} \varepsilon_1^0 \\ \varepsilon_2^0 \\ \varepsilon_6^0 \end{matrix} \right\}.$$

If we carry out a tensile test in the 1-direction on a flat coupon, only N_1 is not zero and $N_2 = N_6 = 0$. From this last equation, we can obtain ε_2^0 and ε_6^0 as a function of ε_1^0. Theories of laminated structures assume that each

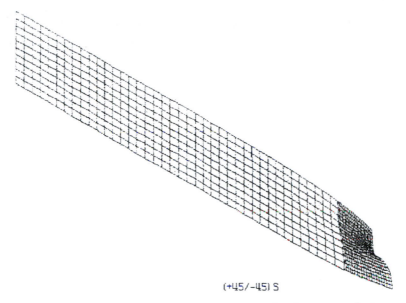

(+45/-45) S

Figure 10.14. Local warping at a free-edge interface of a $[\pm45]_s$ laminate ($\frac{1}{4}$ of the cross section).

ply experiences the same strains, thus putting these values in for each ith ply of the laminate, we can calculate σ_2^i and σ_6^i. These stresses cannot be zero at the edge, which is contrary to the expected result as the specimen sides are not loaded.

This is in violation of the boundary conditions which is the cause of edge effects. The abrupt change of properties at the interfaces of adjacent plies, generates local distortions and the local stress state becomes three-dimensional (figure 10.14).

10.3.2 Numerical analysis of 'free edge' effect

Consider a specimen consisting of four unidirectional layers $[\theta_1, \theta_2]_s$. The specimen has mirror symmetry about $x_3 = 0$ plane. The specimen is subjected to a tensile load applied along the x_1 axis and the stress state is examined in a cross section sufficiently far from the grips for there to be no end effects. A finer examination is made around point A which is the intersection between the θ_1, θ_2 interface and the free edge. For this purpose the meshes were refined in this area and the stability of the solution is assessed by refining the original mesh around what is considered to be the critical zone. Different θ_1, θ_2 can be computed. Radiating meshes around point A allow the singularity to be examined in polar coordinates with point A as the origin (figure 10.15).

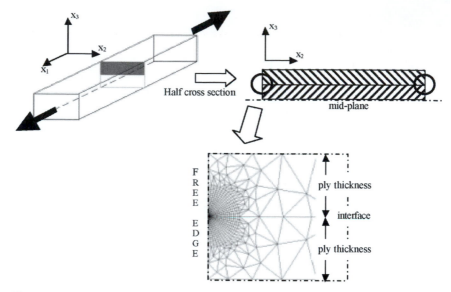

Figure 10.15. Radiating mesh surrounding the free edge interface.

As an example, σ_5 is plotted along the interface up to the edge in figure 10.16. These results, obtained with different mesh sizes, show that with smaller finite elements, the value nearest point A increases sharply, whilst the rest of the curve remains unchanged.

A systematic study for different $[\theta_1, \theta_2]_s$ composites shows that the expression of stress can be written in polar coordinate (r, φ) as the summation of two expressions consisting of a linear and power functions:

$$\sigma = f(\varphi)r^{-\alpha} + g(\varphi).$$

Nevertheless to verify the equilibrium equations, all interfacial stresses are singular, close to the edge. An example is given with the normalized tensile stress σ_1 in figure 10.17.

As we go towards the centre of the specimen, the singular part of the stress rapidly decreases and stresses obey laminate theory. Out-of-plane stresses vanish and become zero. This means that the edge singularity is very localized. Physically speaking, this singular zone represents a small number of cells which include a fibre surrounded by matrix. Thus if we want to consider the effect of the edge when designing laminated structures, a mesh size of about half the ply thickness seems to be a realistic good compromise.

Such a refinement is excessive for the calculation of industrial structures but it is possible on simple specimens and helps in the choice of the stacking sequence as a function of the applied loading on the real structure. Let us consider the following industrial problem, which is to choose between two stacking sequences of identical stiffnesses. Which sequence is the better to

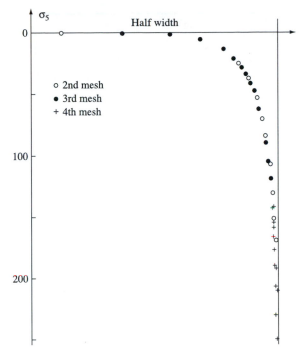

Figure 10.16. Out-of-plane shear stress σ_5, at the interface $+45/-45$ of a $[+45, -45]_s$ composite.

avoid delamination? For this purpose, we are going to compare the $[0, \pm45, 90]_s$ and the $[0, 90, \pm45]_s$ stacking sequences. Finite element calculations give, for each sequence, the results shown in figures 10.18 and 10.19 when a tensile load is applied in the fibre $0°$ ply direction.

If we analyse the results we see that σ_3 stress (pealing stress in mode I) is a maximum and positive in the $[0, \pm45, 90]_s$ laminate. However, the magnitude of this stress is lower and compressive in the $[0, 90, \pm45]_s$ composite. It is obviously preferable to choose the second stacking sequence which is

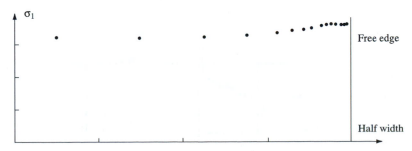

Figure 10.17. Normalized stress σ_1 at the interface $+45/-45$ of a $[+45, -45]_s$ composite.

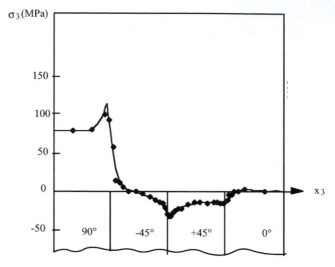

Figure 10.18. Variation of σ_3 at the free edge of a $[0, +45, 90]_s$ composite (1/2 of the edge).

less liable to delaminate as the interface is in compression. To obtain a complete description of the stresses involved, it is of course necessary to make a similar analysis of all stresses.

This example shows that, for a given material, very simple calculations of local stresses allow different possible stacking sequences to be analysed and the best one to avoid delamination chosen. Such investigations have

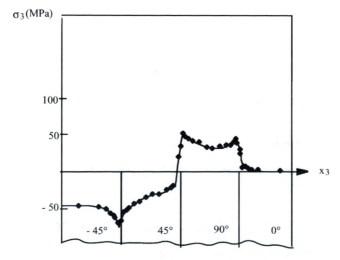

Figure 10.19. Variation σ_3 at the free edge of a $[0, 90, \pm 45]_s$ composite (1/2 of the edge).

also shown that clusters of identical oriented plies are more easily delaminated than a stack of alternating plies at different angles.

10.4   Fibre breakage

This type of damage often corresponds to the last stage of the life of a composite structure. However, laboratory tests on fibres show a large variability of failure properties and for the onset of failure, as described in chapter 2. This explains the brittle characteristics in the fibre direction in carbon and glass fibre composites. In order to mitigate this variability, safety factors play an important role in the design process. Nevertheless, the ability to design components correctly, using advanced composites, requires a detailed understanding of the statistical nature of the fracture. The main problem for the designer is the estimation of the scale effect, namely how the size of the specimen influences the scatter of the material properties such as strength. The question is to extend experimental values, obtained in the laboratory with small gauge lengths to full-scale calculations of structures. Once again we are going to see how a multi-scale approach based on homogenization techniques allows the scale effect, on the failure probability of the composite structure, to be appreciated. The procedure which will be proposed will be applied to the statistics of the fracture stress of a unidirectional T300/914 carbon fibre reinforced composite under a load applied parallel to the fibres. The steps are the following.

1. Obtain, through appropriate experiments, the statistical information concerning the distribution of the defects.
2. Introduce the results in a finite element calculation to precisely describe the damage mechanisms in the material and to calculate the local stress and strain fields.
3. During the calculation, the randomly distributed defects in the finite element mesh are broken according to a local fracture criterion and using homogenization, the macroscopic fracture of the homogenized volume can be estimated.
4. Carry out such a calculation at different scales. The results of one scale is then used as data for the computation at a higher level up to the macro scale of the industrial structure. Meanwhile at each intermediate scale, any new mechanisms or new scatter in the behaviour has to be appreciated. These must be revealed through appropriate experiments.

In this way, the scale effects and the influence of the micromechanical parameters involved in the failure of the material can be estimated and the strength of the laboratory specimen can be extended to real structural components. This method could be a useful tool for the designer, as the use of a probabilistic design for composite structures can achieve potential weight reductions in comparison with the conventional safety factors.

10.4.1   Experimental procedure

Tensile tests and analysis of experimental results

The experimental values of fracture stresses obtained for two gauge lengths are classified in increasing order and the cumulative fracture probability is plotted versus the fracture stress (figure 10.20).

No important scale effect is observed in spite of the inherent experimental differences due to the rigid clamping, which are more important for short test specimens and a more uniform strain and stress field which can be achieved with a longer gauge length.

To appreciate better the influence or not of scale effects in the fibre breaking process, the relationship between two important stochastic parameters which are involved in the fracture behaviour of a unidirectional composite have to be considered. First, the distribution of defects along the fibres which will be estimated using single fibre tests and the multi-fragmentation test. These are discussed in chapter 2. Secondly, the variation of the volume fraction of the fibres which will be estimated by image analysis.

Image analysis

Image analysis is used to estimate the local fibre and resin volume fraction of the material. The initial image can be obtained using an optical microscope which is connected directly to the image analyser through a video camera. The image analyses store the obtained images with high resolution and

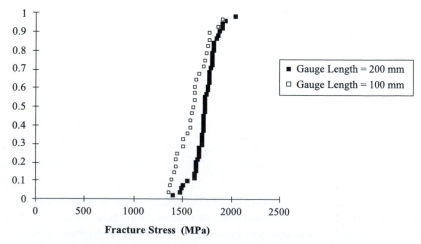

Figure 10.20. Fracture stress distribution of a UD (T300/914) composite specimen.

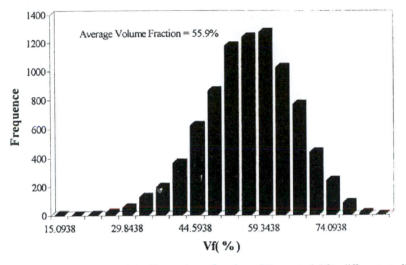

Figure 10.21. Distribution of the fibre volume fraction of the material for different studied specimens.

many grey levels, from which binary images (1 for the fibres and 0 for the matrix) are created by selecting a suitable grey intensity level. Figure 10.21 illustrates the fibre volume fraction distribution. A considerable variation in fibre volume fraction is often found, even though the average volume fraction is close to the nominal volume fraction of the plates. These results will be used as input in the proposed model in order to determine the influence of the local volume fraction on the fracture behaviour of the uni-directional composite.

Multi-fragmentation test

The variability of the strength of a unidirectional composite is mainly due to the distribution of flaws along the fibres. When a load is applied parallel to the fibres, breaks occur randomly at the most critical defects. As the load increases, damage zones arise in the material and coalesce to create a crack, the unstable propagation of which induces the failure of the material. Many authors have investigated the failure of a unidirectional composite [6, 7] using a Weibull strength distribution for the fibres, as explained in chapter 2. This is estimated through single fibre tests and the results applied to analytical solutions of the failure of unidirectional composites taking into account the assumptions of load transfer between fibres and matrix which occur by the shear lag process. Once again we are going to show that a multi-scale approach based on appropriate statistics and finite element calculations can be used to describe local mechanisms. We shall reconsider the Weibull analysis introduced in chapter 2 and apply it to our composite structures.

If we assume the material to be brittle in the fibre direction, 'the weakest link' assumption is made, so that the failure of the material occurs when the fracture stress of the most critical defect is reached. The cumulative fracture probability is given by

$$P_r(\sigma_R < \sigma) = 1 - \exp[-\theta(\sigma)V]$$

where P_r is a random generator, σ_R is the fracture stress of the material and σ is the applied stress. The density of the defects $\theta(\sigma)$ will be estimated by using the multi-fragmentation test, described in chapter 7, which reveals the whole population of defects on a fibre, in contrast to the single-fibre test which only exhibits the most critical defects. The density function of the defects is not an intrinsic parameter of the material because it depends on the sampling method. However, in the multi-fragmentation test, the results obtained reproduce better the state of the fibres in the matrix, and it provides information on the population of defects involved in the fracture process.

If the experimental values of fracture stress σ_R which are obtained are classified in increasing order, the density of defects along the carbon fibre can be fitted by a sigmoidal function of the applied stress:

$$\theta(\sigma) = A\{1 - \exp[-((\sigma - \sigma_s)/\sigma_0)^m]\}$$

where A represents the maximum number of defects per unit length, σ is the applied stress on the fibre, σ_0 is the scale parameter, and σ_s is the threshold stress. For the studied material T300/914, $A = 8.66$, $\sigma_0 = 6318$ and $m = 7.26$.

Using a Taylor series, this function can be approximated by a power law function for the weak values of the stress:

$$\theta(\sigma) = A\left(\frac{\sigma - \sigma_s}{\sigma_0}\right)^m$$

which characterizes the limiting case of the Weibull distribution. Therefore, using the sigmoidal density function, the fracture stress of a fibre with length L can be calculated as a function of the fracture probability. This will be useful in simulating the random variable σ_R for a random generator P_r:

$$\sigma_R = \sigma_0\left(-\ln\left[1 + \frac{\ln(1 - P_r)}{AL}\right]\right)^{1/m}$$

with the condition that $P_r < 1 - \exp[-(AL)]$. According to this condition, the sigmoidal distribution is valid for a fibre length $L \geq 0.5\,\mathrm{mm}$ for the fibre T300. This length corresponds to the mean fragment length, described as the critical length in chapter 7, at the saturation limit. Consequently the distribution function cannot be extrapolated for fibre lengths lower than 0.5 mm.

10.4.2 Finite element simulation

The finite element method is based on the discretization of the continuous medium, which is represented by a mesh with a number of elements. As the number of elements is increased, the numerical solution converges to the 'exact solution'. Assuming that each finite element represents a defect with an associated fracture stress σ_R, the value of the fracture stress depends on the size of the element. Therefore the coupling between finite elements and appropriate statistics requires a precise estimation of the fracture stress as a function of the applied scale. In more detail, as the size of the finite element decreases, the convergence of the stress and strain fields occurs, whilst the probability of zero defect tends to one.

The major problem arises from the extrapolation of the results from the scale where experiments have been carried out to the finite element scale. Furthermore, the multi-fragmentation test requires that fibres of a minimum length of 0.5 mm be used.

It then becomes possible to consider a small part of the whole composite containing only a few fibres. These results can then be applied to lengths of thin layers of unidirectional composite material (intermediate scale), containing different numbers of fibres and different volume fractions of fibres under a load parallel to the fibres. These thin layers play the role of the representative volume element (RVE), the fracture of which is going to be simulated. For this purpose, the fibres are divided into elements the length of which corresponds to the critical length. Each element then corresponds to a defect, the fracture stress of which is randomly assigned according to the fracture probability σ_R using sigmoidal and power law density functions. If the average stress calculated at the Gauss point of a finite element overcomes the associated fracture stress σ_R, the element is broken and the stiffness of this element is weakened (to the matrix stiffness for instance).

If the critical length corresponds to several finite element meshes (the case of a very refined finite elements mesh), all the elements over this distance are associated with the same value of fracture stress. In this way we decouple the introduction of probabilistic information associated with random fibre failure and the finite element size of the mesh. The failure stress of the RVE is defined when instability occurs during the calculation (figure 10.22).

At the next scale level (meso scale) the composite is considered to be an assembly of RVEs. The same procedure as before is adopted with a fracture stress distribution which is estimated through calculations at the previous scale. Step by step the final level simulates the real scale of specimens tested in the laboratory and includes the scatter in failure stress.

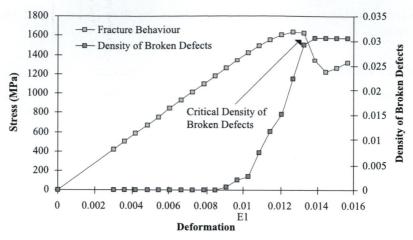

Figure 10.22. Fracture behaviour of a representative volume element.

Figure 10.23 gives a comparison between numerical cumulative fracture distribution and experimental data.

10.4.3   Numerical results

Many simulations, made for elements of different sizes and different distributions of the fracture stress σ_R along the fibres, show that the dispersion of the

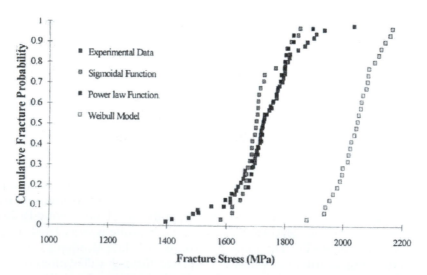

Figure 10.23. Simulated cumulative fracture distribution as a function of different density functions of defects along the fibres; comparison with the experimental data.

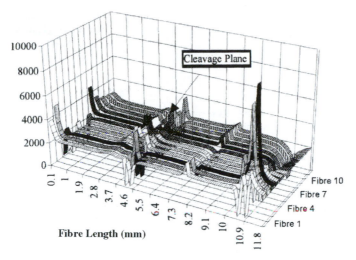

Figure 10.24. Stress field along the fibres close to rupture.

fracture stress decreases with an increasing length of the RVE and the average value and the standard deviation of the macroscopic fracture stresses remain nearly constant when a cluster of about 20 fibres is broken. An increase of the applied load initially causes diffuse damage corresponding to the fracture at the weakest points on fibres, increasing as the controlling defects become less important. At the same time, other defects initiate failure by inducing stress concentrations in the vicinity of the already broken elements. A point is reached, due to the large difference of carrying capacity of the fibres and of the matrix, when the stress concentration around broken fibres causes the failure of their neighbours and clusters of broken fibres develop locally. The stress concentration factors then become too large for any one fibre to stop the propagation of damage and the failure of the material is practically inevitable (figure 10.24). A critical density of broken fibres can be described as characteristic of the fracture behaviour in the fibre direction whatever the size of the specimen. Over a certain size, no important size effect is observed and the behaviour of the material becomes nearly deterministic. We therefore obtain, for a given composite, the size of the element representative of the fracture of a uni-directional composite [8]. This is the basis of the life prediction technique described for carbon fibre reinforced composite pressure vessels described in chapter 11.

Numerical and experimental results show that the material fails by the propagation of a single crack when a limiting of fibre failure density is reached whilst macroscopic behaviour remains almost linearly elastic.

This kind of fracture is considered as brittle and the fracture mechanics criterion can be used based on a critical number or a critical density of broken fibres. In this case the concept of unstable crack propagation can be used and macroscopic failure stress can be calculated. An average value of the distribution of the macroscopic fracture stresses obtained for large-scale composite parts represented by the fracture of UD elements, both in terms of size and in terms of critical bundles of broken fibres, can be used as a failure criterion. In the case of T300/914 composites, the values do not exhibit a large scatter and do not depend on the volume. Such an average strength can be used as a criterion for lifetime calculations.

This absence of scale effects can be shown by the similarity of fracture stresses when testing 100 mm and 200 mm gauge length specimens (already shown in figure 10.20).

10.5   Conclusion

The numerical and experimental analysis of damage in composite materials shows that it is important to consider each individual failure mechanism in the design of structures. Each induces local deteriorations which can be accelerated when they are coupled with other mechanisms. Global criteria are able to predict neither these processes nor their interactions. The essential tool which is necessary to link these physical events which happen at the microscopic scale and the global behaviour of the composite structure, or the fracture of the material, is homogenization. It can be used in a deterministic way when applied to calculating the overall behaviour or a ductile fracture of the composite part. Alternatively it provides a probabilistic approach when it deals with brittle fracture. Such criteria are used by designers to predict lifetimes of composite materials and to better appreciate safety factors which must be applied to composite structures. This allows greater weight reductions to be achieved in comparison with traditional safety factors which generally are overestimated and lack any scientific validation.

10.6   Revision exercises

1. What are the different types of damage occurring in composite materials?
2. Can you describe in-plane damages?
3. Describe inter-ply damage.
4. What is the influence of damage on a stiffness matrix?
5. Mention the main techniques which can be used to characterize damage and its evolution.
6. One of the most critical forms of damage is delamination between plies. This interfacial phenomenon is due to the so-called 'edge effect'. Mention the main factors which can induce this edge effect.

References

[1] Aveston J and Kelly A 1973 Theory of multiple fracture of fibrous composites *J. Materials* **8** 352–362

[2] Kaczmarek H 1993 Ultrasonic detection of the development of transverse cracking under monotonic tensile loading *Composites Sciences and Technology* **46** 67–75

[3] Renard J and Thionnet A 1992 Meso–macro approach to predict the damage evolution of transverse cracking in laminated composites, AMD-vol. 150/AD-vol. 32 *Damage Mechanics in Composites* ASM 1992, ed D H Allen and D C Lagoudas, pp 31–39

[4] Garret K W and Bailey J E 1977 Multiple transverse fracture in 90° cross-ply laminates of glass-fiber reinforced polyester *J. Materials Sci.* **12** 1957–1968

[5] Rotem A and Hashin Z 1975 Failure modes of angle ply laminates *J. Composite Materials* **9** 191–206

[6] Rosen B W 1964 *AIAA J.* **2**(11) 1985–1991

[7] Zweben C 1968 *AIAA J.* **6**(12) 2315–2331

[8] Baxevanakis C, Jeulin D and Renard J 1995 Fracture statistics of a unidirectional composite *Int. J. Fracture* **73** 149–181

Chapter 11

The long-term behaviour of composite materials

11.1 Introduction

The experience in use of conventional materials which has accumulated, sometimes over hundreds of years, is a great advantage for anyone designing a structure. Even on a personal level we quickly learn to avoid dropping ceramics. We know that they withstand high temperatures, but not abrupt changes in temperature and, once hot, they cool slowly. We know that metals can resist being deformed without breaking easily but if they are bent repeatedly they will break. We know that metals get hot quickly but also cool rapidly. This type of information has been analysed and codified so that designers have confidence in using traditional materials. Composites, however, have not existed for so long and most people have had little opportunity to acquire experience of advanced fibre reinforced materials, which in any case are evolving at a much faster rate than other more established materials. It would be very unsatisfactory if it were necessary to wait for experience of the use of composites to accumulate before they were used, not to say dangerous. This is because much of the acquired knowledge with conventional materials has been through accidents and failures, whereas composites, especially high performance composites, are often used in structures for which failure must be avoided at all cost. In addition the behaviour of fibre reinforced composite materials is often very different from that of other types of materials. For example, both ceramics and metals fail by simple crack propagation. This is not the case for many composites which can accumulate damage at the level of individual fibres or at the fibre–matrix interface without any apparent crack development. In addition, delamination of composite laminates can occur, often due to an impact, but it may not be visible at the surface of the composite. These differences can be disconcerting to designers used to dealing with other materials and this lack of experience has slowed the growth of the composite industry.

344

In order to better understand and to predict long-term behaviour of composites it is necessary to have a detailed knowledge of the mechanisms which control their properties and failure. In this chapter we shall examine both the effects of applied loading over periods of years on some types of advanced composite structures and also the effects that the environment can have in degrading the composite materials.

11.2   The effects of long-term mechanical loading on advanced fibre composites

Composite materials reinforced with continuous fibres can be made in many forms. The composite part is usually, if not always, basically two-dimensional even if it is a skin on a core material. The composite can be a stratified structure composed of layers or plies of composite material with the fibres in each ply being oriented in different directions so as to overcome the restrictions of unidirectional composites. Cross-plied composite structures are described as being quasi-isotropic if the fibres are arranged so as to give balanced properties in the plane of the laminate. Approaches to describe failure of these materials are described in chapter 5. Long-term degradation can occur due to delamination of the plies, and elastic fracture mechanics have been applied to describe this behaviour, despite the composite being far from the ideal material normally considered for this type of approach. However, the most difficult structure to test and to analyse so as to understand its long-term behaviour is the unidirectional fibre reinforced composite. This type of material is the essence of what a composite is and it poses many problems which are both theoretical and experimental.

It is often stated that the uses of unidirectional composite materials are limited because of their inherent anisotropy, and as most structures are subjected to multi-axial loadings cross-plied structures are generally used. An important exception, however, is filament-wound structures. We saw in chapter 2 that aramid fibres should not be loaded in any direction except parallel to their axis and in tension, but they are commonly used in filament-wound structures such as rocket motors. This is because filament-wound vessels subjected to internal pressure subject the fibres to tensile loads. Such vessels are increasingly being made but the tests used to periodically control metal pressure vessels are not applicable to composite vessels as the latter do not fail by simple crack propagation. During the winding process the tows of fibres are positioned along geodesic paths. Any force other than a tensile force, for example a shear force, will cause the tow to slide on the mandrel so that the fibres take these paths quite naturally. They are then maintained in these positions by the cured resin. A pressure vessel made by filament winding will therefore be made up of fibre tows arranged along geodesic paths and when the vessel is under pressure the fibres will be subjected only to tensile forces, as in a unidirectional composite. This means that at any given point the composite behaves and can be considered as a unidirectional fibre reinforced material. This observation is vital in attempting to understand and ultimately predict the long-term behaviour of composite pressure vessels.

11.2.1 Lifetime prediction and in-service control of carbon fibre composite pressure vessels

The use of natural gas as a fuel for buses and trucks is increasing rapidly due to the relatively low cost of the fuel and the low numbers of polluting by-products. This technology, which is presently being introduced, most probably foreshadows a future hydrogen economy in which hydrogen will provide much of the energy which is now provided by oil. The pressure vessels used to hold the gas under pressure have to be as light as possible, so high performance composites are an obvious choice. They are usually made of filament-wound high strength carbon fibres in an epoxy matrix. In service these vessels, holding natural gas, are pressurized up to 200 atm (20 Mpa). A typical bus requires nine such vessels, each having a capacity of around 135 litres, as can be seen in figure 11.1.

Figure 11.1. Natural gas carbon fibre reinforced plastic pressure vessels used on buses.

As with all pressure vessels, legislation requires that they are subjected to a proof test after manufacture and then periodically tested, for example every three or five years, to determine whether they can continue to be used until the next control. Standards and techniques exist for testing metal structures. These require the vessel to be pressurized to above (usually 1.5 times) the maximum pressure in service. The reasoning, for metal structures, is that any crack will be blunted by the plastic deformation at its tip so that if the vessel survives the test it will survive subsequent loading to lower pressures. Composite pressure vessels do not fail by the simple propagation of a crack but, in the absence of stress raisers or localized damage, by the accumulation of a more diffuse and global type of damage, of which the most important is the failure of fibres. It has long been known that overloading a composite structure leads to an increase in this type of damage, so the test required for a metal pressure vessel is inappropriate. The result of such a test on a composite structure is that the composite is more damaged and closer to final failure after the test than before. The overload test is therefore discredited. Composite pressure vessels can be inspected visually to determine if any accidental damage such as by an impact has occurred. In the absence of obvious damage it has to be assumed that the vessel is as new. Visual inspection of composite structures is, however, extremely limited in use as an impact and the subsequent local bending induced is most likely to produce delamination on the underface of the surface of the composite impacted and no external damage may be apparent. In addition the vessels should be removed from the vehicle, which introduces the possibility of further damage being produced to the pressure lines.

Acoustic emission has been shown to be a useful technique for monitoring damage in composites and by localization for identifying localized areas of damage. Many types of composite structure have been tested under cyclic and steady loads. In the case of carbon fibre reinforced composites, in which the fibres are disposed so as to be primarily loaded in tension, it is the reinforcements which determine the strength and rigidity of the structure. The failure of the structure, in this case, requires the fibres to fail. The acoustic emission technique can detect such fibre breaks, even though the fibres are of small diameter, usually 7 µm, and as they may occupy 60% of the volume of the structure, the number of fibres in the composite is very great. It is generally assumed that, as carbon fibres do not show any viscoelastic behaviour, once a carbon fibre composite has been loaded in the direction of the fibres no further evolution of the material can be expected. It is true that no discernible overall creep is detectable in the fibre direction if the load remains constant. However, it has long been known that damage in the form of fibre failures can be shown, by acoustic emission, to continue after a constant load has been applied, even in the case of unidirectional specimens loaded in the direction of the fibres. The damage mechanism which accounts for this behaviour has to involve the

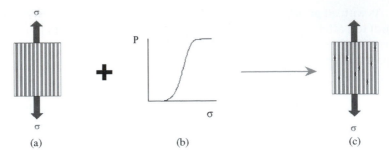

Figure 11.2. (a) The composite structure is described as behaving as a unidirectional composite. (b) The failure of the fibre reinforcements obeys a probability curve as a function of applied stress, leading to a random distribution of fibre breaks (c).

viscoelastic behaviour of the matrix material around sites of broken fibres. Under a constant applied load the rate of fibre failure can be modelled mathematically, and a relationship between damage accumulation and the duration of the loading can be determined. The technique can therefore have the potential to be used to monitor damage accumulation in carbon fibre composite pressure vessels and to allow minimum lifetimes to be predicted.

A typical carbon fibre pressure vessel consists of carbon fibre tows wound onto a liner which serves as a mandrel and provides a gas-proof seal. The fibres are impregnated with epoxy resin which is cured after the filament winding is complete. Both aluminium liners and polyethylene liners are used. They typically have a capacity of 135 litres and measure approximately 1.4 m in length and 0.4 m in diameter. After manufacture, the pressure vessels are subjected to a hydraulic pressure test of 300 atm, which corresponds to 1.5 times the operating pressure.

The development of damage in a carbon fibre composite structure, in which the fibres are disposed so as to transfer maximum strength and rigidity to the structure so that the failure of the composite depends on the failure of the fibres, is likened to the accumulation of damage in a unidirectional composite loaded in the fibre direction. Figure 11.2(a) represents a unidirectional composite. The failure probability as a function of applied stress on the fibres, shown in figure 11.2(b), can be determined experimentally and is described with considerable accuracy by Weibull statistics, which are discussed in greater detail in chapter 2 and which describe the probability of fibre failure as

$$P = 1 - \exp\left[-\int \left(\frac{\sigma - \sigma_{\mathrm{u}}}{\sigma_0}\right)^m V\right] \tag{11.1}$$

where P is the failure probability of a fibre of volume V, σ is the applied stress, σ_{u} is the stress failure threshold and σ_0 and m are material constants.

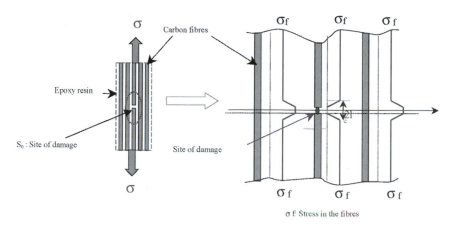

Figure 11.3. Creation of a damage zone around a fibre break: Zweben's model.

The factor m is known as the Weibull modulus and describes the scatter in fibre strength.

Loading the composite in the direction of the fibres produces fibre breaks, as shown in figure 11.2(c), which are randomly distributed in the volume of the composite, in the absence of any stress concentration, as there is a wide range of fibre strengths.

Each fibre break is isolated from other breaks by the shear of the matrix which produces a load transfer along the fibre so that at a critical distance from the break the fibre is able to continue to support the load. This distance is given by

$$l_c = \left[\frac{1}{2[1 + (1/n)]} \ln\left(\frac{1}{V_f}\right) \frac{E_f}{G_m} \right]^{1/2} r_f \qquad (11.2)$$

where r_f and E_f are the radius and Young's modulus of the fibre, G_m is the shear modulus of the matrix which varies as a function of time t due to relaxation, V_f is the fibre volume fraction, and n the number of neighbouring fibres around the fibre break.

This means that each fibre break creates a damage zone around the break, the limit of which is determined by the shear properties of the matrix and the Young's modulus of the fibre, as is illustrated in figure 11.3. In addition and importantly, there is a length of the fibre–matrix interface around the break which is debonded and which increases the load transfer length. The intact fibres neighbouring a broken fibre experience an increase in stress over a short length of each fibre as the total load supported by the plane passing through the fibre break has to remain unchanged. Shear stress relaxation of the matrix means that, with time, the lengths of the damage zones and the lengths of neighbouring intact fibres which experience increased stresses become longer.

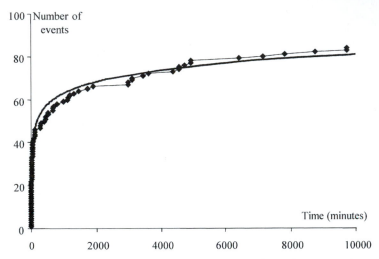

Figure 11.4. A comparison between the experimental and theoretical curves for a carbon fibre composite pressure vessel pressurized to 200 atm.

As the probability of failure increases with fibre length, for a given stress, additional fibres fail which can be detected by acoustic emission and the damage zone increases. When the damage zones begin to interact, damage accumulation accelerates, and this represents the onset of instability which is taken as the definition of useful life of the pressure vessel.

Relaxation of the shear stress leads to an evolution of the size of the damage sites, and an accumulation of fibre failures which can be detected as acoustic emission. The same behaviour is observed with a pressurized filament-wound vessel under a steady pressure and is shown in figure 11.4. This curve has been shown to obey a law of the following form:

$$\frac{\mathrm{d}N}{\mathrm{d}t} = \frac{A}{(t+\tau)^n} \qquad (11.3)$$

where N is the number of acoustic emission events, t is the time, A is a factor which depends only on the applied stress, τ is a time constant, and n is a factor less than unity.

By experimentally determining an upper damage threshold in terms of a maximum number of acoustic emission events which is acceptable, it now becomes possible to use equation (11.3) to create master curves for structures which have exactly the lifetimes required, as shown in the curve of acoustic emission as a function of time in figure 11.5.

In order that a pressure vessel is acceptable its acoustic activity must lie on a curve under the appropriate master curve. This is the basis of a technique proposed and patented jointly by the Ecole des Mines de Paris and Gaz de France for calculating residual lifetimes of pressurized carbon

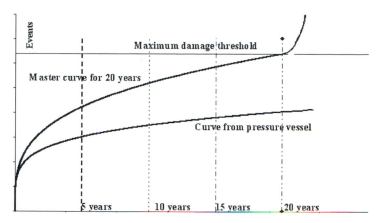

Figure 11.5. The accumulated acoustic emission events curve is compared with a master curve calculated for a life of 20 years. If the pressure vessel is acceptable for further use its curve should be lower.

fibre filament-wound pressure vessels. It is not feasible to obtain all the curve of a pressure vessel; however, at any given time its rate of acoustic activity must be lower than that of the master curve. In this way a comparison of the activity during a short period with the activity predicted by the master curve allows an evaluation of the residual lifetime of the pressure vessel to be made. The vessel is acceptable if the number of emissions during the test is less than or equal to that given by the master curve. If not, it should either be changed or subjected to further examination.

Using equation (11.3) this curve can be extrapolated over a period of 20 years and compared with a master curve giving an exact lifetime of 20 years, as shown in figure 11.5. It is assumed that control tests have to be carried out immediately after manufacture and then every five years. It can be seen that at each control the gradient of the experimental curve is less than that of the master curve for the case of a pressure vessel which is acceptable for further service. The values of A and τ in equation (11.3) for pressure vessels subjected to 200 atm were obtained from the experimental curve and this allowed the master curve to be calculated.

It now becomes possible to draw curves which can be used in practical tests. The numbers of emissions expected from the master curve depends on the length of time the control test is conducted and this can be chosen so as to accommodate the needs of the bus operator. Figure 11.6 shows curves for a control period of 24 hours corresponding to pressure vessels which have been in service for different periods of time.

It can be seen that carbon fibre reinforced composites continue to accumulate damage even when they are subjected to a constant load, or pressure in the case of a pressure vessel. This damage can accumulate until

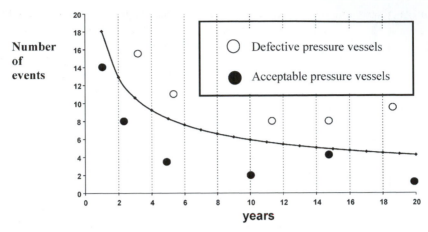

Figure 11.6. During a given control period of 24 hours the master curve indicates the maximum number of emissions which are acceptable (source: PhD thesis, S Blassiau).

it reaches a critical level at which the composite becomes unstable. As this behaviour is seen even in unidirectional composites loaded in the direction of the fibres, the time-dependent mechanism controlling delayed failures must be the relaxation of the matrix around broken fibres. This leads to an increasing influence of the fibre break and an overload effect on the intact fibres neighbouring the break. As the strength of fibres shows a wide scatter, a local increase in stress on intact fibres will inevitably cause further breaks which can be detected by the acoustic emission technique.

The regular rate of emission which is detected reflects the effects of shear stress relaxation around fibre breaks which leads to further failures. The accumulation of damage can therefore be modelled to give the accumulation of damage as a function of time. For a given series of supposedly identical pressure vessels, initial tests can identify the maximum acoustic emission which is acceptable and so master curves can be drawn for pressure vessels with just the lifetimes required. Pressure vessels at manufacture or in service can be controlled periodically by a comparison between their actual acoustic emission count in a given period and the count predicted by the master curve for the same period. If the actual count is less than the limit given by the master curve the pressure vessel is deemed acceptable for further service. If not it should be withdrawn from service.

11.3 Fatigue deterioration of composites

The ageing of composite pressure vessels, as described above, shows that even under steady loads composites will evolve and gradually deteriorate. This is common to everything and so should not come as a surprise. Some and perhaps even most materials can also be progressively damaged by another process called fatigue when they are subjected to cyclic loading. This has been a particular concern of designers when using metals for construction. The failure of metal structures due to metal fatigue has been a major cause for accidents and loss of life from the beginning of the industrial revolution and continues to be a major cause of concern for all metallic structures subjected to varying loads.

The discussion in chapter 7 and figure 7.28 shows the fundamental cause of premature failure of materials at loads which are lower than what would be expected from a simple knowledge of the stresses necessary to provoke failure at the microstructural level. The presence of irregularities or small cracks in a continuous medium produces an amplification of the stress applied to the body of the structure so that at the microscopic level, around such a defect, the local stress can be much higher. This concentration of the stress produces, locally, a rearrangement of the structure, in metals by the movement of dislocations, and repeated cyclic loading leads to the slow growth of the crack until it becomes critical for the loads applied. At this point the crack growth becomes unstable and the structure breaks. This type of behaviour is often depicted for specimens subjected to cyclic stresses, for example with a minimum zero stress taken to different maximum stress, as a function of numbers of cycles to failure. These curves are called S–N curves.

Steel pressure vessels, similar to those made with carbon fibre reinforced plastic (CFRP), discussed above, are used to store gas at high pressure. They are much heavier than the CFRP pressure vessels and so are less suitable for use in transport. Even before the mechanisms producing fatigue were understood, experience showed that steel pressure vessels could explode unexpectedly after some time in use and many did at the beginning of the use of steam power. For this reason such pressure vessels are subjected to periodic proof testing in which they are briefly loaded to one and a half times their maximum load in service. If the pressure vessel survives the proof test it seems intuitively obvious that it will survive the lower pressures that it will experience in service. This procedure can be justified for steel structures as the overload will produce a zone of plastic deformation around any incipient crack. At lower loads this plastic region will block further crack growth so the structure has been stabilized by the proof test. We have seen above that composites do not generally fail by simple crack growth and because of the random nature of defects on the reinforcing fibres the nature of damage accumulation is different from that in metals. The consequence, as has been explained, is that an overload, above the maximum load experienced in service, means that the structure would be more damaged and therefore closer to ultimate failure after such a test than before. The discussion above describes how a composite pressure vessel can be proof tested based on an understanding of the failure processes at the microstructural level of the fibres and surrounding matrix. Nevertheless the possibility of deterioration of the composite when it is subjected to cyclic loads which are lower than those which would normally cause failure in monotonic tests has to be investigated. That is to say, the question as to whether composites suffer from a fatigue process, analogous to that which exists in metals, has to be answered. As the highest performance composites used in structures are made from CFRP, it is this material which will be discussed below.

11.3.1 Fatigue in carbon fibre reinforced plastics

As we are concerned by the processes of damage accumulation in these composites we shall limit our discussion to unidirectional materials. This is because most composites are made up of unidirectional layers superimposed on one another and in some cases entirely made with all the fibres aligned in the same direction. Cross-plied composite laminates can suffer from progressive delamination between the plies which can be provoked by cyclic loads, and this can be treated by an approach using classical fracture mechanics, as with metals, as it involves crack propagation, between the plies, in a region mainly dominated by the matrix. In the case of unidirectional material, the composite structure is highly anisotropic, which makes understanding and testing the material, to determine its fatigue behaviour, extremely difficult.

What is clear, as is discussed in section 7.2 of chapter 7, when a unidirectional composite is loaded parallel to the fibre direction it is the fibres which control behaviour, and failure of the composite must involve breakage of the fibres. We can therefore start by wondering if carbon fibres themselves can fail by a fatigue process. The answer is that they appear not to be susceptible to fatigue failure. This is because the microstructure of the carbon fibres is composed of carbon atoms bonded to each other by very strong covalent forces which can be deformed by an applied load but which, at room temperature, do not allow the movement of the atoms. The consequence is that carbon fibres are perfectly elastic and on cyclic loading follow the same load-elongation curve on loading and unloading even at the microscopic scale. The insensitivity of carbon fibres to fatigue failure was seen reflected in the fatigue behaviour of unidirectional CFRP and early studies, usually in the aircraft industry, showed S–N curves which were nearly horizontal with any decrease in strength falling within the scatter of tensile strengths found with the specimens. Note, however, that it was, not surprisingly, the aircraft industry which was conducting these studies. This industry, gladly, is conservative and concerned with absolute reliability, so they use large safety factors, especially with materials with which they are lacking experience. The maximum loads to which these specimens were subjected were of the order of 25–30% of average breaking loads which, as the earlier high strength carbon fibres had failure strains of less than 1.2%, meant maximum imposed strains of between 0.3 and 0.4%. Glass fibres, on the other hand, fail at more than 3% and unidirectional glass fibre reinforced plastics (GRP) subjected to similar fractions of ultimate failure load were experiencing strains of up to and above 1%. As GRP is a cheaper material it found more rapid use in a wider range of industries than CFRP and was often subjected to greater strains than 1%. Under these conditions GRP was found to be deteriorated under cyclic loads—not by crack propagation normal to the fibres but rather by the disintegration of the matrix. The mechanisms involved were seen to be failure of the resin between the fibres. This observation underlines that the behaviour of a composite material depends on all it constituents and that even a unidirectional composite loaded in the fibre direction can be deteriorated if the matrix is damaged. It is clear that the low strain amplitudes traditionally studied in the aircraft industry do not subject the matrix or the fibre–matrix interfaces to sufficiently large strains for them to be damaged. However, improvements in carbon fibre properties now mean that many CFRP structures are made with fibres having a strain to failure approaching or exceeding 2%, so it is necessary to see if the larger strains that the composites could experience in other areas of use could lead to fatigue failure. This type of information is an absolute necessity if CFRP is to penetrate general engineering applications and be accepted by designers.

What is clear is that CFRP is better than GRP in resisting fatigue failure when cycled to the same fractions of tensile breaking load for each type of composite, as can be seen from figure 11.7 based on data from Curtis. The GRP, however, has a much lower elastic modulus and larger failure strain. The principal characteristic of fatigue in both these composites is longitudinal splitting parallel to the fibres which is induced by inherent anisotropy of the material. As with the model described above for damage accumulation in the CFRP pressure vessels, damage is initiated at failure points of weaker fibres which break during initial loading. The shear stresses around the breaks cause the resin and the interfaces to degrade and promote cracks which propagate along the fibre–matrix interfaces. The cracks affect the ability of the material to redistribute load as it becomes further damaged and so more fibres break which leads to further resin damage and ultimate failure. Curtis concluded that the fatigue sensitivity of unidirectional composite materials depends on the susceptibility of the material to the initiation and growth of longitudinal splits. Final failures have a brush-like appearance.

With improvements in carbon fibre strength and strains to failure there has developed an increasing interest in using CFRP in major engineering projects in which the composites could be strained to greater levels than those considered in figure 11.7. More recent studies have examined the effects of loading so as to produce larger cyclic strains. It can be expected

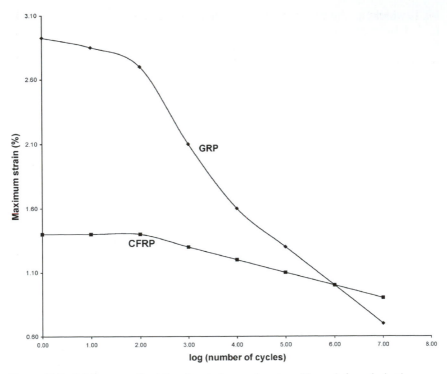

Figure 11.7. S–N curves of unidirectional glass and carbon fibre reinforced plastic.

that as CFRP is subjected to larger strain amplitudes the matrix and the fibre–matrix interfaces will be damaged and that the earlier conclusion that such composites are insensitive to fatigue processes would be found not to hold.

The behaviour of the unidirectional CFRP subjected to large strain amplitudes has been seen to depend on the loading level with lifetimes reducing as the load amplitude increases. The load ratio, R, is defined as the ratio of the minimum stress to the maximum stress. Let us consider three load ratios, $R = 0.5$, $R = 0.3$ and $R = 0.1$. For the same stress level (σ failure %), there is an increase in the number of cycles to failure with increasing load ratio R (reduction of cyclic load amplitude). This evolution is clearly seen from the experimental curves shown in figure 11.8. The arrows represent results obtained from specimens which did not fail.

The representation on a linear scale seems to indicate a tendency towards a threshold under which failure of the specimen would not occur. This observation agrees with the literature that CFRP does not suffer from fatigue when tested with small strain cycles but that larger cyclic strains can have a damaging effect. For this study a non-failure threshold was

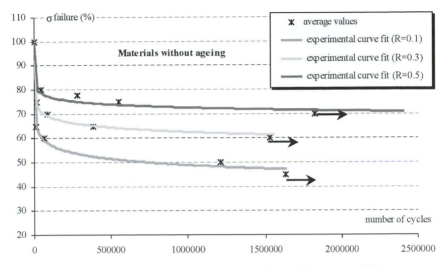

Figure 11.8. Typical S–N curves obtained with similar unidirectional CFRP specimens subjected to large cyclic strains for three load ratios. Linear scale.

defined for the specimens reaching 1.5 million cycles. The tests were stopped after this number of cycles and the results indicated by an arrow on the curve. The threshold has been confirmed, for $R = 0.5$ and a maximum applied stress of 70% of the nominal strength, with a specimen tested for more than 2.4×10^6 cycles without failure. Figure 11.9 shows the appearance of a specimen after failure.

These results indicate that unidirectional CFRP composites do fail by a fatigue process involving the degradation of the matrix and of the fibre–matrix interfaces. Any modification to the matrix or to the interfaces could be expected to modify the fatigue lifetimes and this is seen to be the case when the composite has been exposed to a wet environment. The absorption of water can have major effects on the behaviour of composites as it can destroy the interfaces and also make molecular movement in the resin easier. It can be seen from figure 11.10 that this is the case as the fatigue threshold levels for specimens aged in water have been reduced compared with those found with the specimens in their original dry state as shown in figure 11.8.

When the fracture surfaces of the fatigued specimens are examined by scanning electron microscopy two types of failure process can be observed in the as-received specimens, as shown in figure 11.11. These are matrix failure at an angle of 45° compared with the loading direction (fibre direction) and fibre–matrix interfacial debonding. Figure 11.11 shows a micro-crack which has initiated from a region of interfacial debonding and this seems to be the origin of greater cracking of the matrix. The

Figure 11.9. The brush-like appearance of a unidirectional CFRP specimen broken during a fatigue test.

micro-cracks are oriented at 45° compared with the loading direction. The growth and coalescence of micro-cracks led to the general cracking of the matrix. Although local debonding of the interface is seen in the as-received dry specimens, sufficient bonding remains between the fibres to hold them together and the resin fracture surface is characteristic of a brittle material. Debonding in the aged specimens is much more obvious so, although both matrix cracking and debonding are seen in fatigued aged specimens, debonded fibres are much more in evidence in these specimens, as can be seen in figure 11.12.

It is evident therefore that unidirectional CFRP is not protected from fatigue processes if the load and strain amplitudes are sufficiently large. This is because the matrix and fibre–matrix interfaces degrade due to the

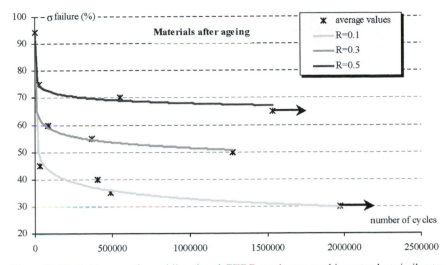

Figure 11.10. S–N curves for unidirectional CFRP specimens aged in water but similar to those used in their dry form to obtain the results given in figure 11.8. Linear scale. The curves show lower fatigue threshold levels for the aged specimens.

cycling, and the onset of matrix cracking leads to further degradation. Taken to an extreme the composite is reduced to its constituent fibres with the matrix reduced to a powder. The effect of water is seen to help in the breakdown of the matrix.

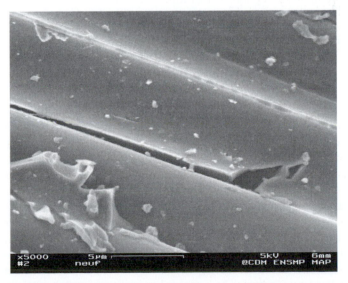

Figure 11.11. Fatigue failure in as-received unidirectional CFRP can be seen to involve both interfacial debonding and matrix cracking.

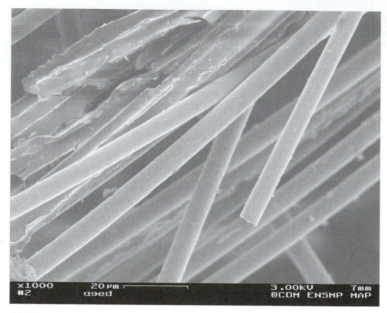

Figure 11.12. Aged specimens showing debonded carbon fibres.

The fatigue behaviour of unidirectional CFRP is therefore seen to be essentially similar to that of GRP and in both composites it is the failure of the matrix and interfaces which limit the fatigue lives.

11.4   Environmental ageing

We have seen, above, that the effect of CFRP absorbing water is to reduce its fatigue performance. Water is everywhere and composites will inevitably absorb water, from the air which holds water, and is measured as relative humidity, or from being in direct contact with liquid water by being immersed in it, or as a composite pipe, by transporting it. Although composites may be in contact with many forms of corrosive liquid it is water which presents an all-pervading concern as absorbed water has the potential of breaking the interfacial bonds between the fibres and the matrix on which the composite depends for its mechanical properties.

Fibre reinforced resins are often used, not only for their good mechanical properties and light weight, as well as their ability to be made into useful and sometimes complicated forms, but also because of their chemical inertness. Tanks for the storage and transport of corrosive chemicals are often made of composites. The preferred choice of materials in environments for which even stainless steel would be quickly corroded, such as in deep-seated oil drilling, is also moving towards fibre reinforced resins. Never-the-less composites do have an Achilles heel—the fibre–matrix bond. We have seen in chapter 7 that the properties of composites depend critically on the transfer of loads between the matrix and the fibre, which in turn depends on the quality of the adhesion between the components. This adhesion is often controlled by rather weak atomic bonds which can be broken by the arrival of water molecules in the vicinity of the fibre–matrix interface. Dramatic falls in composite properties can be expected if the bond is broken in this way, even if the fibres and the bulk matrix material are not modified. However, in some cases either the reinforcements or the matrix, or both, can also be deteriorated by water. Although composites may be exposed to many different types of environment, some of which may cause degradation, it is most often water absorption which can be the most damaging and which must be understood.

The fibre–matrix interfaces in resin matrix composites provide preferential channels for the ingress of outside agents which can modify the interactions between the fibres and the matrix. This means that water absorption can proceed much more quickly in the direction parallel to the fibres than in other directions, so we shall see that not only are mechanical properties of composites dependent on fibre arrangements but also the uptake of water and therefore the rate of ageing of the composites. The characteristics of the reinforcements and of the matrix materials can also be altered by the absorption of agents or by the temperature at which the materials are used.

Such deterioration is usually irreversible and strikes at the basic mechanism of fibre reinforcement as it removes the ability of the matrix to transfer load to the fibres and can reduce the performance of the component parts making up the composite. This is made worse as the water at the interface acts as a lubricant. Applications for which the composite must be immersed in water obviously raise the question of its effects, but water is also in the air, in the form of humidity. Resin matrix composites will absorb water to a lesser or greater extent and although measures such as painting can hinder its penetration, eventually the composite will either come to an equilibrium with its environment or will be degraded by it. It is therefore necessary to understand and to be able to calculate the rate of diffusion of water in these materials as a function of water concentration and temperature, as the process is thermoactivated. Water also has the effect of a lubricant at the molecular level on the bulk structure of the resin which is revealed most strikingly by a fall in glass transition temperature, T_g, as absorption proceeds. In some resin systems this can lead to dramatic falls in T_g so that at the beginning of exposure the composite may be well below the T_g but as water is absorbed the T_g falls below the ambient temperature and the physical behaviour of the matrix is as a consequence greatly changed. This effect can be totally or partially reversible.

Water can also provoke the hydrolysis of the resin leading to a deterioration of the matrix and composite. In the case of thermoplastic matrix composites this can lead to an increase in crystallinity and progressive embrittlement of the matrix. In some high temperature resin systems, oxidation of the polymer matrix can also be a source of degradation.

In the case of metal–matrix composites it is most often the interaction between the fibres and matrix during composite manufacture which can be the cause of deterioration or galvanic corrosion between the components, when water is present, if the two types of material possess markedly differing electron energy levels.

Ceramic matrix composites rely on the interface to act as a mechanical fuse which fails when a crack approaches and so blunts further propagation. Even when the crack passes around the fibre the failure of the interface over a substantial distance provides an important mechanism for absorbing energy and so contributes to the overall tenacity of the composite. The interface in such materials is often controlled by a carbon layer deposited on the fibre surface or by multilayers of carbon with other materials. Oxygen in the air, however, can remove the carbon when the composite experiences high temperatures, usually over 700 °C, and the interface is replaced by brittle silica which has the effect of dramatically reducing tenacity. In addition the fibres themselves evolve at high temperatures, so they may lose strength and begin to creep.

11.4.1   Water diffusion

The diffusion laws have generally been applied to the modelling of simple water uptake, and it has been shown that the simple Fick's law can be applied to carbon fibre reinforced epoxy resin subjected to humid environments. In this model the water is considered to remain in a single free phase driven to penetrate the resin by the water concentration gradient. Other studies have indicated, however, that non-Fickian processes do occur which can complicate the understanding of the role of the water and which may lead to irreversible changes in mechanical properties. Water penetration can cause swelling and plastification of the resin, and ingress can occur by capillary action along the fibre–matrix interface. Composite structures consisting of several layers of differing types of fibre lay-up, often with a thick resin coat or gel coat, can suffer from water being trapped in a particular layer or at the interface between layers. This effect may be irreversible because of osmosis, and blistering may result. This is particularly a problem with glass fibre reinforced boats for which considerable effort is made, by a judicious choice of resin materials and manufacturing techniques, to avoid this type of damage. In addition to changes in mechanical properties, modifications to other physical characteristics may be observed, such as changes in dielectric properties.

Diffusion models

Fick's law and single-phase diffusion. Fick's law is applied to simple single free-phase diffusion of water into a material. The law assumes that the rate of water absorption $(\mathrm{d}M/\mathrm{d}t)$ is determined by the water concentration gradient $(\mathrm{d}C/\mathrm{d}x)$, and is related by the coefficient of diffusion, D. This means that we can write

$$\frac{\mathrm{d}M}{\mathrm{d}t} = -\left(D_x \frac{\partial C}{\partial x} + D_y \frac{\partial C}{\partial y} + D_z \frac{\partial C}{\partial z}\right). \tag{11.4}$$

In general, water can be absorbed in all three orthogonal directions which may have different absorption coefficients. Water concentration is the amount of water in a unit volume of the material. The negative sign in equation (11.4) is necessary as the increase in mass M has to be positive and the concentration gradient is negative.

Composite materials are most often used as large plates with little thickness. This is the case for even big structures such as an aircraft wing for which the composite may be used for the skin but the interior would be of a honeycomb structure. This means that water absorption in many composite structures is in one direction normal to the plan of the fibres. If we call this direction x, we can now see that equation (11.4) can be written as

$$\frac{\mathrm{d}M}{\mathrm{d}t} = -D_x \frac{\mathrm{d}C}{\mathrm{d}x}. \tag{11.5}$$

If the rate of water diffusion is written as F then

$$F = \frac{\mathrm{d}M}{\mathrm{d}t} = -D_x \frac{\mathrm{d}C}{\mathrm{d}x}. \tag{11.6}$$

If we consider an elementary section of the material, of unit surface area and thickness δx, as shown in figure 11.13, the rate of change of water concentration will be the amount of water absorbed divided by the volume of the element (δx). So

$$\frac{\mathrm{d}C}{\mathrm{d}t} = \frac{F_x - F_{(x+\delta x)}}{\delta x}. \tag{11.7}$$

The value of the water concentration varies through the thickness of the material as absorption takes place so, if the concentration at the one face is C, the concentration at the other face must be $C + (\mathrm{d}C/\mathrm{d}x)\delta x$.

From equation (11.5) we see that the difference between the rate of water entering and leaving is

$$-D_x \frac{\mathrm{d}C}{\mathrm{d}x} - \left[-D_x \frac{\mathrm{d}C}{\mathrm{d}x} - \frac{\mathrm{d}}{\mathrm{d}x}\left(D_x \frac{\mathrm{d}C}{\mathrm{d}x}\right)\delta x\right].$$

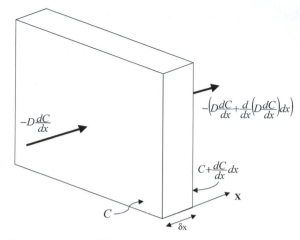

Figure 11.13. The rate of water diffusing through one face of an elementary section of the composite and leaving through the opposite face.

As D_x is considered constant, the increase in weight is $D_x(\mathrm{d}^2C/\mathrm{d}x^2)$, which from equation (11.7) gives

$$\frac{\mathrm{d}C}{\mathrm{d}t} = D_x \frac{\mathrm{d}^2 y}{\mathrm{d}x^2}. \tag{11.8}$$

In the general case where the absorption is in three directions and, as we have seen above, the coefficients of diffusion are different in the three orthogonal directions, we must write

$$\frac{\mathrm{d}C}{\mathrm{d}t} = D_x \frac{\mathrm{d}^2 C}{\mathrm{d}x^2} + D_y \frac{\mathrm{d}^2 C}{\mathrm{d}y^2} + D_z \frac{\mathrm{d}^2 C}{\mathrm{d}z^2}. \tag{11.9}$$

Most composite structures are, however, basically two-dimensional as they are made of layers of fibres, often in the form of unidirectional lamina or woven cloth. This means that for many composite structures the thickness of the material is much less than the lateral dimensions, so the problem can be reduced to the penetration of water at right angles to the plane of the fibres. This can cause difficulties in interpreting laboratory specimens subjected to ageing experiments in a humid environment as such specimens are often of small size, and the surface areas of the edges can be a significant fraction of the total surface area of the material. This is exacerbated by the diffusion coefficient being greater in the direction of the fibres. We have all seen examples of the effects of water uptake in the fibre direction leading to accelerated deterioration of the material. Wooden posts placed in the ground rot from the bottom up because of the diffusion of water up the wood along the grain, or fibre, direction. However, most artificial composite structures are big enough for the effects of water uptake in the fibre direction to be neglected and the interpretation of laboratory results can be misleading.

In practice it is the variation in weight as the composite plate absorbs water which is measured. The relationship between the water uptake M_t and the change of water concentration C_t at a depth x in the composite material is given by

$$M_t = 0.5 \int_{x=-h/2}^{x=h/2} C_t \, dx. \tag{11.10}$$

The weight gain, due to water absorption obeying Fick's law by a composite plate for which unidirectional absorption can be assumed, was determined by Crank. If the plate is of thickness h, the mass M_t of water absorbed in time t across a unit surface area as a fraction of the mass of water absorbed at saturation, M_m, is given by

$$\frac{M_t}{M_m} = 1 - \frac{8}{\pi^2} \sum_{n=1}^{\infty} \frac{1}{(2n+1)^2} \exp[-D(2n+1)^2 \pi^2 (t/h^2)]. \tag{11.11}$$

This equation has been simplified (1), to give

$$\frac{M_t}{M_m} = 1 - \exp\{-7.3[D(t/h^2)]^{0.75}\} \tag{11.12}$$

and they showed that, for $\sqrt{Dt}/h < 0.2$, the initial absorption is given by

$$\frac{M_t}{M_m} = \frac{4}{h} \left(\frac{Dt}{\pi} \right)^{1/2} \tag{11.13}$$

and for $\sqrt{Dt}/h > 0.2$

$$\frac{M_t}{M_m} = 1 - \frac{8}{\pi^2} \exp\left(-\frac{Dt\pi^2}{h^2} \right). \tag{11.14}$$

The initial linear relationship, of M_t/M_m as a function of the square root of time, is to be noted, and so water uptake is usually plotted as M_t/M_m versus the square root of time. It is considered that the first molecular layer of the specimen reaches equilibrium with the environment instantly so that the water concentration at this point is always equal to the equilibrium concentration C_m. The value of C_m will vary with the relative humidity of the environment. Water uptake by the body of the composite is described by equation (11.11) so that water molecules take increasing time to arrive at increasing different depths in the material. A composite, or other material, which absorbs water and which is subjected to varying degrees of humidity in the environment will experience a variation of water content at the surface and the centre of the material may only begin to experience water diffusion after a delay.

If we consider a large, initially dry, composite plate of thickness h placed in constant humid conditions at a constant temperature, water will diffuse

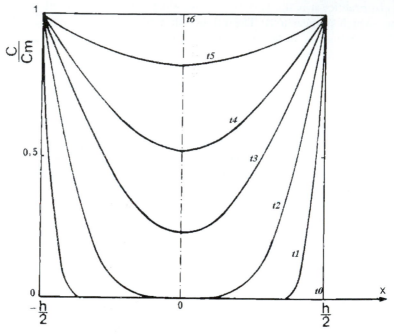

Figure 11.14. Water concentration profiles as diffusion proceeds through both faces of a large composite specimen of thickness h, for increasing lengths of time ($t_0 < t_1 < t_2 < t_3 < t_4 < t_5 < t_6$).

through both surfaces in the direction, normal to the surfaces, which we shall designate as the x direction, with the origin being at the centre of the plate. The surfaces are considered to reach saturation instantaneously but the diffusion takes time so that, as figure 11.14 shows, the water concentration will vary within the material as a function of time. The equation for the change of water concentration at a distance x from the centre of the plaque is given by

$$\frac{C_t}{C_m} = \left[1 - \frac{4}{\pi} \sum_{n=0}^{\infty} \frac{(-1)^n}{2n+1} \cos\left(\frac{(2n+1)}{h} \pi x \right) \exp\left(-\frac{Dt}{h^2} \pi^2 (2n+1)^2 \right) \right].$$

(11.15)

Figure 11.14 shows in graphic form the change of water concentration profile as a function of time t, which is given by equation (11.15).

It can be seen that the water concentration is a maximum at C_m at the surfaces from the beginning of the diffusion but that water has to diffuse to the centre and that this takes time. Saturation is reached when the water concentration profile is uniform across the thickness of the plate which in the case of thick specimens could take years. Figure 11.15 shows the results

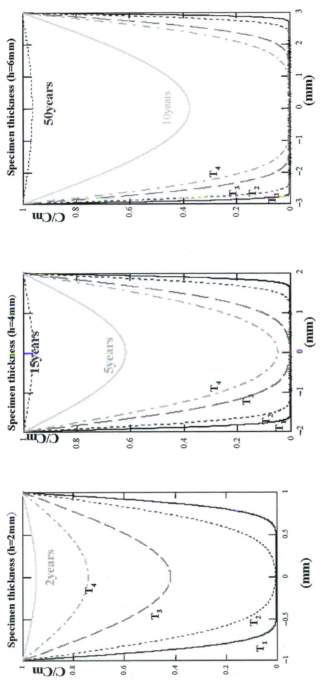

Figure 11.15. The change of water concentrations in three similar composite plates but of different thicknesses ($T_1 = 1$ week, $T_2 = 1$ month, $T_3 = 6$ months, $T_4 = 1$ year).

Table 11.1. The time for water to arrive at the centre of similar composite specimens of different thicknesses and for them to become saturated.

Composite thickness	1 mm	2 mm	4 mm	6 mm
Time for water to reach centre of composite	6 days	35 days	200 days	400 days
Time for the composite to become saturated	6 months	2.5 years	18 years	50 years

of calculations, using equations (11.13) and (11.14), for a typical composite having a coefficient of diffusion of $D = 4 \times 10^{-8} \, \text{mm}^2 \, \text{s}^{-1}$. It can be seen that for a thickness of 2 mm the composite is nearly completely saturated after two years' immersion but that it would take 15 years with specimen twice as thick and 50 years for a specimen three times as thick. The times for water to arrive at the centre of the composite specimens and also for them to become fully saturated are given in table 11.1.

The increase in weight of the composite, due to water absorption, is given by the increase in water concentration multiplied by the volume of the material. If Fick's law is obeyed, the form of the curve of weight against square root of time, multiplied by $\sqrt{D}/h$, is as shown in figure 11.16.

It is only when the composite has reached saturation under the prevailing humidity conditions that a uniform water concentration across the specimen exists. This should be a warning for conducting experiments to reveal the

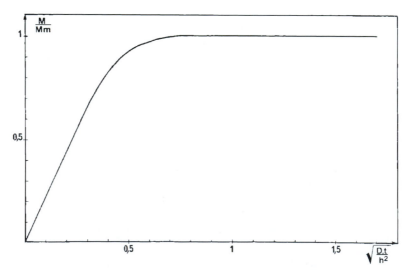

Figure 11.16. Weight increase due to water absorption which obeys Fick's law which supposes that the water is in a single free state and is able to circulate freely in the composite.

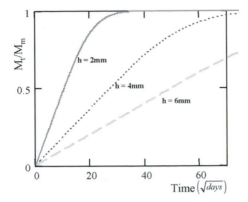

Figure 11.17. The initial increase in weight due to Fickian absorption of water by three composite plates of different thickness.

effect of water uptake on composite behaviour, since if the specimen has not reached saturation a complex diffusion state exists which most probably will give unreliable results. The effects of thickness on the initial increase in weight of the composite is shown in figure 11.17, using the same data as were used to draw figure 11.15. It can be seen how, even though all three specimens if left long enough would arrive at the same level of saturation, this is not obvious from the initial slopes of the curves and this can lead to misinterpretation of what is occurring. The time to saturation as a function of composite plate thickness can be seen to increase exponentially from figure 11.18.

Figure 11.19 shows classical Fickian behaviour of a glass fibre reinforced epoxy resin at two temperatures and at different rates of relative humidity. It can be seen that the saturation levels are independent of temperature but do depend on relative humidity.

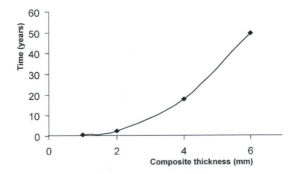

Figure 11.18. Time to saturation for similar composite plates of different thickness.

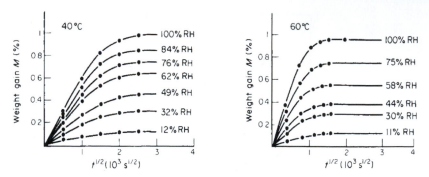

Figure 11.19. Absorption curves obtained under different conditions of temperature and relative humidity with glass fibre reinforced DGEBA epoxy resin cured with a diamine hardener, showing classical Fickian absorption (source: PhD thesis, Ph Bonniau).

The coefficient of diffusivity, D, which is directly related to the slope of the curve, is therefore seen as being dependent only on the temperature, and it can be surmised that it obeys an Arrhenius type law such as

$$D = D_0 \exp\left[\frac{-E}{RT}\right] \tag{11.16}$$

where R is the gas constant and is equal to $8.314\,\mathrm{J\,mol^{-1}\,K^{-1}}$, E is the activation energy which, depending on the resin used, is generally in the range 40–80 kJ/mol and T is the temperature in Kelvin. Figure 11.20 shows the Arrhenius curve for a DGEBA epoxy resin system cured with a diamine hardener and reinforced with glass fibres. Figure 11.21 shows how

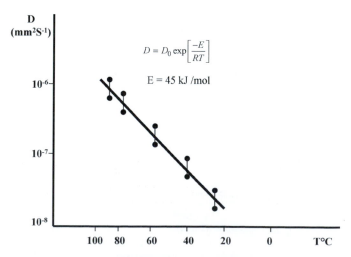

Figure 11.20. The coefficient of diffusivity, D, as a function of $-\log T$ (source: PhD thesis, Ph Bonniau).

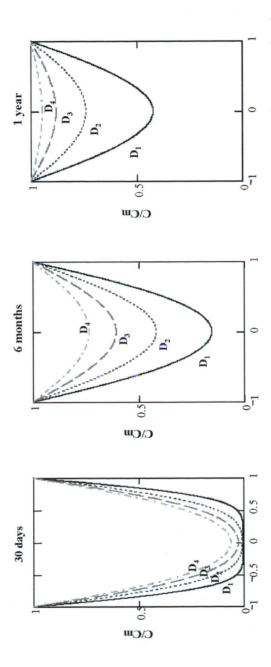

Figure 11.21. The effect of an increasing value of the diffusion coefficient can be seen to become very apparent after some weeks of water absorption. ($D_1 = 2 \times 10^{-8}$ mm^2 s^{-1}, $D_2 = 4 \times 10^{-8}$ mm^2 s^{-1}, $D_3 = 6 \times 10^{-8}$ mm^2 s^{-1}, $D_4 = 8 \times 10^{-8}$ mm^2 s^{-1}).

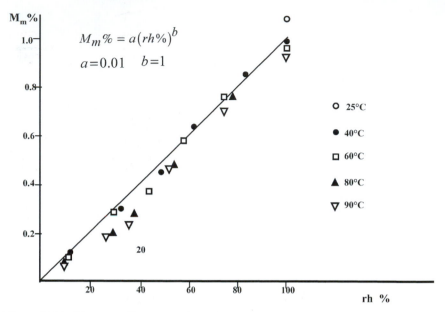

Figure 11.22. The saturation level of a glass fibre reinforced DGEBA epoxy resin as a function of relative humidity.

changes to the coefficient of diffusion due to changes in the temperature have a marked effect on the rate of water uptake.

The saturation level of the same composite system as a function of relative humidity is shown in figure 11.22. It can be seen that the saturation level increases linearly with relative humidity and that the temperature has little effect.

In the case where water can penetrate into the material in several directions and the coefficient of diffusivity varies with the direction, edge effects can become important and should be considered. For a unidirectional specimen, two diffusion coefficients must be considered accounting for penetration normal $(D_\perp)$ and parallel $(D_\parallel)$ to the fibres, and a third co-efficient, D_r for the unreinforced resin, also exists.

It has proved possible to derive the relationships between the diffusion coefficients of the composite and that of the resin by analogy to thermal and electrical diffusion, taking into account the fibre volume fraction V_f and assuming a square or hexagonal fibre-packing arrangement. In the case of a square array of fibres having a diameter d and distance between centres a, the volume fraction is given by

$$V_f = \frac{\pi d^2}{4a^2}.$$

(11.17)

The relationships are therefore

$$D_{\parallel} = (1 - V_{\mathrm{f}})D_{\mathrm{r}} \tag{11.18}$$

$$D_{\perp} = [1 - 2(V_{\mathrm{f}}/\pi)^{1/2}]D_{\mathrm{r}}. \tag{11.19}$$

The Langmuir model of diffusion.   In cases where the simple Fickian model proves inadequate, such as when two plateaux in the absorption curve are observed or if it is found that the absorption depends on the specimen thickness, it is often useful to consider a more complex model of the Langmuir type. In applying this model it is often assumed that the diffusion coefficient remains independent of water concentration as in the Fickian model. However, the water was considered to be in two phases, one free to diffuse and the other trapped and so not free to move in the absorbing medium.

The analysis of diffusion involving chemical reaction, reversible or otherwise, has been treated by Crank. In addition to the two parameters D and M_{∞} used in the simple free-phase diffusion model, the two-phase model introduces two other parameters: α, which is the probability of a trapped water molecule being released and γ, the probability of a free water molecule being trapped.

In passing to a two-phase model it is assumed that the absorbed water is in two states: the free phase having a water concentration $n(t)$ water molecules per unit volume can diffuse in the resin, but can be linked or trapped in the resin with a probability γ, and the combined phase with concentration $N(t)$ able to break free with a probability α. At saturation

$$n_{\mathrm{sat}}\gamma = N_{\mathrm{sat}}\alpha \tag{11.20}$$

that is to say, there is an equilibrium between those water molecules being trapped and those being freed. By analogy with equation (11.6) we can write

$$D\frac{\partial^2 n}{\partial x^2} = \frac{\partial n}{\partial t} - \frac{\partial N}{\partial t} \tag{11.21}$$

and

$$\frac{\partial N}{\partial t} = \gamma n - \alpha N. \tag{11.22}$$

The analysis of this situation results in the following simplified equation which describes the rate of water uptake in a plate of thickness $2h$. For $[\alpha, \gamma \ll (\pi^2 D)/h^2]$:

$$\frac{M_t}{M_m} = 1 - \frac{\gamma}{\gamma + \alpha}\mathrm{e}^{-\alpha t} - \frac{8}{\pi^2}\frac{\alpha}{\gamma + \alpha}\mathrm{e}^{-\gamma t}\sum_{i=1}^{i=2n-1}\frac{1}{i^2}\exp[-(\pi i/2h)^2 Dt]. \tag{11.23}$$

It should be noted that the Langmuir model reduces to the simple Fickian model where $\gamma = 0$. As for the Fickian model the initial water uptake of a plate, as a function of time, is linear and is given by equation (11.24):

$$\frac{M_t}{M_m} = \left(\frac{\alpha}{\alpha + \gamma}\right) \frac{4}{\pi} \sqrt{\frac{dt}{h^2}} \qquad (11.24)$$

with $\alpha/(\alpha + \gamma)$ being the proportion of free water in the composite. If we put $\gamma = 0$ we find the expression for the initial part of the curve given by Fick's law. The linear part of the curve occurs in a time given by

$$t \leq \frac{0.7h^2}{\pi^2 D}. \qquad (11.25)$$

For a time of the order of $5h^2/\pi^2 D$ an initial levelling out of the absorption curve is observed at which point

$$\frac{M'}{M_m} = \frac{\alpha}{\alpha + \gamma} \qquad (11.26)$$

where M' is the levelling off point.

For $t \gg h^2/(\pi^2 D)$, the composite slowly approaches saturation and we obtain

$$\frac{M_t}{M_m} = 1 - \frac{\gamma}{\alpha + \gamma} e^{-\alpha t} \qquad (11.27)$$

and assuming that both α and $\gamma \ll (\pi^2 D)/h^2$ it can be shown that

$$-\left(\frac{dM_t}{dt}\right)^{-1} \frac{d^2 y}{dx^2} \approx \text{constant} \equiv \alpha \qquad (11.28)$$

and

$$e^{-\alpha t}\left[\alpha\left(\frac{dM_t}{dt}\right)^{-1} M_t + 1\right] \approx \text{constant} \equiv 1 + \frac{\alpha}{\gamma}. \qquad (11.29)$$

The four variables in the Langmuir two phase model (D, M_m, α and δ) can be determined by using equations (11.23), (11.28) and (11.29) and the experimental values taken from the absorption curve. The calculation is very sensitive to inaccuracies in determining the experimental values so that great care should be taken in their determination.

Figure 11.23 shows two-phase water absorption, as described by the Langmuir model and equation (11.23) for three specimens of DGEBA resin cured with a dicyandiamide hardener and subjected to three different humid environments.

The result of this type of two-phase diffusion is that the composite eventually comes to an equilibrium with its surroundings so that it becomes saturated. However, this takes longer than with single-phase diffusion and the curve of weight gain as a function of the square root of time can show a pseudo-plateau. This pseudo-plateau is not necessarily

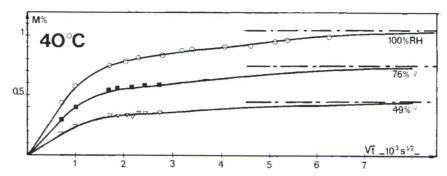

Figure 11.23. Water absorption curves for an epoxy resin reinforced with glass fibres obeying the two phase model (source: PhD thesis, Ph Bonniau).

obvious as, to be so, the thickness of the plate should be small or the probabilities α and γ should be small compared with D.

As the parameters D and M_∞ are still invoked, the model involves four variables and, given the heterogeneity of a composite material, there exist numerous possible interpretations for the physical mechanisms involved in absorption. One is that water can diffuse freely and can interact chemically with the molecular structure of the resin or coupling agent; alternatively, water can be mechanically trapped, absorbed in the matrix or onto the fibre surface. As the interfacial zone is probably very different from the unreinforced resin, different diffusion behaviour could also be expected, even without considering capillary infiltration. The physical processes involved in water absorption by the composite are certainly more complex than the processes considered in the Langmuir model but because of its additional flexibility the model is applicable to a wider range of cases.

Physicochemical aspects of absorption in resin matrix composites. Most polymers absorb water to a greater or lesser degree so that, even in the absence of defects or preferential routes for water uptake such as are provided by poor fibre–matrix adhesion, resin–matrix composites will absorb water in humid environments. Water diffusion occurs in most resins by hydrogen bonds being formed between the water and the molecular structure of the resin.

Epoxy resins are the most widely used resins in high-performance composite structures. They are mixed with a hardener, which can be of several types, to produce a crosslinked structure which provides many sites for hydrogen bonding for the water molecules such as hydroxyl groups (O−H), phenol groups (O−C), amine groups (N−O) and sulphone groups (O−S). Figure 11.24 shows the sites available on a commonly used epoxy consisting of a bisphenol A epoxy resin crosslinked with a dicyandiamide hardener. The hydrolysis of epoxy resins is in itself a cause of degradation and is not by any means limited to this class of matrix materials.

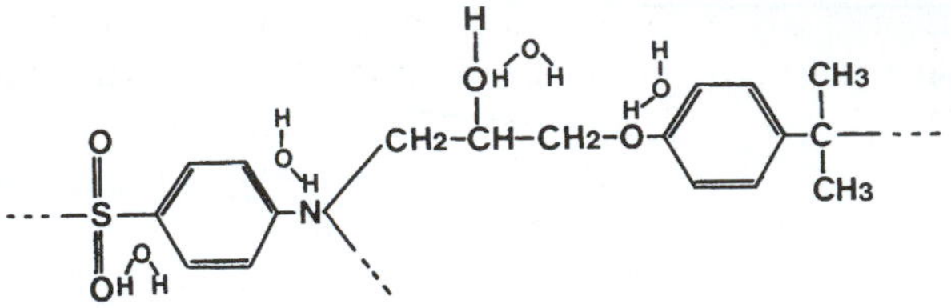

Figure 11.24. Sites available for water molecules to be combined by hydrogen bonding onto the molecular structure of a DGEBA epoxy resin crosslinked with a dicyandiamide hardener.

Figure 11.25 shows a more general example of water molecules becoming attached to the molecular chains of the resin and also forming secondary crosslinking which produces enhanced swelling of the matrix.

Carbon fibres are unaffected by water but others, such as those of glass, can be damaged by prolonged exposure to water which arrives at the interface. Glasses are made up of silica in which are dispersed metallic oxides including those of the alkali metals. These latter non-silicate constituents represent micro-heterogeneities which are hydroscopic and hydrolysable. The absorption of water by glass is therefore characterized by the

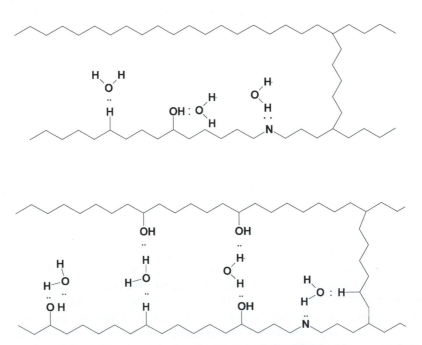

Figure 11.25. Water molecules can adhere to the main chain molecules (shown as a zigzag) and also produce secondary crosslinking which further enhances matrix swelling (source: PhD thesis, L Adamczak).

hydration of these oxides. The most common form of glass fibre is made of E-glass, which contains only small amounts of alkaline-metal oxides ($Na_2 + K_2O < 2wt\%$) and so is resistant to damage by water. However, the presence of water at the glass-fibre surface may lower its surface energy and promote crack growth. Water trapped at the interface may also allow components of the resin to go into solution and if an acidic environment is formed the fibre can be degraded.

Organic fibres such as the aramid family can absorb considerable quantities of water and although this does not lead to a marked deterioration of fibre properties the resulting swelling of the fibre may be a cause of composite degradation. It will certainly change the dielectric properties of the composite which in some applications such as in radomes for which electromagnetic transparency is important may present a problem.

The size, or coating, put onto many fibres to protect them from abrasion and to ensure bonding with the matrix also serves to protect them and the composite from damage from water absorption by eliminating sites at which the water can accumulate. However, this is not always adequate.

Influence of the resin matrix on composite degradation. One of the most widely employed composite materials is composed of glass fibres embedded in an epoxy matrix. Degradation of this type of composite depends greatly on the type of hardener used to crosslink the resin. Figure 11.19 shows the classical Fickian behaviour of glass fibre reinforced bisphenol A epoxy cross-linked with a diamine hardener. It has been shown that this behaviour leads to an initial weight gain is a function of the square root of time. The composite reaches an equilibrium with its environment and saturates and the saturation level is a function of relative humidity and its independent of temperature. This behaviour indicates that the water is absorbed by the resin and is linked to it by hydrogen bonds. The saturated composite reveals a somewhat increased tenacity due to a softening of the matrix which is accompanied by a fall in glass transition temperature. Damage does occur under severe conditions (90 °C, 100% relative humidity for two weeks) and appears as a progressive whitening of the material with no detectable additional weight gain. This is probably due to microcracking which begins at the surface.

A change to a dicyandiamide hardener produces, however, a significant change in water-uptake kinetics, and this behaviour can only be explained by the two-phase model of absorption. The water absorption curves shown in figure 11.23 were obtained with this type of composite. The saturation is attained after considerably longer exposure than with the diamine-hardened resin but after equilibrium is reached both composites behave in a similar manner.

The use of an anhydride as the crosslinking agent, which is extremely common, produces a composite which, in a humid environment which is

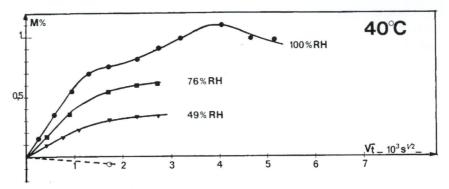

Figure 11.26. The use of an anhydride crosslinking agent with an epoxy resin is undesirable, as any excess anhydride will combine with warm water to give an acid and result in progressive degradation of the composite.

warmer than about $40\,^\circ$C, degrades quite rapidly, as can be seen from figure 11.26. The dotted line in the figure represents the weight of the composite after drying and shows that it loses weight. Immersed in warm water, this composite system shows an initial weight gain due to the absorbed water but this reaches a maximum, after which the weight falls due to leaching of the resin system into the water. Saturation is not achieved and severe weight loss after drying is observed. An anhydride is an acid with the water removed, so that the combination of an anhydride left over from the crosslinking process and warm water produces an acid which can quickly degrade the composite.

The correct choice of hardener in these epoxy composites is therefore extremely important if they are to be used in warm humid conditions.

As mentioned above, the glass transition temperature T_g of organic–matrix composites is lowered with water uptake. In the case of thermosetting resins such as the epoxies, this decrease can be as much as $30\,^\circ$C. It can be even more startling in the case of glass fibre reinforced polyamide 66, which is a thermoplastic. Polyamide 66, better known as Nylon 66, has a glass transition temperature when dry of around $70\,^\circ$C but this falls to around $0\,^\circ$C when the polymer is saturated with water, as shown in figure 11.27.

The effect is largely reversible and hydrolysis is not a major problem with this matrix material, which is found reinforced with short glass fibres in many industrial applications such as in several applications under the bonnet of a car. Such applications also use short glass fibre reinforced saturated polyester, which is also a thermoplastic. Hydrolysis can be a major problem with polyester, as can be seen from figure 11.28 which shows the change of behaviour as hydrolysis occurs at $90\,^\circ$C and 78% relative humidity in the material. The T_g of PET does not fall as with PA66, however, even below the glass transition temperature the behaviour of the composite is modified

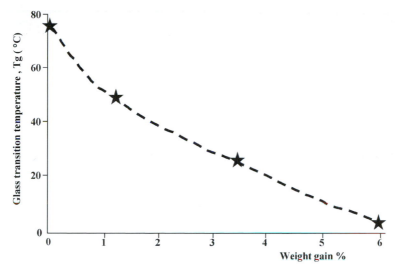

Figure 11.27. The glass transition temperature, T_g, of PA66 falls dramatically with water uptake.

by absorbed water, as can be seen from figure 11.29 which shows the effects of water at 50 °C.

When dry, polyester is deformable but with increasing exposure to water it becomes ever more brittle. The hydrolysis of the polymer cuts the

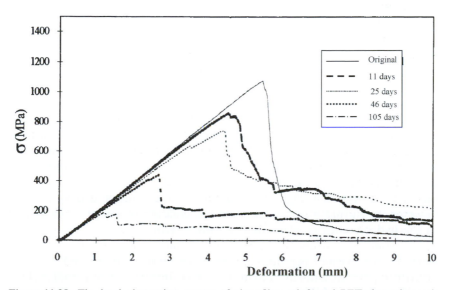

Figure 11.28. The load–elongation curves of glass fibre reinforced PET show dramatic changes as hydrolysis occurs at 90 °C and 78% relative humidity.

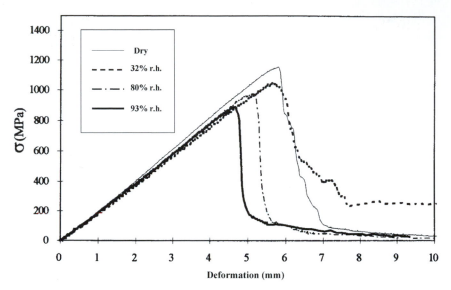

Figure 11.29. The load–elongation curves of glass reinforced PET aged at 50 °C at different values of relative humidity (source: PhD thesis, L Adamczak).

macromolecules into ever shorter lengths, which increases the ability of the molecules to move and so enables the increase in crystallinity. This leads to the embrittlement of the polymer which will spontaneously crack if left in hot water for long enough. By such time the crystallinity has increased significantly.

Families of resins able to withstand higher temperatures than the epoxy resins have been developed and in general are less susceptible to water uptake. Prolonged exposure of these resins to hot water can, however, lead to irreversible changes. Such a resin is polystyrylpyridine (PSP) which is usually reinforced with carbon fibres and destined for aerospace applications. This resin can operate at 250 °C for 1000 hours and can even withstand several hours at 400 °C. However, prolonged exposure to humid conditions, for example at 70 °C and 100% relative humidity, reveals complex behaviour. Initial water uptake appears to be Fickian. Infrared spectroscopy shows that the water is linked to the polar PSP functions by hydrogen bonding, with apparent saturation being reached at this stage. Longer exposures lead, however, to further water uptake and the creation of microcracks as well as the hydrolysis of ethylene double bonds left over after crosslinking of the polymer. The effects of water on other high temperature resins, such as PMR-15 and Avimid-N, which are rated to be used above 300 °C, when dry, are to produce a fall in T_g of around 80 °C during hydrothermal cycling.

Degradation of this type of resin matrix can also occur during thermal cycling, particularly if water is present in microcracks. The carbon fibre

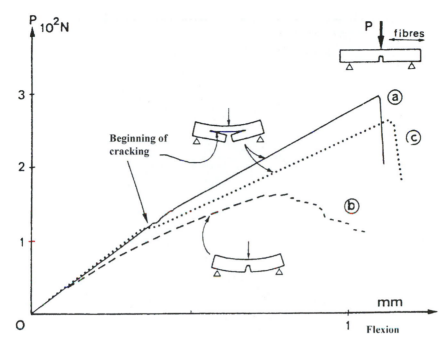

Figure 11.30. Loading in bending of a notched specimen: (a) new and dry, (b) after ageing in boiling water for 96 hours, (c) after ageing and redrying (source: PhD thesis, B Devimille).

reinforced PSP composites show increased water absorption after cycling from 70 °C to 150 °C due to the water being vaporized and forcing the microcracks further open. Cycling to 250 °C leads to a more marked fall in properties due to greater damage exacerbated by oxidation of the resin.

Mechanical degradation due to water absorption in resin matrix composites. As has been mentioned above, water penetration into resin–matrix composites can lead to changes in mechanical properties. A softening of the matrix is commonly observed which in some cases can be beneficial, inducing a greater toughness. However, changes are usually associated with a fall in properties such as strength or modulus. These changes can appear to be reversible as, upon drying, the composite may regain most of its lost properties. Figure 11.30 illustrates this for unidirectional glass reinforced epoxy subjected to three-point bending.

It can be seen that drying the specimen almost reversed the effects seen at saturation. A second cycle in humid conditions revealed that water penetration occurred much more quickly and property loss was much more rapid than in the first cycle. This behaviour is due to the destruction of the chemical bonding at the interface by the arrival of water. Figure 11.31 shows fracture surfaces of the three types of specimens. The original specimen shows fibres

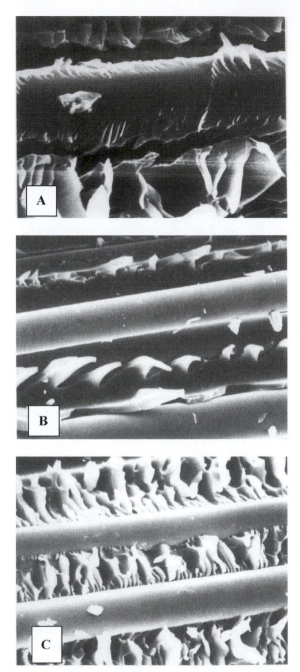

Figure 11.31. Fracture surfaces of a composite (A) before, (B) after ageing in boiling water, and (C) with subsequent drying. The glass fibres have diameters of 13 μm (source: PhD thesis, B Devimille).

which are covered with resin but the other two show smooth fibres bare of resin as the interfacial bond has been destroyed. This leads to greater fibre pullout on failure.

Upon drying, load transfer between fibre and matrix is assured by frictional forces which may largely compensate for the loss of chemical bonding. A second exposure to humidity results in a rapid water uptake as the water can quickly penetrate the composite along the fibre–matrix interfaces and, acting as a lubricant, rapidly results in loss of mechanical properties.

11.5   Corrosion of metal matrix composites

Fibre or particle reinforced metals are made in much smaller volumes than resin matrix composites but are of interest for applications that require higher operating temperatures than are attainable with the latter. In addition MMCs do not suffer from outgassing of water vapour or solvents as do resin composites which cause major problems in space applications. They can be made with very low coefficients of thermal expansion by balancing the properties of the reinforcements and those of the matrix. Such composites can also provide a means of increasing the stiffness of light alloys and of making abrasion-resistant materials. However, the manufacture of MMCs usually involves a stage when the reinforcements are in contact with the molten matrix material. This can lead to chemical interactions and the degradation of the reinforcements. Once the composite has been made a major difficulty can be the galvanic coupling of materials which may have very different electron energy levels. Corrosion in the presence of water and air is an electrochemical process involving anodic (oxidation) and cathodic (reduction) reactions. This can be written as the dissolution of the metal M by the anodic reaction, given as:

$$M \longrightarrow M^{n+} + ne^- \qquad (11.30)$$

For the reaction to proceed the electrons released by the anodic reaction must be consumed by a cathodic reaction. In an aqueous environment such reactions can be by oxygen reduction and proton reduction, such that

$$O_2 + 4e^- + 2H_2O \longrightarrow 4OH^- \qquad (11.31)$$

and

$$2H^+ + 2e^- \rightarrow H_2. \qquad (11.32)$$

The former reaction requires the presence of oxygen whereas the latter can occur in both aerated environments and in the absence of oxygen. In some cases the oxidation process can lead to the creation of a passive layer on the metal, which can eventually inhibit further corrosion, although this protection can be breached if aggressive anions, particularly chlorides, are present. This can result in pitting of the metal.

Corrosion by galvanic processes is the major concern in carbon fibre reinforced aluminium and magnesium. The fibres, which are electrical conductors, act as relatively inert electrodes for both proton and oxygen reduction. Attempts to limit this behaviour by coating the carbon fibres by a variety of materials, such as Ni, SiC, B_4C, TiC and TiB_2, increase the costs of manufacture but work by increasing the electrical resistance between the fibre and the matrix. The protection of the carbon fibres used in such composites may be found necessary as contact between the reinforcements and the molten aluminium leads to the formation of aluminium carbide (Al_4C_3) and falls in fibre strength. In addition the carbide which is formed hydrolyses in damp environments producing methane (CH_4) and aluminium hydroxide. Resin coatings of polyurethane, chlorinated rubber and epoxy on the outside of the composite have been shown to be effective in preventing this type of corrosion for some marine applications.

Silicon carbide as particles, fibres or whiskers is generally less prone to cause corrosion than carbon fibres but does suffer from galvanic corrosion in aerated chloride-rich environments, but this can also be controlled using resin coatings.

Processing can be the cause of fibre strength loss in boron fibre reinforced aluminium for which the fibres can be coated with SiC or B_4C. Silicon carbide fibres made by CVD onto a core must be protected by sacrificial layers, which are usually silicon- or carbon-rich, sometimes with a layer of TiB_2, when they are used to reinforce titanium.

There are no reports of significant corrosion occurring between aluminium- and alumina-based reinforcements although there are suggestions that degradation can occur between such fibres and magnesium matrices, but the mechanisms involved are not understood.

11.6   Ceramic matrix composites

This class of materials, CMCs, offers the possibility of structural materials capable of operating at temperatures above 1200 °C in air. The relationship between the fibres and the matrix material is not that found in most composites. The primary role of the fibres is to provide a mechanism to impede crack propagation in the matrix. This is because, although ceramics can withstand high temperatures, they are hard and brittle polycrystalline structures which usually contain many defects which can initiate crack propagation due to thermal or mechanical shock. Ceramics do not possess mechanisms for accommodating such shocks, so, once a crack is initiated, it propagates catastrophically and causes failure of the structure. Introducing fibres into the ceramic provides mechanisms which can absorb the energy of crack propagation. These mechanisms are crack blunting by debonding of the fibre–matrix interface, fibre bridging of cracks in the matrix and fibre pullout. These processes rely on there being an interface acting as a mechanical fuse which fails when a crack approaches. In this way CMCs can accumulate multicracking without the structure failing so that they can be used as structural materials whereas the unreinforced ceramic cannot.

The most widely studied family of CMCs is composed of beta silicon carbide-based fibres with diameters of between 8 and 15 µm embedded in a SiC matrix. The fibre–matrix interface is usually controlled by making the surface of the fibre rich in carbon, but BN coatings and multilayers of SiC and C have been developed. At room temperature the stress field ahead of a crack in the matrix induces the debonding of the interface with the resulting consequence of an increase in tenacity. An observation of the failure surface reveals extensive fibre pullout. Use of the CMC, with carbon interfaces, in air at temperatures around 700 °C, reveals a dramatic fall in tenacity and no fibre pullout. Under these conditions the carbon interphase is oxidized by the oxygen in the air reaching the interfacial region along the cracks or at surfaces where the interface is revealed. Oxidation proceeds along the interface, creating tubular channels and resulting in the oxidation of the fibre surface to create silica. This silica is produced by the oxidation of the nanometric sized grains of SiC in the fibres. More recent fibres than those originally used have near stoichiometric compositions and as a result have larger SiC grains which are less prone to oxidation so that overall resistance to degradation is improved. The result of the formation of silica at the interface is to create a strong and brittle bond between the matrix and the fibres so that, instead of debonding when a crack reaches the vicinity of a fibre, it passes through it, causing it to break with no fibre pullout. As the mechanisms for absorbing energy have been removed from the composite its tenacity is greatly reduced. The formation of the silica can block the interfacial channels created by the oxidation of the carbon so limiting the degradation. Raising the CMC to higher temperatures above 1200 °C produces a softening of the silica at the fibre–matrix interface and tenacity is increased as debonding and fibre pullout become possible again.

These composites are produced for applications at very high temperatures in air. The first generation of SiC-based fibres began to lose strength and creep above 1100 °C. This was due to the presence of a disordered phase in the fibre structure which allowed creep to occur easily. Temperatures above 1200 °C produced the decomposition of this phase and of the fibre. The reduction of this phase has led to fibres with larger grain sizes which are much more stable. The latest generation of SiC fibres have compositions close to stoichiometry and can retain their properties to around 1400 °C in short-term tests. Eventually these fibres will suffer from oxidation of the SiC and such composites are limited in upper temperature use by this phenomenon. The development of CMCs made from fibres and matrix materials which are both oxides offers a solution to the limitations of the SiC-based fibres. Most oxide fibres are based on alumina, often associated with amorphous silica, and such fibres suffer from rapid creep at temperatures above 1000 °C and lose their strength by grain boundary weakening above 1100 °C. The latest type of such fibres produced by 3M under the name Nextel 720 has a two-phase composition of alpha alumina and mullite and shows good resistance to creep with an upper limit of around 1500 °C. However, the fibre is susceptible to degradation above 1200 °C in the presence of alkaline elements which form silicate phases having low melting points and which consequently promote elemental diffusion in the fibres leading to grain growth and loss in strength.

11.7   Conclusions

Materials and structures age due to the stress experienced in service or due to interaction with the environment, and designers need to know at what rate properties will change. Experience has provided many of the answers to these questions for conventional materials, often through accidents, so that a body of knowledge exists for them which allows safety factors to be built into designs. In addition, a great deal of scientific knowledge of damage processes has been accumulated for conventional materials which allows explanations of observed behaviour to be given. The hydraulic pressure testing of metal pressure vessels is an example in which a brief overload of the vessel allows it to be proof tested. This is justified as it is known that the metal would ultimately break by crack propagation initiated as some small defect, crack or irregularity, which analysis has shown produces local stress concentrations in the material. The stress concentration also induces local plastic deformation. Pressurizing the vessel to a higher pressure than it will experience in service will induce plastic deformation at any crack tip so that, if a crack does not propagate during the proof test, the plastic region ahead of the crack will impede any further propagation at lower pressures. Composites do not break in the same way, so it is unwise to apply the same proof testing method for composite pressure vessels. Indeed a knowledge of failure processes in composites reveals that such a test would produce a vessel which was closer to failure than before the test. The damage processes in these vessels are at a microscopic level, so a visual inspection will not reveal any accumulated damage. The only solution is a thorough understanding of the damage processes governing the evolution of the composite material as, in most cases, allowing experience to be accumulated through accidents is not an option.

The same reasoning applies to the fatigue failure of composites. Fatigue processes in metals are a major concern of designers. It is due to the growth of defects, due to microstructural changes in the metal in the regions of stress concentrations, which eventually leads to the development of a crack big enough to induce failure under the loads applied to the metal structures. Composites are less prone to fail by a fatigue process because of their inherent inhomogeneous structures which impede crack propagation, but the degradation of one of the phases or of the fibre–matrix interface will eventually cause irreversible damage. The behaviour of the composite will depend on the interaction of these phases during the cyclic loading, so although fatigue damage can occur it is different from that found in metals and must be understood as such.

Environmental damage is a concern whether it is the rusting of steel, the oxidation of aluminium or water absorption by composites. Again composites show themselves as being chemically resistant in many environments and the material of choice for many applications for which conventional materials would not be suitable. Water absorption is, however, often the principal concern when using composites. The properties of organic matrices are modified by water, which acts as a lubricant at the molecular level and this can be reversible, leading to a fall in T_g, or irreversible, due to hydrolysis, depending on the nature of the induced changes. Debonding of the fibre–matrix interface is, however, always damaging and irreversible in organic matrix composites. This is a possibility as the water molecules can break the hydrogen bonds adhering the matrix to the fibres. Interactions between the fibres and the matrix can be the cause of degradation in metal matrix composites and these most often occur during manufacture when the metal is molten or can be induced by electro-corrosion between the fibres and the matrix. Ceramic matrix composites rely on the fibre–matrix interface to provide a mechanical fuse so as to impede the propagation of cracks in the ceramic matrix. These materials are most often destined for high temperature use and unfortunately the interfaces can be modified so that, instead of preventing crack propagation, the original interphase is replaced, through oxidation, by a brittle phase which does not stop the cracks. The result can be a dramatic fall in toughness for the composite.

In all of the above examples it is an understanding of the role of the interface or interphase between the matrix material and the reinforcing fibres which is seen as the key to the understanding of the long-term behaviour of composite structures even though, in some cases, the properties of the individual components can change.

11.8 Revision exercises

1. Why do fibres wound onto a mandrel follow geodesic paths? Why does this mean that aramid fibres can be used for pressure vessels whereas they cannot be used for the wing of a plane?

2. What are the principal failure processes in a composite structure? Carbon fibres are perfectly elastic at room temperature but if a unidirectional carbon fibre composite is loaded parallel to the fibres the composite continues to be damaged. What processes are involved in producing this damage? How can this lead to delayed failure?

3. Why can proof testing procedures developed for metal pressure vessels not be used with composite pressure vessels? If a composite pressure vessel is subjected to an overload pressure how does this induce further damage?

4. Explain why a visual inspection is not a reliable way of determining damage in a composite material. How could this type of damage be detected?

5. Explain the acoustic emission technique for detecting damage. By what means are the damage processes detected? Why are emission detected if a unidirectional composite is taken above the previously highest load applied in the direction of the fibres?

6. Explain how a filament-wound carbon fibre pressure vessel can be considered to behave at the microscopic level like a unidirectional composite.

7. The effect of a fibre break in a composite is experienced by neighbouring intact fibres but the effect is limited in distance. Explain why this is the case. Why does the effect evolve with time and how can this lead to delayed fibre failure?

8. A pressure vessel which has been in service every day for five years is held under a constant pressure for 24 hours and 12 failure events are recorded during this period. A comparison with a master curve drawn for the same type of pressure curve predicts 40 events in the same period. What does this mean for the continuing service life of the vessel? What would it mean if the master curve gave a value of 8 events?

9. Composite materials are particularly resistant to fatigue crack propagation. Explain why this is so. Explain what occurs at the tip of a crack at right angles to the fibre direction when the composite is loaded parallel to the fibres. How does this effect the local stress field?

10. Explain why, as the properties of carbon fibres are improved, the possibility of fatigue degradation increases so as to become analogous to that found in glass fibre reinforced composites.

11. What is the driving force for water diffusion into a material?

12. What are the parameters which need to be obtained so as to predict water absorption according to the Fickian model? What are the assumptions in this model?

13. If water absorption obeys Fick's law what does this mean if specimens of various thicknesses are considered? If the specimen thickness is doubled does this only double the time to reach saturation?

14. What are the effects on water absorption of varying the temperature and relative humidity of the environment on a composite which obeys Fick's law of water uptake?

15. A more complex model takes into account the interaction of water with the material. How many parameters are necessary to define this model and what do they represent?
16. What effects can be expected to be seen on the properties of the matrix material? Which of these are reversible?
17. If a composite is found to have lost strength after absorbing water but regains most of this loss on drying, explain why this does not mean that the processes which occurred during water uptake are reversible.
18. If a particular thermoplastic matrix, such as saturated polyester composite, absorbs water when immersed in very hot water it is found to become increasingly brittle. What is happening to the matrix and is it reversible?
19. Water absorption has a dramatic effect on Nylon but it is largely reversible. Explain what is observed as water is absorbed.
20. Explain how the choice of hardener can determine whether a composite is stable or not in a humid environment. Why are anhydride-cured epoxies susceptible to degradation in the presence of warm water?
21. Carbon fibre reinforced aluminium has often been considered as a highly desirable composite material, based on the mechanical properties which can be predicted, but such materials are still not used. Explain why this is so.
22. Explain how the tenacity of ceramic matrix composites is controlled by the fibre–matrix interface. How are the required characteristics for the interface obtained in many SiC–SiC composites? How can this be degraded when the composite is heated in air?
23. Why have oxide–oxide composites not replaced SiC-based composites for use at high temperature?

References

Bonniau P and Bunsell A R 1981 A comparative study of water absorption theories applied to glass epoxy composites *J. Comp. Mat.* **15** 273–293
Cantor B, Dunne F P E and Store I C 2004 *Metal and Ceramic Matrix Composites* (Bristol: Institute of Physics Publishing)
Crank J 1956 *The Mathematics of Diffusion* (Oxford: Oxford University Press)
Curtis P T 1991 Tensile fatigue mechanisms in unidirectional polymer matrix composite materials *Int. J. Fatigue* **13** 377–382
Harris B 2003 *Fatigue in Composites* (Cambridge: Woodhead)
Springer G S 1981 *Environmental Effects on Composite Materials* (Stanford, Conn: Technomic)
Warren R 1972 *Ceramic Matrix Composites* (New York: Blackie–Chapman Hall)
Zweben C and Rosen B W 1970 A statistical theory of material strength with application to composite materials *J. Mech. Phy. Sol.* **18** 189–206

Index